HOUGHTON MIFFLIN
The Mathematics Experience

Senior Authors

Mary Ann Haubner
Mount St. Joseph College
Cincinnati, Ohio

Edward Rathmell
University of Northern Iowa
Cedar Falls, Iowa

Douglas Super
Vancouver School Board
Vancouver, Canada

Senior Consulting Author

Lelon R. Capps
University of Kansas
Lawrence, Kansas

Authors

Harold Asturias
Norwood Street School
Los Angeles, California

Harry Bohan
Sam Houston State University
Huntsville, Texas

William L. Cole
Michigan State University
East Lansing, Michigan

Portia C. Elliott
University of Massachusetts
Amherst, Massachusetts

Francis J. Gardella
East Brunswick Public Schools
East Brunswick, New Jersey

Ana María Golán
Santa Ana Unified School District
Santa Ana, California

Edwin McClintock
Florida International University
Miami, Florida

Jean M. Shaw
University of Mississippi
University, Mississippi

Charles Thompson
University of Louisville
Louisville, Kentucky

Leland Webb
California State University
Bakersfield, California

Barbara Elder Weller
Montclair Public Schools
Montclair, New Jersey

Alma Wright
Trotter Elementary School
Roxbury, Massachusetts

Judith S. Zawojewski
National-Louis University
Evanston, Illinois

Houghton Mifflin Company   Boston
Atlanta   Dallas   Geneva, Illinois
Palo Alto   Princeton   Toronto

Critical Readers

Dennis Anderson
Canyon Park Junior High School
Bothell, Washington

Mary Buck
C. R. Anderson Middle School
Helena, Montana

Lorraine Cooke
Jefferson Elementary School
Bath, South Carolina

Janice Grashel
Instructional Skills Coordinator
Lawrence, Kansas

Lee Hoagland
Westlawn Elementary School
Mobile, Alabama

Charlotte Hughes
Marbrook Elementary School
Wilmington, Delaware

Terell Kaiser
Valley Elementary School
East Grand Forks, Minnesota

Rebecca Kirkland
Highland Elementary School
Dothan, Alabama

Charlene Little
Whitney Young Middle School
Detroit, Michigan

Rebecca Manning
Brant Elementary School
Brant, New York

Mark Medina
King Elementary School
Colorado Springs, Colorado

Michael Monaghan
Emerson Elementary School
Wichita, Kansas

Jill Moore
Yolanda Elementary School
Springfield, Oregon

Betty Pugh
Paris Intermediate School
Paris, Arkansas

Peter Scarano
Elijah Elementary School
Clinton, Iowa

Kathryn Scott
Sandpiper Elementary School
Scottsdale, Arizona

Calvin Shilt
Shroder Middle School
Cincinnati, Ohio

Jeanie Sisson
Red Oak Elementary School
Oklahoma City, Oklahoma

Susan Stonebraker
South Central Elementary School
Canonsburg, Pennsylvania

Donald J. Sweeney
Silverhill School
Silverhill, Alabama

Robert Tate
Thomas Middle School
Philadelphia, Pennsylvania

Ann Watson
Rockefeller Elementary School
Little Rock, Arkansas

Carol Wood
Elma Elementary School
Elma, New York

Multicultural Reviewers

Gail Christopher
Americans All
Chicago, Illinois

Jane Horii
Former Master Teacher
San Francisco Unified
School District
San Francisco, California

Christella D. Moody
Eastern Michigan University
Ypsilanti, Michigan

1994 Impression

Copyright © 1992 by Houghton Mifflin Company. All rights reserved.

No part of this work may be reproduced or transmitted in any form or by any means, electronic or mechanical, including photocopying and recording, or by any information storage or retrieval system without the prior written permission of Houghton Mifflin Company unless such copying is expressly permitted by federal copyright law. Address inquiries to School Permissions, Houghton Mifflin Company, 222 Berkeley Street, Boston, MA 02116.

Printed in U.S.A.

ISBN: 0-395-49409-5

FGHIJ-D-9876543

Field Test Teachers

Adam Artis
Trotter Elementary School
Roxbury, Massachusetts

Julie Book
Edison Elementary School
Waterloo, Iowa

Barbara Costa
St. Teresa School
Cincinnati, Ohio

Carolyn Donahue
Patrick J. Kennedy School
East Boston, Massachusetts

Ruben P. Guzman
William Howard Taft
Middle School
Brighton, Massachusetts

Ann G. Hill
Nishuane School
Montclair, New Jersey

Michael R. Johnson
Wayne Van Horn School
Bakersfield, California

Terry Kawas
Forrestdale School
Rumson, New Jersey

Mary E. Leydon
William Howard Taft
Middle School
Brighton, Massachusetts

Emma Louie
Wayne Van Horn School
Bakersfield, California

Kelly J. Martin
Waverly Middle School
Lansing, Michigan

Trudy Olson
Irving Elementary School
Waterloo, Iowa

Elaine Randolph-Jacobs
Patrick J. Kennedy School
East Boston, Massachusetts

Janet Reinhart
Patrick J. Kennedy School
East Boston, Massachusetts

William J. Rudder
William Howard Taft
Middle School
Brighton, Massachusetts

Mary Sue Salzarulo
St. Teresa School
Cincinnati, Ohio

Louise Scanlon-Oberg
Trotter Elementary School
Roxbury, Massachusetts

Sally Schneider
St. Raphael School
Louisville, Kentucky

Kathleen Schweer
St. Teresa School
Cincinnati, Ohio

Henry Smith
William Howard Taft
Middle School
Brighton, Massachusetts

Jo Ann Smithmeyer
St. Teresa School
Cincinnati, Ohio

Frances M. Stuart
Patrick J. Kennedy School
East Boston, Massachusetts

Kathleen Harris Sullivan
William Howard Taft
Middle School
Brighton, Massachusetts

Susan Thompson
Sam Houston Elementary
School
Huntsville, Texas

Willard Vredenburg
South Miami Middle School
Miami, Florida

Robert Walsh
Patrick J. Kennedy School
East Boston, Massachusetts

Julie Weseman
Blackhawk Elementary School
Waterloo, Iowa

Polly Wing
Trotter Elementary School
Roxbury, Massachusetts

Contents

Discovery: Addition and Subtraction of
Whole Numbers and Decimals …… xii
Discovery: Multiplication of Whole Numbers and Decimals …… xiv

1 Number Sense and Numeration: Whole Numbers and Decimals

1 Working with Data
2 Let's Talk Math
  What Is Number Sense?
4 Place Value: Whole Numbers
6 Comparing and Ordering Whole Numbers
8 Rounding Whole Numbers
10 Rounding Greater Numbers
12 Let's Talk Math
  When Do We Estimate?
14 Greater Numbers

16 Problem Solving Strategy: Using Patterns
18 *Midchapter Checkup*
20 Decimals: Tenths and Hundredths
22 Decimals: Thousandths
24 Rounding Decimals
26 Comparing and Ordering Decimals
28 Let's Talk Math
  Problem Solving: Reading for Understanding

30 Chapter Checkup
32 Extra Practice
33 Enrichment: Independent and Cooperative Activities
34 Technology

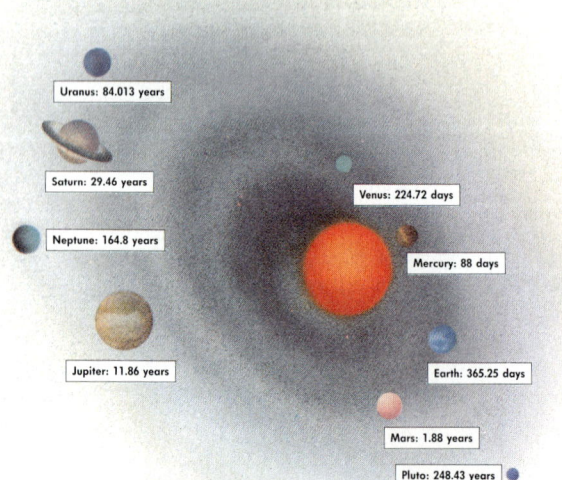

2 Addition and Subtraction: Whole Numbers, Decimals

- 35 Working with Data
- 36 Properties of Addition
- 38 Adding Whole Numbers
- 40 Front-End Estimation
- 42 Subtracting Whole Numbers
- 44 Subtracting Across Zeros
- 46 Solving Equations Using Mental Math
- 48 *Midchapter Checkup*
- 50 Let's Talk Math

  Problem Solving:
  Choosing Mental Math, Pencil and Paper, or Calculator

- 52 Adding Decimals
- 54 Subtracting Decimals
- 56 Making Change
- 58 Problem Solving:
  Four-Part Process

- 60 *Chapter Checkup*
- 62 *Extra Practice*
- 63 *Enrichment: Independent and Cooperative Activities*
- 64 *Cumulative Review*
- 65 *Problem Solving Review*
- 66 *Technology*

3 Multiplication: Whole Numbers and Decimals

- 67 Working with Data
- 68 Properties of Multiplication
- 70 One-Digit Multipliers
- 72 Mental Math: Multiplying by Multiples of 10, 100, and 1,000
- 74 Estimating Products
- 76 Two- and Three-Digit Multipliers
- 78 Zeros in the Multiplier
- 80 Let's Talk Math

  Problem Solving:
  Representing a Problem

- 82 *Midchapter Checkup*
- 84 Exploring Multiplication of Decimals
- 86 Multiplying Decimals
- 88 Multiplying Whole Numbers and Decimals
- 90 Problem Solving:
  Too Much, Too Little Information

- 92 *Chapter Checkup*
- 94 *Extra Practice*
- 95 *Enrichment: Independent and Cooperative Activities*
- 96 *Cumulative Review*
- 97 *Problem Solving Review*
- 98 *Technology*

v

4 Division: Whole Numbers and Decimals

- 99 Working with Data
- 100 Exploring Inverse Operations
- 102 One-Digit Divisors
- 104 Dividing by Multiples of Ten
- 106 Two- and Three-Digit Divisors
- 108 Zeros in the Quotient
- 110 Problem Solving: Multistep Problems
- *112 Midchapter Checkup*
- 114 Dividing a Decimal by a Whole Number
- 116 Exploring Multiplying and Dividing by Powers of Ten
- 118 Exploring Dividing by Decimals
- 120 Dividing By Decimals
- 122 Remainders in Decimal Division
- 124 Averages
- 126 Estimating Quotients
- 128 Let's Talk Math
- Problem Solving: Looking Back
- 130 Creative Problem Solving
- *132 Chapter Checkup*
- *134 Extra Practice*
- *135 Enrichment: Independent and Cooperative Activities*
- *136 Cumulative Review*
- *137 Problem Solving Review*
- *138 Technology*

5 Measurement

- 139 Working with Data
- 140 Exploring Measurement
- 142 Customary Units of Length
- 144 Customary Units of Capacity and Weight
- 146 Fahrenheit Temperature
- 148 Let's Talk Math
- Problem Solving: Writing Problems
- *150 Midchapter Checkup*
- 152 Metric Units of Length
- 154 Perimeter
- 156 Metric Units of Capacity
- 158 Metric Units of Mass
- 160 Problem Solving: Elapsed Time
- 162 Creative Problem Solving
- 164 Mixed Review
- *166 Chapter Checkup*
- *168 Extra Practice*
- *169 Enrichment: Independent and Cooperative Activities*
- *170 Cumulative Review*
- *171 Problem Solving Review*
- *172 Technology*

6 Number Theory, Fraction Concepts

- 173 Working with Data
- 174 Least Common Multiples
- 176 Exploring Divisibility
- 178 Prime and Composite Numbers
- 180 Exponents
- 182 Problem Solving Strategy: Making a Table
- *184 Midchapter Checkup*
- 186 Let's Talk Math

   Non-numerical Graphing
- 188 Meaning of Fractions
- 190 Equivalent Fractions
- 192 Greatest Common Factor
- 194 Lowest Terms
- 196 Mixed Numbers
- 198 Comparing and Ordering Fractions and Mixed Numbers
- 200 Fractions and Decimals
- 202 Exploring Writing Fractions as Decimals
- 204 Problem Solving Strategy: Drawing a Diagram
- 206 Mixed Review
- *208 Chapter Checkup*
- *210 Extra Practice*
- *211 Enrichment: Independent and Cooperative Activities*
- *212 Cumulative Review*
- *213 Problem Solving Review*
- *214 Technology*

7 Fractions: Addition and Subtraction

- 215 Working with Data
- 216 Adding and Subtracting Fractions: Same Denominators
- 218 Adding and Subtracting Mixed Numbers: Same Denominators
- 220 Renaming Before Subtracting: Same Denominators
- 222 Exploring Fractions and Measurement
- 224 Problem Solving Strategy: Guess and Check
- *226 Midchapter Checkup*
- 228 Adding and Subtracting Fractions: Different Denominators
- 230 Adding and Subtracting Fractions: Any Denominator
- 232 Adding and Subtracting Mixed Numbers: Any Denominator
- 234 Renaming Before Subtracting: Any Denominator
- 236 Problem Solving Strategy: Using a Simpler Problem
- 238 Creative Problem Solving
- *240 Chapter Checkup*
- *242 Extra Practice*
- *243 Enrichment: Independent and Cooperative Activities*
- *244 Cumulative Review*
- *245 Problem Solving Review*
- *246 Technology*

8 Fractions: Multiplication and Division

- 247 Working with Data
- 248 Multiplying Fractions: Using Arrays
- 250 Multiplying Fractions
- 252 Multiplying Mixed Numbers
- 254 Multiplying Fractions: Using the Shortcut
- 255 Reciprocals
- 256 Problem Solving Strategy:

    Using Estimation
- 258 *Midchapter Checkup*
- 260 Exploring Dividing Fractions
- 262 Dividing Fractions
- 264 Dividing Fractions and Whole Numbers
- 266 Dividing Mixed Numbers
- 268 Problem Solving Strategy:

    Using Equations
- 270 *Chapter Checkup*
- 272 *Extra Practice*
- 273 *Enrichment: Independent and Cooperative Activities*
- 274 *Cumulative Review*
- 275 *Problem Solving Review*
- 276 *Technology*

9 Geometry and Measurement

- 277 Working with Data
- 278 Exploring Basic Figures
- 280 Angles
- 282 Triangles
- 284 Problem Solving:

    Venn Diagrams
- 286 Polygons
- 288 Exploring Quadrilaterals
- 290 Problem Solving Strategy:

    Using Generalizations
- 292 *Midchapter Checkup*
- 294 Exploring Slides, Flips, and Turns
- 296 Congruent and Similar Figures
- 298 Symmetry
- 300 Circle and Circumference
- 302 Exploring Area
- 304 Area of Rectangles and Triangles
- 306 Area of a Circle
- 308 Creative Problem Solving
- 310 Mixed Review
- 312 *Chapter Checkup*
- 314 *Extra Practice*
- 315 *Enrichment: Independent and Cooperative Activities*
- 316 *Cumulative Review*
- 317 *Problem Solving Review*
- 318 *Technology*

10 Ratio, Proportion, and Percent

319 Working with Data
320 Ratios
322 Rates and Unit Rates
324 Exploring Proportional Thinking
326 Meaning of Proportion
328 Solving Proportions
330 Problem Solving:
Scale Drawings
332 Midchapter Checkup
334 Meaning of Percent
336 Percent and Decimals
338 Fractions and Percents
340 Exploring Percent
342 Finding a Percent of a Number

344 Let's Talk Math
When Do We Use Percent?
346 Problem Solving Strategy:
Using Percent
348 Mixed Review

350 Chapter Checkup
352 Extra Practice
*353 Enrichment:
Independent and Cooperative Activities*
354 Cumulative Review
355 Problem Solving Review
356 Technology

11 Statistics and Probability

357 Working with Data
358 Exploring Statistics
360 Mean and Median
362 Problem Solving:
Interpreting Graphs
364 Making Double Bar Graphs
366 Making Circle Graphs
368 Making Line Graphs
370 Let's Talk Math
Problem Solving:
Misleading Graphs
372 Midchapter Checkup
374 Sampling and Predicting
376 Measuring Chance
378 Equally Likely

380 Exploring Probability
382 Independent Events
384 Sample Space
386 Exploring Combinations and Permutations
388 Creative Problem Solving
390 Mixed Review

392 Chapter Checkup
394 Extra Practice
*395 Enrichment:
Independent and Cooperative Activities*
396 Cumulative Review
397 Problem Solving Review
398 Technology

12 Geometry and Measurement

399 Working with Data
400 Let's Talk Math
    Geometry and Bridges
402 Exploring Prisms and Pyramids
404 Exploring Surface Area
406 Exploring Constructions
408 Spheres, Cylinders, Cones
410 Problem Solving:
    Visual Perception
412 *Midchapter Checkup*
414 Volume: Rectangular Prisms
416 Volume: Cylinders
418 Problem Solving Strategy:
    Using Formulas
420 Creative Problem Solving
422 Mixed Review

424 *Chapter Checkup*
426 *Extra Practice*
427 *Enrichment: Independent and Cooperative Activities*
428 *Cumulative Review*
429 *Problem Solving Review*
430 *Technology*

13 Integers, Coordinate Graphing

431 Working with Data
432 Integers
434 Comparing and Ordering Integers
436 Exploring Adding Integers
438 Adding Integers
440 Problem Solving Strategy:
    Using Patterns
442 *Midchapter Checkup*
444 Exploring Subtracting Integers
446 Subtracting Integers
448 Coordinate Graphs and Ordered Pairs
450 Problem Solving Strategy:
    Working Backward
452 Creative Problem Solving

454 *Chapter Checkup*
456 *Extra Practice*
457 *Enrichment: Independent and Cooperative Activities*
458 *Cumulative Review*
459 *Problem Solving Review*
460 *Technology*

x

14 Pre-Algebra

461 Working with Data
462 Writing Expressions
464 Order of Operations
466 Evaluating Expressions
468 Exploring Equations: Addition and Subtraction
470 Solving Equations: Multiplication and Division
472 *Problem Solving Strategy:* Writing and Solving Equations
474 *Midchapter Checkup*
476 Exploring Functions
478 Exploring Function Rules
480 *Problem Solving Strategy:* Guess and Check
482 *Mixed Review*

484 *Chapter Checkup*
486 *Extra Practice*
487 *Enrichment: Independent and Cooperative Activities*
488 *Cumulative Review*
489 *Problem Solving Review*
490 *Technology*

491 Games

501 Calculator

504 Glossary

510 Table of Measures

511 Index

520 Credits

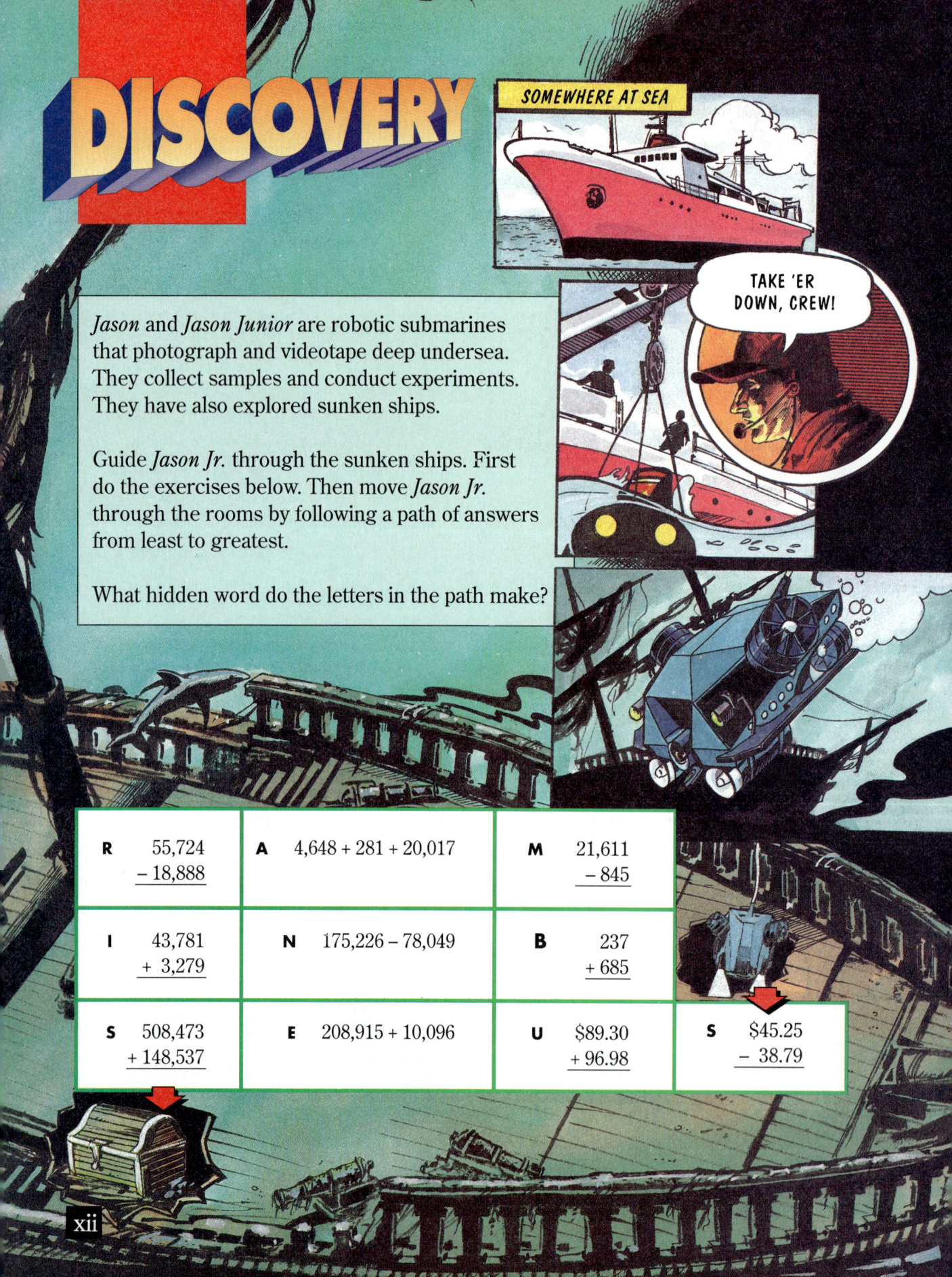

An electronic network connects *Jason* undersea with students visiting science museums.

The students can watch live video transmissions sent by satellite from *Jason's* undersea cameras. They may also use remote control to "fly" *Jason* as it explores deep beneath a distant sea.

Find the answers.

A. 300
   − 274

B. $6.00
   − .59

C. 3,020
   − 1,465

D. 30,000
   − 26,721

Find the path of answers that add up to one of the answers in Exercises A–D.

START

1. 15.52 − 2.8
2. 1.45 + 3.16
3. 0.178 + 0.045
4. 6 + 4.08 + 0.045
5. 6.009 + 7.02
6. 2.5 − 1.8
7. 1.4 + 7.25
8. 0.21 − 0.19
9. 15.52 − 2.8
10. 6.005 − 1.25
11. 23.48
    15.6
    + 0.759
12. 3 − 2.76
13. 4.1 − 1.307
14. 1.45 + 3.16

xiii

DISCOVERY

Complete Exercises A–R. Use your answers to copy and complete the sentences below.

Scuba divers can dive `E minus 500` feet undersea.

Jason and *Jason Junior* can explore `G` feet undersea.

When *Jason* explored sunken 1812 warships in Lake Ontario, about `H` students watched the live broadcasts at their local museums.

The satellite used to relay TV signals from *Jason* in Lake Ontario to students at museums was located `Q` miles above the equator.

Jason can collect `I` pounds of samples from the ocean floor.

Jason is only 7 feet long but it weighs `E plus I` pounds.

A. 724 × 8	**B.** $34.71 × 9	**C.** 6,502 × 8	**D.** $76.42 × 3	**E.** 90 × 7
F. $6.00 × 8	**G.** 400 × 50	**H.** 500 × 500	**I.** 88 × 20	**J.** 3,089 × 96
K. $4.06 × 548	**L.** 559 × 155	**M.** 432 × 608	**N.** 5,507 × 402	**O.** 4,005 × 205
P. 403 × 403		**Q.** 446 × 50		**R.** 45 × 23

Table of Numbers

	O	P	Q	R	S	T	U	V	W
A	50	32	73	15	61	92	28	86	47
B	568	240	809	485	714	123	926	679	391
C	82	325	146	65	508	237	78	444	99
D	6,385	3,672	5,127	8,996	2,001	9,131	1,518	4,280	7,324
E	2,430	767	6,420	3,518	942	5,291	895	4,713	1,727
F	0.7	0.3	0.9	0.1	0.5	0.8	0.2	0.6	0.4
G	0.48	0.99	0.10	0.66	0.83	0.26	0.50	0.72	0.36
H	5.4	3.1	1.6	4.3	3.9	3.2	2.8	2.5	7.7
I	3	2.18	3.5	2.19	7.42	1.38	4.2	4.02	5.08
J	$\frac{4}{5}$	$\frac{1}{5}$	$\frac{2}{2}$	$\frac{1}{10}$	$\frac{7}{10}$	$\frac{2}{5}$	$\frac{3}{10}$	$\frac{1}{2}$	$\frac{3}{5}$
K	$1\frac{1}{2}$	$3\frac{1}{4}$	$4\frac{1}{3}$	$2\frac{1}{4}$	$5\frac{5}{6}$	$1\frac{1}{3}$	$3\frac{1}{2}$	$2\frac{2}{3}$	$1\frac{3}{4}$
L	$3\frac{1}{8}$	$6\frac{1}{2}$	$4\frac{3}{4}$	$5\frac{7}{8}$	$7\frac{2}{4}$	$1\frac{1}{4}$	$7\frac{7}{16}$	$2\frac{2}{4}$	$8\frac{4}{8}$
M	$\frac{1}{4}$	3.5	$3\frac{5}{10}$	2.8	5.2	$8\frac{3}{4}$	0.75	0.15	4.25
N	5%	25%	100%	30%	10%	75%	1%	50%	80%

Ideas for using this table for estimation, mental math, and for calculator activities are found periodically in the Two Minute Math sections of the Teacher's Edition.

Number Sense and Numeration: Whole Numbers and Decimals 1

DID YOU KNOW....?
At 1 mile per second, it would take about 31.7 years to travel 1 billion miles.

Planet	Average Distance from Sun	Travel Time at 1 mi/s*
Mercury	36,000,000 mi	1.1 years
Venus	67,000,000 mi	2.1 years
Earth	93,000,000 mi	2.9 years
Mars	142,000,000 mi	4.5 years
Jupiter	484,000,000 mi	15.3 years
Saturn	887,000,000 mi	28.1 years
Uranus	1,784,000,000 mi	56.5 years
Neptune	2,794,000,000 mi	88.5 years
Pluto	3,674,000,000 mi	116.4 years

*Traveling in a straight line

USING DATA
Collect
Organize
Describe
Predict

Do you think a jet plane goes slower or faster than 1 mi/sec?

Use a number line to draw a diagram that represents the distances of the planets from the sun.

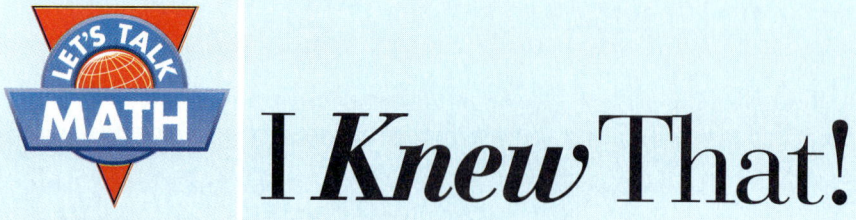

I *Knew* That!

What Is Number Sense?

READ ABOUT IT

You need to have number sense to understand the many ways we use numbers in our daily lives.

If you have number sense:

You know that 40 is a large number if it tells the number of points you score in a basketball game, but → that 40 is a small number if it tells how many you got right out of 100 on a math test.

You know that if you like pizza you would rather have $\frac{1}{3}$ of a pizza than $\frac{1}{4}$, and → that if you eat $\frac{3}{4}$ of a pizza, you have eaten more than half of it.

You know that a quarter of a dollar means 25 cents, but → that "quarter after two" means 15 minutes after 2:00.

You know the pencil you use weighs much less than one pound, but → the bike you ride weighs a lot more than one pound.

You know that 400, $\frac{400}{1}$, and 400.0 all represent the same number, 400, and → that $200 + 200 + 0$, forty tens, and 20×20 also represent 400.

You know that if you buy 3 pairs of socks for $1.87 per pair, you'll owe about $6, and → that your change from a $10 bill will be more than $4.

TALK ABOUT IT

Choose the best answer.

1. $100 could be
   a. the price of a new car
   b. your weekly allowance
   c. the price of a new bicycle

2. $\frac{3}{4} + \frac{3}{4}$ is
   a. greater than 1
   b. equal to 1
   c. less than 1

3. The amount of water needed to fill a bathtub is about
   a. 50 gal
   b. 1 gal
   c. 1 qt

4. The sum of the weights of all the students in your class is closest to
   a. 25,000 lb
   b. 2,500 lb
   c. 250 lb

Use your number sense. Write *true* or *false*.
The number **1,292**

5. is closer to 1,000 than 2,000.

6. is the number of seconds in 10 minutes.

7. is about 700 years in the past.

8. is less than the number of pounds in a ton.

9. could be the sum of two scores in a basketball game.

10. is about the number of pennies in $13.

11. is how many $100 bills it takes to make $1 million.

WRITE ABOUT IT

12. Describe number sense in your own words. Then give an example that proves you have number sense.

CHAPTER 1   3

Place Value: Whole Numbers

Do you know that 506,235 people walked the entire length of the Mississippi River in less than an hour?

Larger numbers like 506,235 have commas to separate the digits into groups of three. Each group of three digits is called a **period**.

Thousands			Ones		
Hundreds	Tens	Ones	Hundreds	Tens	Ones
5	0	6	2	3	5

River Walk in Mud Island Park, Memphis, Tennessee is a 5-block long scale model of the Mississippi River.

We read this number as *five hundred six thousand, two hundred thirty-five.*

This number can be written in different ways.

- **standard form**   506,235
- **short word form**   506 thousand, 235
- **expanded form**   500,000 + 6,000 + 200 + 30 + 5

GUIDED PRACTICE

Use **892,403** to answer.

1. Write in short word form.
2. Write in expanded form.
3. Which is the correct value of the 9—9,000 or 90,000? Explain your answer.
4. Write the number that is 10 thousand less.

PRACTICE

Write in standard form.

5. four thousand, sixty-eight
6. 25 thousand, 9 hundred fifteen

4   LESSON 1-2

MATH AND SOCIAL STUDIES

7. eighty-six thousand, fifty
8. 96 thousand, nine
9. 70 thousand
10. a thousand thousands

Write in expanded form.
11. 796
12. 5,485
13. 63,240
14. 98,000

Write the value of the underlined digit.
15. 4,$\underline{7}$03
16. 3$\underline{5}$,892
17. 7$\underline{8}$4,400
18. $\underline{8}$0,322
19. 670,$\underline{2}$13
20. $\underline{7}$19,230
21. 54,4$\underline{8}$1
22. 8$\underline{0}$7,671

Use the number **198,715**. Write the number that is:
23. six hundred less
24. 15 thousand more
25. 100 thousand less
26. twenty less
27. 500 thousand more
28. 20 thousand less

PROBLEM SOLVING

Complete each fact about the Mud Island River Walk by choosing a number from the box. Use the hint in parentheses to help you.

29. Mud Island Park opened in ▦. (ten thousand less than 11,982)

30. Concerts are given in the Mud Island ▦-seat outdoor theater. (between 5,000 and 6,000)

31. The Mississippi River Museum has Native American pottery that is more than ▦ years old. (greater than 9,000; less than 15,000)

150,000   1,982   219
10,000   1,746   5,064   1,812
6,000   500,000

32. Each concrete section of the River Walk weighs almost ▦ lb. (Sum of the four digits is 6.)

33. The actual Mississippi River flows at a rate of ▦ gal/min. (5 in ten thousands' place)

34. During the great earthquake of ▦ the Mississippi River flowed backward and formed Reelfoot Lake. (sum of the digits is 12)

35. The River Walk, constructed of molded concrete sections, has ▦ sections in all. (close to 1,800; even digit in tens' place)

Comparing and Ordering Whole Numbers

If you know how to compare three-digit numbers and you know the names of the periods, you can compare greater numbers.

When you use something you already know to figure out something you don't know, you make a MATH CONNECTION.

Look at the figures for car sales. Which company sold more cars, Great Wheels Motors or Awesome Auto?

Compare periods.

AUTOTOWN NEWS

Car Sales

Great Wheels Motors	357,918
Awesome Auto Company	365,077
Rex's Wrecks, Inc.	92,064
Galaxy Motors	114,321

Look at the thousands' periods.	Compare.
357,918	365 thousand > 357 thousand *(is greater than)*
365,077	so, 365,077 > 357,918.

Awesome Auto Company sold more cars last year.

THINK ALOUD Give an example where you need to look at both the thousands' period and the ones' period to compare two six-digit numbers.

GUIDED PRACTICE

Compare. Write >, <, or =.

1. 3,129 ▩ 3,192
2. 49,320 ▩ 49,302
3. 899,551 ▩ 899,596
4. 210,991 ▩ 21 thousand, 991
5. 759,023 ▩ 759 thousand, twenty

Order the numbers from least to greatest.

6. 123,540; 12,354; 12,540; 133,054
7. 3,523; 3,532; 3,502; 3,520; 3,539

PRACTICE

Write >, <, or =.

8. 1,951 ▩ 1,915
9. 19,510 ▩ 21,990
10. 18,506 ▩ 18,650
11. 38,551 ▩ 38,515
12. 590,930 ▩ 590,939
13. 989,009 ▩ 989,900

14. 7,008 ▨ 7 thousand

15. 600,000 ▨ 600 thousand

16. 139 thousand, six hundred fifteen ▨ 139,650

Order the numbers from greatest to least.

17. 802,000; 828,000; 816,000

18. 16 thousand; 60 thousand; 600 thousand

19. eighteen thousand; 8 thousand; 880 thousand; 80,000

20. 58,054; 150,000; 500,000; 51,000; 58,980

NUMBER SENSE Answer *true* or *false*.

21. $654,321 > 456,321 + 200,000$

22. $456,231 - 1,000 < 466,231$

23. $954,321 - 20,000 = 754,321$

24. $465,119 + 3,000 = 3,000$ less than $495,119$

PROBLEM SOLVING

Use the picture to answer.

25. List the cars and their prices from most to least expensive.

26. **IN YOUR WORDS** If the digits in the thousands' periods are reversed, does the order of most to least expensive change? Why or why not?

27. Susan bought a Mervette for $1,000 off the regular price. How much did the car cost?

28. **IN YOUR WORDS** If every car price goes up $500, does the order of most to least expensive change? Explain your answer.

CHAPTER 1 7

Rounding Whole Numbers

The Daily News
1,000 Volunteers Build Sand Castle

There were actually 1,142 volunteers. Sometimes a rounded number is used when an exact number isn't needed.

To round 1,142 to the nearest thousand, think:
- Which two thousands does 1,142 come between?
- Which number is halfway between 1,000 and 2,000?
- Is 1,142 greater than or less than the halfway number, 1,500?

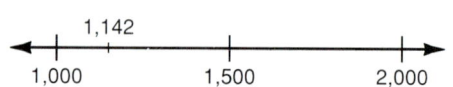

1,142 is less than 1,500, so we round it to 1,000.

THINK ALOUD Describe the steps you would follow to round 3,897 to the nearest thousand.

Other examples:

650 rounds to 700

971 rounds to 1,000

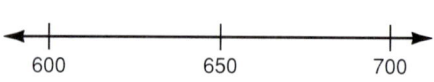
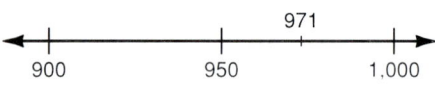

GUIDED PRACTICE

Use **5,842** to answer.

1. Which two thousands is the number between?

2. Which number is halfway between the two thousands?

3. Round 5,842 to the nearest thousand.

Round to the nearest ten.
4. 43
5. 25

Round to the nearest hundred.
6. 579
7. 846

Sleeping Beauty's Castle

PRACTICE

Round to the nearest ten.

8. 39 **9.** 91 **10.** 76 **11.** 34 **12.** 79 **13.** 95

Round to the nearest hundred.

14. 147 **15.** 293 **16.** 337 **17.** 949
18. 699 **19.** 708 **20.** 477 **21.** 557

Round to the nearest thousand.

22. 8,532 **23.** 3,804 **24.** 5,001 **25.** 4,996
26. 9,099 **27.** 7,460 **28.** 2,489 **29.** 6,812
30. 3,089 **31.** 2,879 **32.** 599 **33.** 19,887

NUMBER SENSE Write two different numbers that round to

34. 500 **35.** 80 **36.** 900 **37.** 4,000 **38.** 8,000

PROBLEM SOLVING

Use the newspaper article to answer.

39. IN YOUR WORDS Give examples of two rounded numbers. Explain why rounded numbers are used.

40. If each volunteer carries one 8-lb pail of sand per minute, about how much sand could 1,000 people move per minute?

Lost City of Atlantis

Sand Sculptors International (SSI) set a world's record in 1985 with its 29½-foot-high Sleeping Beauty's Castle. They broke their own record in 1986 by creating the 52½-foot, 49,000-ton

Lost City of Atlantis. The SSI team holds records for the longest, tallest, oldest and largest sand sculptures. Some of its sand sculptures have drawn as many as 400,000 visitors.

CHAPTER 1 9

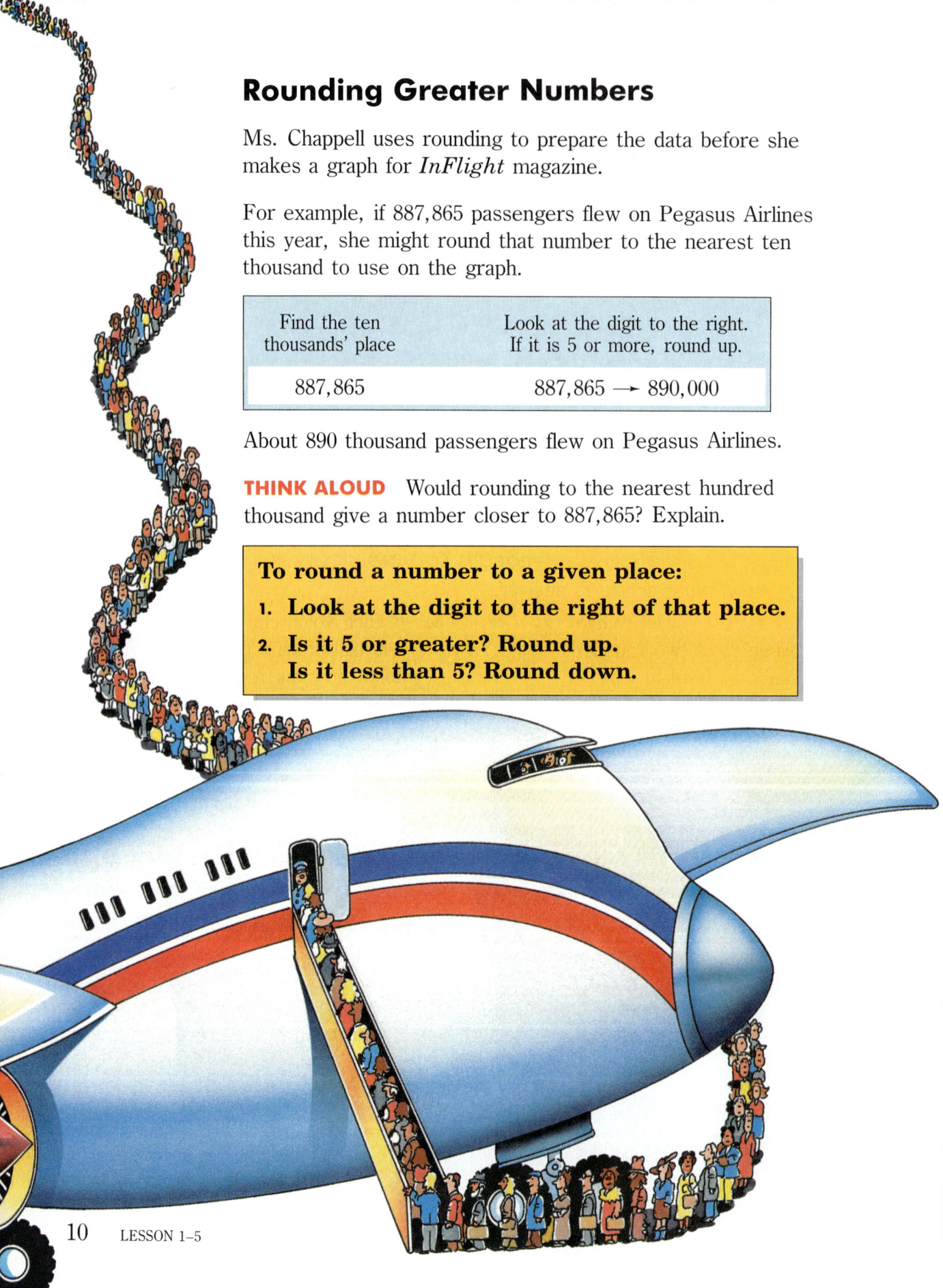

Rounding Greater Numbers

Ms. Chappell uses rounding to prepare the data before she makes a graph for *InFlight* magazine.

For example, if 887,865 passengers flew on Pegasus Airlines this year, she might round that number to the nearest ten thousand to use on the graph.

Find the ten thousands' place	Look at the digit to the right. If it is 5 or more, round up.
887,865	887,865 → 890,000

About 890 thousand passengers flew on Pegasus Airlines.

THINK ALOUD Would rounding to the nearest hundred thousand give a number closer to 887,865? Explain.

To round a number to a given place:
1. **Look at the digit to the right of that place.**
2. **Is it 5 or greater? Round up.
   Is it less than 5? Round down.**

10   LESSON 1–5

GUIDED PRACTICE

Round **791,579** to the nearest

1. ten thousand
2. hundred
3. thousand
4. hundred thousand

PRACTICE

Round to the nearest ten thousand.

5. 16,487
6. 73,495
7. 123,599
8. 28,422
9. 94,578
10. 45,238
11. 129,007
12. 751,631
13. 454,701
14. 619,220

Round to the nearest hundred thousand.

15. 442,798
16. 511,005
17. 705,999
18. 287,024
19. 511,715
20. 302,113
21. 872,432
22. 998,128

Round to the place of the underlined digit.

23. 1<u>9</u>,675
24. <u>5</u>82,129
25. 8<u>5</u>6
26. 73,8<u>9</u>5
27. <u>5</u>03,062
28. 8,5<u>6</u>8
29. <u>8</u>48,075
30. 665,1<u>2</u>1

NUMBER SENSE Find the greatest and the least whole numbers that can round to:

31. 500
32. 3,000
33. 20,000
34. 25,000
35. 50,000

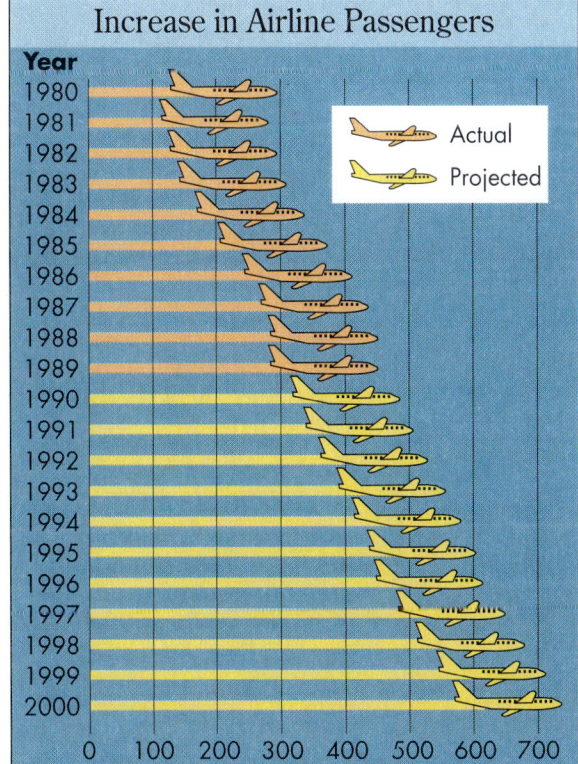

PROBLEM SOLVING

Use the graph to answer the questions.

36. What kind of information is displayed on the graph?

37. Between which two years did the number of passengers decrease?

38. In 1980 there were about 300 million passengers. By which year is that number of passengers expected to double?

39. **IN YOUR WORDS** Why do you think this graph was made? Who might be interested in this information?

CHAPTER 1   **11**

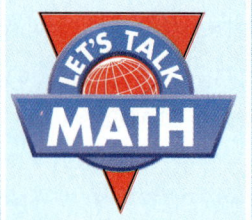

Better Than a Guess

When Do We Estimate?

READ ABOUT IT

Think about all the ways you use numbers every day. Sometimes the numbers you use are exact, like

- the number of people in your family, or
- the number of pages in the book you are reading.

Sometimes (more often than you think!) the measures you use are not exact. They are **estimates**.

We use estimates when . . .

an estimate is just as good as an exact number.
- How many miles apart are New York and San Francisco?
- How many people live in your town?

there is no way of getting an exact number.
- How many pennies are in circulation today?
- How many people are watching this baseball game on TV?

we could get the exact number but it is too difficult.
- How many people pass through an airport in one day?
- How many steps do you walk in one week?

we want to check if an exact computation is reasonable.
- $195 \times 19 = 3{,}705 \Rightarrow 200 \times 20 = 4{,}000$
  Yes, the answer is reasonable.

LESSON 1-6

🟥 TALK ABOUT IT

Work with your group. Look at the numbers used below. Write *exact* or *estimate*.

1. That newspaper has 76 pages.
2. I am 11 years old.
3. We're counting on an audience of 200 people for our school play.
4. Each passenger car on the train seats 100 people.
5. My plane leaves at 9:00 A.M. and arrives at 12:20 P.M.

The numbers are missing from the situations described below. For each situation, would you need to know an exact number or would an estimate be enough? Explain.

6. I need ▨ cups of flour for the cookie recipe.
7. You go to the supermarket with $▨.
8. Before leaving on a long car trip, you decide you need to cover ▨ miles a day.

Use the travel brochure below.

9. Which numbers used in the brochure are exact? estimates?

🟥 WRITE ABOUT IT

10. Work with your group to create a travel brochure about a place you would like to visit or have visited. Use exact and estimated numbers. Exchange with another group and identify the exact and estimated numbers.

Greater Numbers

VOYAGER Views Solar System

NASA's *Voyager 1* spacecraft continued to send photos home. Photos showing most of the planets, taken from almost 4 billion miles away, took more than $5\frac{1}{2}$ hours to be transmitted to Earth. The signals traveled at 186,000 miles per second.

1. Write 4 billion in standard form. Use the place-value chart to help you.

2. How many millions are in 4 billion?

3. *Voyager 1* was launched in 1977. How many years ago was this?

Billions			Millions			Thousands			Ones		
Hundreds	Tens	Ones	Hundreds	Tens	Ones	Hundreds	Tens	Ones	Hundreds	Tens	Ones

Cosmic Tug of War

"The Great Attractor," an enormous gravitation source, has been discovered 150 million light-years from Earth. This huge magnet-like structure has enough force to pull entire galaxies toward it at a rate that is not yet able to be measured.

4. Which is greater, 115 million or 150 million?

5. Write these numbers from least to greatest: 500 million; 5,000,000; 15,000,000; 150 million.

Solar Blasts

Solar Max, an Earth-orbiting satellite, photographs clouds of gas called solar flares. These 10-to-100-billion-ton masses of energy erupt from the Sun at speeds of 22,000 to more than 2 million mi/h.

6. How can you represent the number, 2 million, without using the digit **2**.

7. Is 2 million closer to 10 times 22,000 or 100 times 22,000?

8. The Sun is 93 million mi from Earth. Write a number that is 13 million less than 93 million.

9. Is the standard form for 100 billion 1,000,000,000; 100,000,000,000; or 100,000,000?

Cheap Heat

The self-heating $91-million Boston, Massachusetts Transportation Building saves more than $400,000 yearly in heating costs. The heat, generated by people, lights, and office machines, is circulated by heat pumps and stored in huge, basement water tanks. Solar collectors provide hot water.

10. Would the savings in 10 years be more or less than $1 million?

11. Is $91 million greater or less than $9,100,000?

Ecology Facts

Did you know . . .

Los Angeles residents drive 142 million miles, the distance from Earth to Mars, every day.

12. How much is 100 million more than 142 million?

13. Is 142 million a rounded or an exact number? Explain.

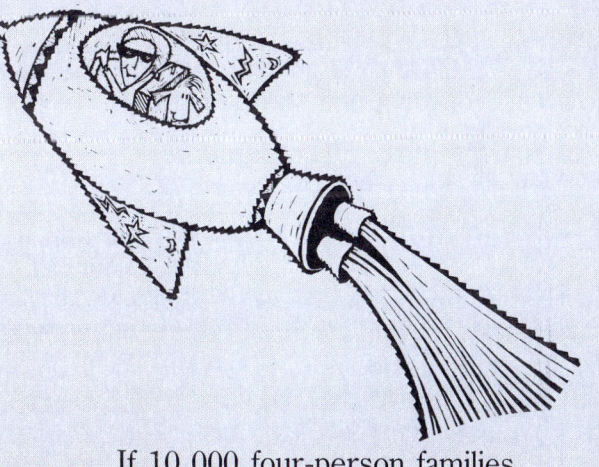

If 10,000 four-person families used "low-flow" shower heads, about 140 million gallons of water would be saved.

14. How much water would be saved if 100,000 families did the same? 1 million?

Problem Solving Strategy: Using Patterns

Did you know that cells grow in patterns? A starfish cell is an example of this. One cell divides into two cells. Each of those two cells divides into two cells so that there are four cells. This pattern of cell growth continues.

List the number of cells after each division to help you see the pattern.

1, 2, 4, 8, 16, 32, 64, . . .

- How is one number related to the next?
- How can you use the pattern to predict the next number in the pattern?

Here is another type of pattern.

5, 3, 7, 5, 9, 7, 11 . . .

Find the relationship between the numbers to help you see the pattern.

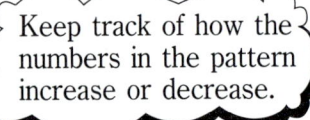

5, 3, 7, 5, 9, 7, 11 . . .
−2 +4 −2 +4 −2 +4

Keep track of how the numbers in the pattern increase or decrease.

- How can you describe the pattern?
- What are the next three numbers?

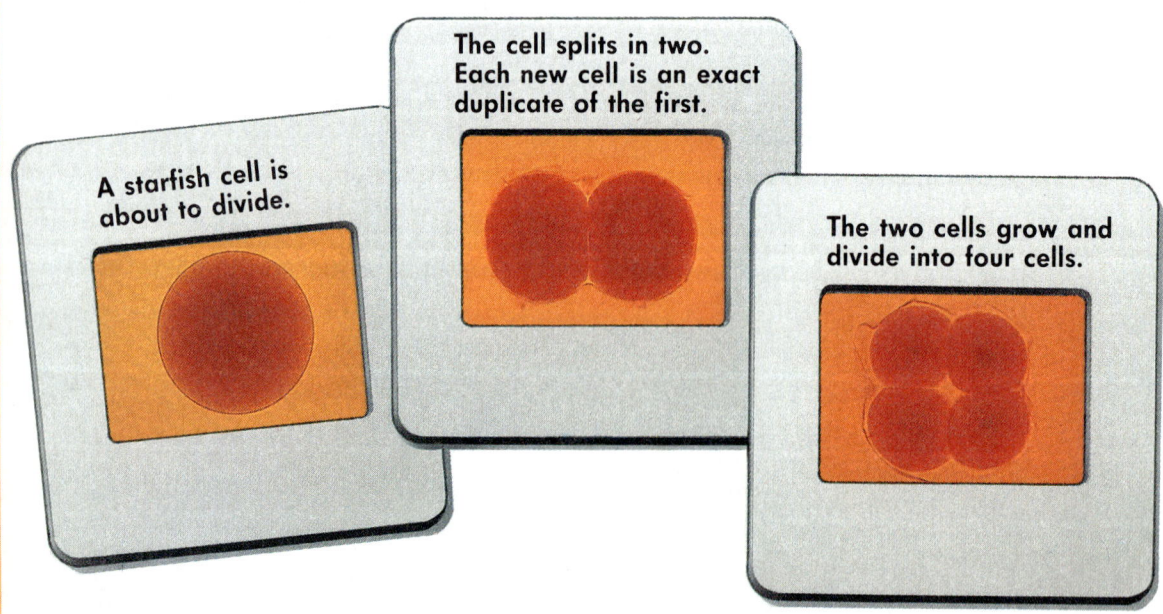

A starfish cell is about to divide.

The cell splits in two. Each new cell is an exact duplicate of the first.

The two cells grow and divide into four cells.

GUIDED PRACTICE

Look at the pattern. Answer the question.

1. 6, 8, 12, 14, 18, 20, 24
    a. Is the difference between each pair of numbers the same?
    b. Explain the relationship between 6 and 8? 8 and 12?
    c. Write the next three numbers in the pattern.

Continue the pattern.

2. 4, 5, 10, 11, 16, ▨, ▨

3. □, ○, ●, □, ○, ▨, ▨

PRACTICE

Continue the pattern.

4. 29, 26, 23, 20, ▨, ▨

5. 16, 21, 26, 31, ▨, ▨

6. 99, 89, 79, 69, ▨, ▨

7. 6, 11, 16, 21, ▨, ▨

8. ↑, →, ↓, ▨, ▨

9. ○, ⊖, ⊕, ▨, ▨

10. 18, 27, 36, 45, ▨, ▨

11. aa, ba, bb, cb, cc, ▨, ▨

12. 2, 5, 8, 11, ▨, ▨

13. 05789, 57890, 78905, ▨, ▨

14. 101, 110, 105, 114, 109, ▨, ▨

15. 7, 13, 11, 17, 15, 21, ▨, ▨

16. ⊤, ⊤⊤, ⊥, ⊥⊥, ▨, ▨

17. △, △·, ·△·, ▨, ▨ (in circles)

IN YOUR WORDS Continue the pattern two different ways. Explain your pattern.

18. 2, 2, 4, 6, ▨, ▨

19. 1, 1, 2, 2, ▨, ▨

20. Ann, Art, Cara, Carl, ▨, ▨

21. 5, 10, 15, ▨, ▨

Describe the pattern.

22. 5.01, 5.001, 5.0001

23. •, ••, ••• (triangular dot arrangements)

24. ⌐, ⌐⌐, ⌐⌐⌐, ⌐⌐⌐⌐

25. 1, 1, 2, 3, 5, 8, 13

CREATE YOUR OWN Work with a partner to make a pattern that

26. increases by an even number

27. uses two or more geometric figures

28. decreases, then increases

29. can continue in two or more ways

MIDCHAPTER CHECKUP

LANGUAGE & VOCABULARY

Explain how each of the following ways of writing a numeral differ.
1. standard form
2. short word form
3. expanded form

QUICK QUIZ

Use 24,806,450 to answer. *(pages 4–5, 14–15)*
1. Write in short word form.
2. Write in expanded form.
3. What is the value of the 8?

Compare. Write >, <, or =. *(pages 6–7)*
4. 26,529 ■ 26,925
5. 417,072 ■ 417 thousand, seventy

Order the numbers from least to greatest. *(pages 6–7)*
6. 46,803; 46,830; 46,308; 46,380; 46,813

Round 84,356 to the nearest *(pages 8–11)*
7. hundred.
8. ten thousand.
9. thousand.

Solve. *(pages 16–17)*
10. Look at the pattern.
    3, 5, 4, 6, 5, ■, ■, ■ . . .
    a. What is the difference between pairs of numbers?
    b. What would you need to do to find the next three numbers in the pattern?
    c. What are the next three numbers in the pattern?

18   MIDCHAPTER CHECKUP

LEARNING LOG

Write the answers in your learning log.

1. How would you explain our place value system to someone who is not familiar with it?

2. The numbers between 3,001 and 3,499 are all the numbers that round to 3,000. Explain why this statement is true or false.

MATH AMERICA

DID YOU KNOW . . . ? The distance across the United States is about 3,000 mi. How does the distance across your state compare with the distance across the United States?

BONUS

CALCULATOR Make organized lists to help you answer these questions.

1. How many different 1-digit numbers could be made using only the digits 1, 2, or 3?

2. How many different 2-digit numbers could be made using only the digits 1, 2, or 3?

3. How many different 3-digit numbers could be made using only the digits 1, 2, or 3?

Look for a pattern in your answers. Use the pattern to find how many different 12-digit numbers can be made using only these digits:

4. 1, 2, or 3

5. the digits 1, 2, 3, or 4

Decimals: Tenths and Hundredths

It takes about 0.1 s to snap your fingers.
Your friend across the room will hear it in about 0.01 s.

You can use models to show decimals.

 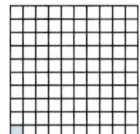 

0.1
one tenth
1 tenth

0.01
one hundredth
1 hundredth

1.05
one and five hundredths
1 and 5 hundredths

Equivalent decimals name the same amount.

 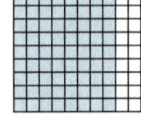

0.8   =   0.80
8 tenths   =   80 hundredths

THINK ALOUD Does writing a zero to the right of the decimal places change the value of the decimal? Why or why not?

GUIDED PRACTICE

THINK ALOUD Use decimal models to show how you would represent these decimals.

1. 0.3
2. 5.90
3. $12.09

PRACTICE

Write the decimals.

4. 5 and 5 tenths
5. forty hundredths
6. nineteen and one tenth
7. five and six tenths
8. fifty and one tenth
9. 3 and 5 hundredths

Write a decimal for each letter on the number line.

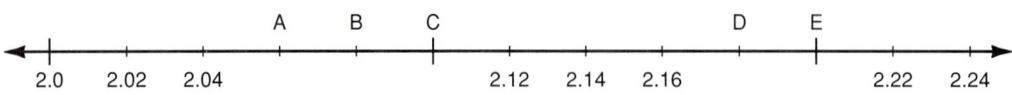

10. A
11. E
12. C
13. B
14. D

LESSON 1-9

Match with an equivalent decimal.

15. 13.4
16. three and forty hundredths
17. 40.40
18. 4.00
19. 0.40
20. two and thirteen hundredths

a. 3.4
b. 4 and 0 tenths
c. 2 and 13 hundredths
d. 4 tenths
e. forty and four tenths
f. thirteen and forty hundredths

NUMBER SENSE Use the digits **0, 3, 6, 4** once each in an exercise. Write a decimal that is

21. 10 more than 20.46
22. less than 4
23. close to 600
24. between 4 and 6
25. 0.1 less than 63.14
26. greater than 40

PROBLEM SOLVING

Mach numbers are used to compare speeds to the speed of sound. Use the Mach scale graph to solve.

27. The high-speed boat, *Spirit of Australia* reached a top speed of 345.23 mi/h. Did it reach Mach 1? Explain.

28. A Boeing 747 cruises at about 60 mi/h less than the speed of sound in the stratosphere. What is the 747's cruising speed.

29. At approximately what rate of speed must an aircraft travel to reach Mach 2?

30. An experimental jet reached a speed of Mach 1.5. About how many miles per hour did it travel?

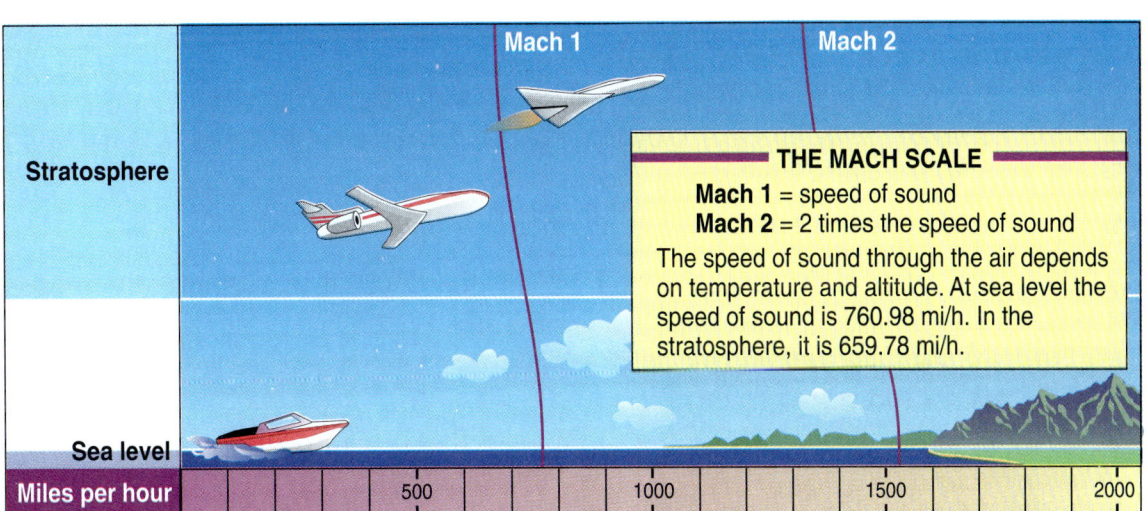

THE MACH SCALE
Mach 1 = speed of sound
Mach 2 = 2 times the speed of sound
The speed of sound through the air depends on temperature and altitude. At sea level the speed of sound is 760.98 mi/h. In the stratosphere, it is 659.78 mi/h.

Decimals: Thousandths

You would be more than 84 years old before you had your first birthday if you lived on Uranus! It takes Uranus 84.013 Earth years to revolve once around the sun.

Uranus: 84.013 years

The decimal tells you that it takes 84 full years plus 0.013 of a year for Uranus to complete its orbit.

A place-value chart can help you read decimals.

Saturn: 29.46 years

Hundreds	Tens	Ones	.	Tenths	Hundredths	Thousandths
	8	4	.	0	1	3
	1	6	.	0	4	1
		0	.	3	2	5

Neptune: 164.8 years

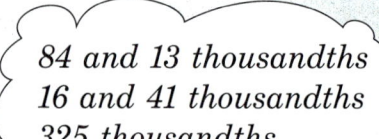

84 and 13 thousandths
16 and 41 thousandths
325 thousandths

Jupiter: 11.86 years

You can write 84.013 in different ways.

- **standard form**   84.013
- **short word form**   84 and 13 thousandths
- **expanded form**   80 + 4 + 0.01 + 0.003

GUIDED PRACTICE

Use **76.025** to answer.

1. Write in short word form.
2. Write in expanded form.
3. Write the value of the 2; the 5.
4. Write the decimal that is 7 tenths more.

PRACTICE

Write in standard form.

5. 99 thousandths
6. 15 thousandths
7. 9 thousandths
8. 5 and 5 hundredths
9. forty thousandths
10. 39 and 3 thousandths

Write the value of the underlined digit.

11. 0.00<u>5</u> **12.** <u>6</u>0.01 **13.** 32.0<u>8</u>3 **14.** 116.9<u>8</u> **15.** 90.28<u>6</u>

16. 23.<u>8</u>67 **17.** 54<u>3</u>.007 **18.** 98.83<u>7</u> **19.** 449.8<u>0</u>1 **20.** 43.86<u>5</u>

Write in short word form.

21. 1.5 **22.** 7.392 **23.** 0.4 **24.** 0.07

25. 2.98 **26.** 9.65 **27.** 8.08 **28.** 6.03

29. 36.02 **30.** 16.16 **31.** 507.082 **32.** 0.20

Write in expanded form.

33. 5.025 **34.** 0.88 **35.** 39.201 **36.** 5,009 **37.** 7.899

Venus: 224.72 days

Mercury: 88 days

Earth: 365.25 days

Mars: 1.88 years

Pluto: 248.43 years

PROBLEM SOLVING

The picture shows the planets' orbit time around the sun in Earth time. Solve.

38. About how many birthdays would you have in an Earth year if you were living on Mercury?

39. If you lived on Mars, how many Mars years old would you be?

40. CREATE YOUR OWN Write two statements about the data.

41. IN YOUR WORDS If another planet were in our solar system beyond Pluto, do you think its orbit time would be greater than or less than that of Pluto? Explain.

Critical Thinking

Use the numbers below. Work with a partner to group or classify the numbers in at least five different ways. (Example: evens, odds)

129.02	3,679,003	340,897	56	1,682.0245	89.6
5	87,301,488	$3.67	78,642	5.0008	17

Rounding Decimals

In Switzerland you can cross the Alps by train through Simplon Tunnel, 19.8 km long, or by car through St. Gotthard Tunnel. St. Gotthard Tunnel, 16.41 km long, is the world's longest road tunnel.

Distances like these are usually rounded to the nearest tenth or whole kilometer.

If you know how to round whole numbers, you can make a MATH CONNECTION to round decimals.

16.4̲1 → 16.4

19̲.8 → 20

To round a decimal to a given place:
1. **Look at the digit to the right.**
2. **If it is 5 or more, round up.
   If it is less than 5, round down.**

GUIDED PRACTICE

Round **567.983** to the nearest

1. tenth
2. hundred
3. ten
4. hundredth

PRACTICE

Round to the nearest tenth.

5. 3.54
6. 16.152
7. $5.58
8. $19.29
9. 23.888
10. 5.06
11. 9.64
12. 52.073

Round to the nearest hundredth.

13. 34.567
14. 56.713
15. 9.005
16. 44.901
17. 189.508
18. 66.666
19. 125.057
20. 21.999

Round to the nearest whole number.

21. 58.213
22. 134.009
23. 98.6
24. 404.005
25. 431.789
26. $13.76
27. 9.999
28. $899.79

Round to the place of the underlined digit.

29. 14.9<u>6</u>　　**30.** <u>9</u>.234　　**31.** $5.<u>2</u>2　　**32.** 0.1<u>4</u>9

33. 9<u>0</u>.7　　**34.** 394.8<u>2</u>9　　**35.** 125.55<u>5</u>2　　**36.** 17.0<u>5</u>78

NUMBER SENSE Use the digits, **4, 3, 0,** and **6** once each. Write a decimal that rounds to

37. 35　　**38.** 34　　**39.** 6.3　　**40.** 0.35

PROBLEM SOLVING

Use the information in the picture to solve.

41. Which trip costs about $1,000 more than another trip? about $100 less than another trip?

42. What other information would you need to know in order to estimate the total cost of any of the trips?

43. If you had $1,000, which 3 trips could you take and still have about $50 left over?

44. Today is October 13. Your trip to Dallas is 3 weeks from tomorrow. What date will that be?

Comparing and Ordering Decimals

Who had the faster time in the Giant Slalom event, Andrea Lawrence or Vreni Schneider?

Compare the two decimals to find out.

Line up the decimal points	Write a zero.	Compare the decimal places.
2:06.8	2:06.80	49 hundredths < 80 hundredths
2:06.49	2:06.49	2:06.49 < 2:06.80.

Schneider's time was faster than Lawrence's time.

CRITICAL THINKING Which would break the record set by Schneider in 1988, a time of 2:06.87 or of 2:06.21? Explain.

GUIDED PRACTICE

Compare. Write <, >, or =.

1. 16.12 ▆ 16.125
2. 98.099 ▆ 98.901
3. 11 ▆ 10.989

Order from least to greatest.

4. 6, 6.04, 6.008, 66.002
5. 123.1, 124, 12.3, 1.233, 0.123

PRACTICE

Compare. Write <, >, or =.

6. 0.7 ▆ 0.39
7. 0.80 ▆ 0.8
8. 0.65 ▆ 0.6
9. 1.23 ▆ 1.32
10. 0.9 ▆ 0.900
11. 0.01 ▆ 0.001
12. 0.290 ▆ 0.30
13. 8 ▆ 0.8
14. 0.870 ▆ 0.87
15. 40.41 ▆ 4.1
16. 3.30 ▆ 2.838
17. 77.2 ▆ 77.25
18. 0.31 ▆ 31 thousandths
19. 8 and 4 thousandths ▆ 8.004
20. 17.354 ▆ 17.3504
21. 10.4 million ▆ 4.10 million

Write the missing numbers in the pattern.

22. 3.62, 3.72, 3.82, ▆, ▆, ▆
23. ▆, ▆, 0.007, 0.010, 0.013
24. ▆, 3.66, 3.67, 3.68, ▆, ▆
25. ▆, ▆, ▆, 3.270, 3.274, 3.278

Order from greatest to least.

26. 97.38, 98.73, 89.73

27. 0.008, 0.009, 0.07

28. 0.089, 0.090, 0.009

29. 1.9, 1.99, 1.099, 1.909

NUMBER SENSE Write three decimal numbers between the given numbers.

30. 65.3 and 65.8 **31.** 4.72 and 4.758 **32.** 0 and 1 **33.** 13.05 and 13.06

PROBLEM SOLVING

Use the information in the chart to solve.

34. What time would a skier need to get to break the record among these gold medal winners?

35. CRITICAL THINKING Give two reasons why the winning times in 1976-1984 are less than those in the other years.

36. IN YOUR WORDS Can you predict what the gold medal winners' times will be in 1992 and 1996? Explain why or why not.

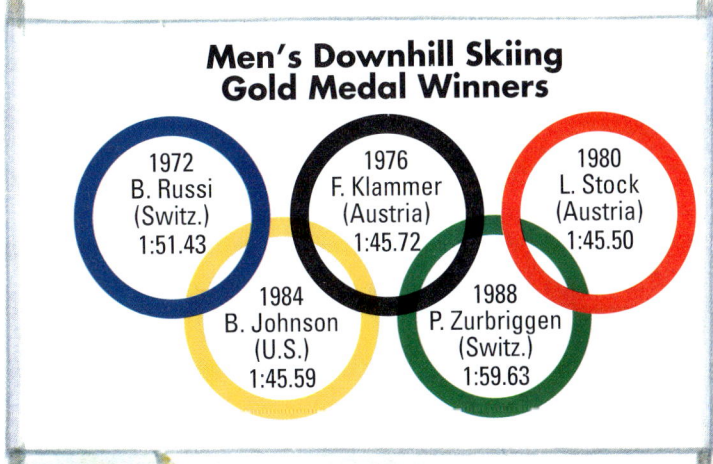

Men's Downhill Skiing Gold Medal Winners

1972 B. Russi (Switz.) 1:51.43
1976 F. Klammer (Austria) 1:45.72
1980 L. Stock (Austria) 1:45.50
1984 B. Johnson (U.S.) 1:45.59
1988 P. Zurbriggen (Switz.) 1:59.63

Critical Thinking

Work with a partner to create a new number system with different symbols. Use the new system to:

- write the numbers one through twenty.
- write 1,000; 9,005; 30.4
- compare two of those numbers.

Discuss with the class any problems you may have had and how you solved them.

READING BETWEEN THE LINES

Problem Solving:
Reading for Understanding

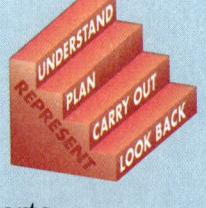

READ AND TALK ABOUT IT

You may need to read word problems and charts several times, each time for a different purpose.

First, read quickly to get the topic.
- What is the topic?

Nani and her friends are making a quilt to raffle off at the school's Aloha Week Fair. To finish the quilt, they need 3 spools of quilting thread, 1 spool of regular thread, 2 packets of needles, and 4 sheets of tracing paper.

Then, read carefully to answer questions.
- How many types of thread do they need?
- How many spools of each type are needed?
- Where can you find the price of needles?

Quilting Materials
Cotton Cloth
    $2.75 per yard
Quilting Thread
    $1.15 per spool
Regular Thread
    $.75 per spool
Needles
    $1.25 per packet
Graph Paper
    $.10 per sheet
Tracing Paper
    $.05 per sheet

Read again to answer these questions.

1. What information is not needed to find the cost of the supplies Nani and her friends need?
   a. the cost of each type of thread
   b. the cost of a packet of needles
   c. the cost of a sheet of graph paper

2. Which cannot be determined with the information given?
   a. the cost of the items listed
   b. the total cost of the quilt
   c. the cost of 2 packets of needles

3. How would you find the cost of the thread needed to finish the quilt?

4. How would you find the cost of 3 yd of cotton cloth and 1 packet of needles?

5. **IN YOUR WORDS** Why might you need to read a passage again once you have found an answer?

READ AND WRITE ABOUT IT

Read the problem as often as necessary to answer each question.

At the fair, the quilt Nani and her friends made was the raffle's grand prize. Nani's class also gave away 25 travel posters that were donated by a local travel agent. At $1.00 each, they sold 200 raffle tickets, 75 shell necklaces, and 300 tropical fruit salads. How much money did the class raise at the fair?

6. What is the problem about?
   a. how many people attended
   b. how much money was raised
   c. how long the fair lasted

7. What information is not needed?
   a. 25 posters were donated.
   b. 75 necklaces were sold.
   c. Each item cost $1.00.

8. Which question do you need to answer to help solve the problem?
   a. How many items were made?
   b. How many items were sold?
   c. Which item sold the most?

9. Decide what operation you will use to solve the problem. Then solve it.

Now that you have read both problems in this lesson at least twice, answer these questions.

10. When you scanned each problem quickly the first time, for what kinds of things were you looking?

11. How did your reading change when you read each passage again more carefully? What did you look for then?

12. Explain why both kinds of reading are important to your understanding of a problem.

CHAPTER 1   29

CHAPTER CHECKUP

LANGUAGE & VOCABULARY

Write the short word form of two decimals equivalent to 4 tenths.

TEST ✓

CONCEPTS

Write *true* or *false*. If false, explain why. *(pages 4–5, 20–23, 24–25)*

1. The number 52.206 is equal to 52 + 0.2 + 0.6.
2. The number 29.3 when rounded to the nearest whole number is 30.
3. In 6,287,454, the digit 8 is in the thousands' period.

SKILLS

Write in standard form. *(pages 4–5, 14–15, 20–23)*

4. 52 thousand, 60
5. 4 billion, 6 thousand, 512
6. 5 and 6 hundredths
7. seventy thousandths

Write the value of the underlined digit. *(pages 4–5, 14–15, 20–23)*

8. 7<u>6</u>,439
9. <u>3</u>,407,328
10. 326.<u>4</u>78
11. 53.08<u>2</u>

Compare. Write >, <, or =. *(pages 6–7, 26–27)*

12. 42,052 ■ 42,502
13. 0.3 ■ 0.25
14. 5.307 ■ 5.46

Order from least to greatest. *(pages 6–7, 26–27)*

15. 8,509; 8,590; 8,950; 8,195; 8,059
16. 2.357; 23.75; 0.2375; 235; 2.53

Round 349,748 to the nearest *(pages 8–11)*

17. ten thousand.
18. hundred thousand.
19. thousand.

Round 83.256 to the nearest *(pages 24–25)*

20. hundredth.
21. whole number.
22. tenth.

PROBLEM SOLVING

Look at the pattern. Answer the questions. *(pages 16–17)*

23. 1, 2, 4, 7, 11,

   a. What is the difference between the first and the second numbers? between the second and third numbers?
   b. How would you determine the pattern used in this list of numbers?
   c. What are the next three numbers in the pattern?

Solve. *(pages 8–11)*

24. During the election for mayor, Ms. Juarez received 349,748 votes. The newspaper reported this number to the nearest ten thousand. What number did the newspaper use?

LEARNING LOG

Write the answer in your learning log.

1. One of your friends says that numbers with 3 decimal places are always greater than numbers with 2 decimal places. Is this correct? Explain why or why not.

2. How would you explain the kinds of patterns that can be found when you look closely at decimal place value?

EXTRA PRACTICE

Write in standard form. *(pages 4–5, 14–15, 20–23)*

1. five thousand, seventy-six
2. 235 thousand, eight
3. 25 million, 346 thousand
4. 2 billion, 32 thousand, 927
5. 18 and 4 hundredths
6. two and three tenths
7. 5 and 24 thousandths

Write the value of the underlined digit.
(pages 4–5, 14–15, 20–23)

8. $3\underline{7}8{,}493$
9. $43{,}0\underline{2}9$
10. $8{,}\underline{9}15{,}374{,}208$
11. $2{,}15\underline{6}{,}000{,}319$
12. $0.0\underline{2}5$
13. $3.\underline{1}4$
14. $51.0\underline{3}$
15. $1\underline{7}.465$
16. $1.30\underline{8}$

Compare. Write >, <, or =. *(pages 6–7 and 26–27)*

17. 4,025 ▒ 4,052
18. 37,426 ▒ 37,246
19. 215,368 ▒ 215,386
20. 3.026 ▒ 3.30
21. 7.50 ▒ 7.5
22. 12.5 ▒ 12.54

Order from least to greatest. *(pages 6–7 and 26–27)*

23. 28,176; 28,671
24. 7,803; 7,583; 6,237; 7,327; 6,240
25. 52.4; 5.42; 0.542; 40.25
26. 3.007; 30.07; 3.07; 3; 3.077

Round to the place of the underlined digit.
(pages 8–11 and 24–25)

27. $5\underline{8}2$
28. $4{,}\underline{3}51$
29. $\underline{2}{,}164$
30. $7{,}\underline{9}16$
31. $4\underline{6}2{,}379$
32. $\underline{8}71{,}025$
33. $29\underline{9}{,}642$
34. $6\underline{4}{,}057$
35. $1.3\underline{6}8$
36. $0.\underline{4}29$
37. $\underline{7}.512$
38. $18.2\underline{7}1$

Solve. *(pages 16–17 and 28–29)*

39. The first three racers who signed up were given numbers 13, 24, and 35. If this pattern continues, what numbers would the next two racers have?
40. There were 23,350 tickets sold for today's baseball game. There were 5 thousand more tickets than that sold for last week's game. How many was that?

ENRICHMENT

Base to Base

Our number system is a base 10 system. We use ten different digits. If you had only five digits—0, 1, 2, 3, and 4—you could write numbers using a base 5 system.

Base 10	1	2	3	4	5	6	7	8	9	10
Base 5	1	2	3	4	10	11	12	13	14	20

10 base five (10_5) = 1 five + 0 ones, or 5.

13 base five (13_5) = 1 five + 3 ones, or 8.

Work with a partner. Write the first 25 counting numbers in base 5. Then try to answer these questions.

- How did you write 25 in base 5?
- You can write any number in base 5 without starting from 1. Try 85 and 155. Explain how finding the number of 125s, 25s, and 5s in each helps you.
- Complete the equation.

    $1,000_5 = \boxed{}_{10}$

In Ancient Times

Work with a partner to find out about the Roman number system. Make a list of the Roman number symbols and their meanings.

Suppose we still used Roman numbers. Draw (or use pictures cut from magazine ads) a store scene where the items are priced with Roman numbers.

Ask another group to tell you the cost of the items in our number system.

Check This

If you have $200 in a checking account at a bank, you can pay for things by writing checks up to that amount.

Locate three items in catalogs or in ads that you would like to buy. Write the checks you would need to pay for these items. (Do not include sales tax for this activity.) Do you have any money left in your checking account? How much?

TECHNOLOGY

CLOSE, CLOSER, CLOSEST

This team game will help you review decimal comparisons and prepare you to use the computer game "Decimal Hunt".

Make three card sets of the digits 0 to 9 (30 cards in all). Make some decimal point cards. Shuffle the digit cards, then draw four cards.

Use the four cards you drew and a decimal point card to form a target decimal.

An example:

Now each player draws four other digit cards. The player who can make a decimal closest to the target number earns 1 point. The first player to earn 5 points wins.

DECIMAL SHIFT

How can you use a calculator to change *one* decimal digit? Here's a way to change the 9 in 17.94 to a 6:

1. Enter 17.94
2. Enter ⊟ 0.3 ⊟.

The display will show 17.64. Now try these decimal shifts:

a. In 8.376, change the 7 to a 4.
b. In 24.704, change the 0 to a 5.
c. In 40.484, change all 4s to 2s.

HOW OLD?

Here are the estimated ages of the Ruiz children (ignoring leap years):

- Anna is 3,285 days old.
- Benito is 832 weeks old.
- Clara is 105,120 hours old.
- Dan is 5,256,000 minutes old.

Find their ages in years, then list the four children in order from the oldest to the youngest.

Addition and Subtraction of Whole Numbers and Decimals 2

DID YOU KNOW…?
"Happy Birthday to You" is the most frequently sung song in the world. At least 10,000,000 people have a birthday each day!

In the United States, a gold record is awarded when a "single" has sold 1 million copies. In some other countries, different numbers of copies must be sold to win a gold record.

Copies of a "Single" Needed to be Sold To Win a Gold Record	
COUNTRY	NUMBER OF COPIES
Australia	50,000
Canada	75,000
France	500,000
Italy	1,000,000
Poland	250,000
Spain	100,000
United Kingdom	500,000

Make a bar graph or a picture graph using the data in the table.

Add the United States to your graph. In which other country is the number of copies needed to be sold the same as in the United States?

Use an almanac to find out which musical group has won the most gold records in the United States.

USING DATA
Collect
Organize
Describe
Predict

Properties of Addition

The shortest route from the West Entrance to Old Faithful Geyser is 14 mi plus 16 mi. The trip back is the reverse, 16 mi plus 14 mi. Both trips are 30 mi.

Knowing the properties of addition can help you add.

Commutative Property
Changing the order of the addends does not change the sum.

$$14 + 16 = 16 + 14$$

In general:

$$a + b = b + a$$

Associative Property
Changing the grouping of the addends does not change the sum.

$$(14 + 16) + 9 = 14 + (16 + 9)$$

In general:

$$(a + b) + c = a + (b + c)$$

Zero Property
The sum of zero and any other number is that number.

$$14 + 0 = 14$$

In general:

$$a + 0 = a$$

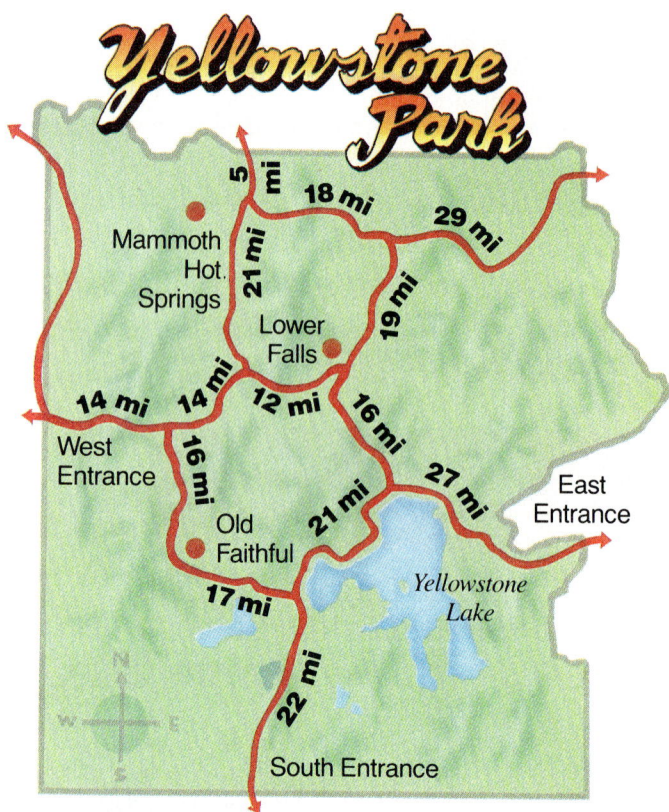

Scale: 1 inch = approximately 20 mi

Old Faithful and Mammoth Hot Springs are two of the main attractions in Yellowstone National Park located in Wyoming.

GUIDED PRACTICE

Give an example of the property. Use the digits **0, 3, 4,** or **5**.

1. Associative
2. Zero
3. Commutative

Complete. Name the property.

4. $23 + (3 + 2) = (23 + \boxed{}) + 2$
5. $a + \boxed{} = b + a$

PRACTICE

Complete. Name the property.

6. $\boxed{} + 7 = 7 + 5$
7. $9 + 16 = \boxed{} + 9$
8. $13 + \boxed{} = 13$
9. $(66 + 19) + 5 = \boxed{} + (19 + 5)$
10. $45 + 163 = 163 + \boxed{}$
11. $275 + 3 = 3 + \boxed{}$
12. $\boxed{} + (32 + 4) = (33 + 32) + 4$
13. $0 + \boxed{} = 1{,}578$
14. $a + \boxed{} = a$
15. $(\boxed{} + b) + c = a + (b + c)$

16. **NUMBER SENSE** Explain how these addends can be grouped to make mental math easier: $36 + 45 + 4 + 15$.

Find the answer if $a = 6$; $b = 4$; $c = 2$.

17. $(a + b) + c = \boxed{}$
18. $a + (b + c) = \boxed{}$
19. $a - (b - c) = \boxed{}$
20. $(a - b) - c = \boxed{}$

21. **CRITICAL THINKING** Use your answers to Exercises 19 and 20. What can you conclude about using the Associative Property with subtraction?

PROBLEM SOLVING

Solve. Use the map of Yellowstone National Park.

22. **IN YOUR WORDS** What is the shortest route from the South Entrance to Mammoth Hot Springs? Explain your answer.

23. Old Faithful Geyser erupts every 73 min on the average. If the last eruption was at 3:30 P.M., what is a reasonable time to expect the next eruption?

24. Lower Falls is about twice as high as Niagara Falls. If Niagara Falls is about 160 ft, about how high is Lower Falls?

25. Use the map scale to estimate the distance you would travel if you hiked the entire northern boundary of Yellowstone.

CHAPTER 2

Adding Whole Numbers

You can travel from China to Sweden, Moscow, Mexico, and Paris, then back to China and never leave the state of Maine!

To find the total distance of the trip, use the map. Find the sum of the mileage between the towns.

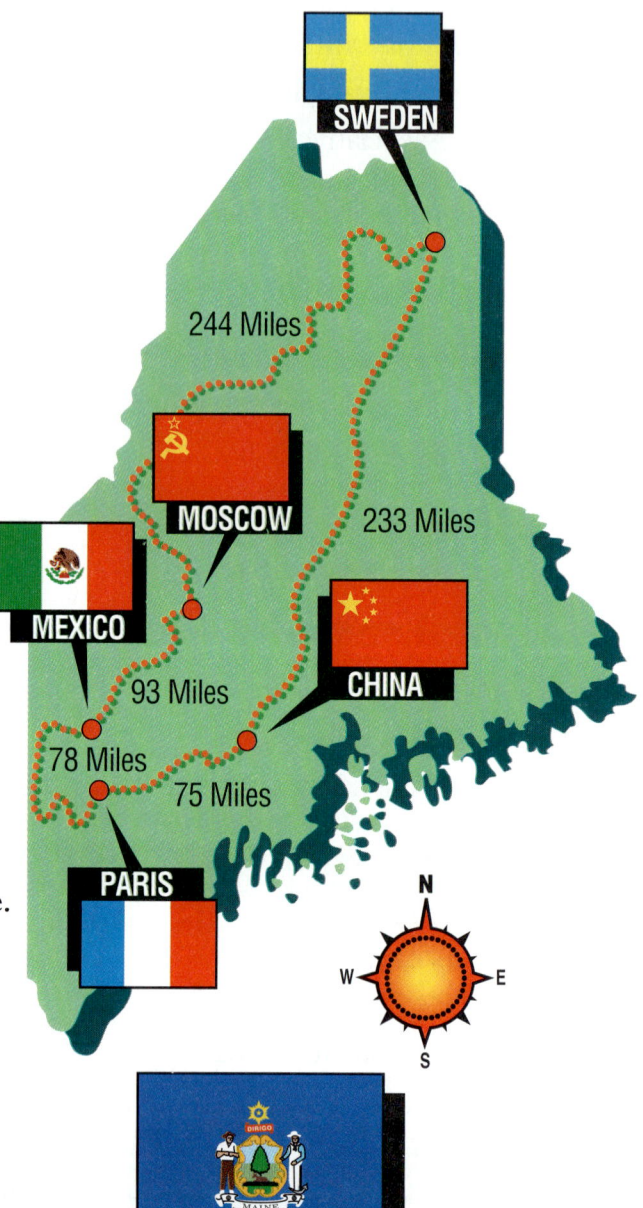

Add the ones and rename.	Add the tens and rename.	Add the hundreds.
2 23**3** 24**4** 9**3** 7**8** + 7**5** **3**	**3 2** 2**3**3 2**4**4 **9**3 **7**8 + **7**5 **2 3**	3 2 **2**33 **2**44 93 78 + 75 **7**23

The trip will cover about 723 mi.

Estimate to be sure your answer is reasonable.

200 + 200 + 100 + 100 + 100 = 700

723 is close to 700.

Other examples:

```
  1 2 2 2
  $136.98           1 1
     6.59        40,980
 + 677.65      + 367,721
  $821.22       408,701
```

THINK ALOUD Why is it important to have the places of the addends lined up correctly?

GUIDED PRACTICE

Add. Estimate to check that your answer is reasonable.

1. 394
   + 64

2. 495
   + 306

3. $15.96
   + 3.64

4. 53,697
   + 219,968

5. 2,798
      14
   + 67,215

PRACTICE

Add. Use a calculator, mental math, or paper and pencil.

6. 67 + 32	**7.** 175 + 213	**8.** $1.97 + 6.09	**9.** 495 + 289	**10.** 944 + 437
11. 2,784 4,435 + 5,187	**12.** 158,428 342 + 5,779	**13.** 344,984 8,086 + 1,254	**14.** 1,672 365 19 + 25,348	**15.** 8,793 47 129,395 + 917,165

16. 760 + 639 + 489 + 905

17. $2,674 + $87 + $393,512 + $456

ESTIMATE Round to the nearest hundred, then add.

18. 121 + 325 + 499

19. 384 + 289 + 5,207

20. 1,249 + 1,568 + 96

Round to the nearest thousand, then add.

21. 964 + 1,410 + 15,132 + 3,839

22. $3,012 + $16,905 + $98,003 + $551

PROBLEM SOLVING

Solve these problems about the state of Maine.

23. Acadia National Park, the oldest park east of the Mississippi River, was established in 1916. Is the park more than 70 years old?

24. At a rate of 10 mi per day, about how long does it take to hike the 280-mi length of the Maine portion of the Appalachian Trail?

25. Twenty-one bathtubs, 167 canoes, 329 sailboats, and 35 homemade rafts took part in the 8-mi Great Kennebec River Race. How many watercraft were in the race?

26. Portland is south of Bangor. Augusta is north of Portland. Bangor is north of Augusta but south of Caribou. List the four cities in order from south to north.

Mental Math

Group by tens. Find the sum.

1. 6 + 4 + 2 + 8

2. 1 + 5 + 9 + 5

3. 21 + 5 + 17 + 4 + 3

4. 2 + 33 + 7 + 28

5. 11 + 53 + 9 + 7

6. 22 + 14 + 6 + 68

Front-End Estimation

You have $7 left to spend, and you want to go on each of the rides once more. Do you have enough money?

Crazy Coaster $2.10

Monster Mix $1.95   **Twister** $1.50

Bumper Cars $1.25   **Mega Wheel** $1.30

You can use **front-end estimation** to get a quick estimate.

	Add the front digits.	Group the cents by dollars.
$6	$2.10 1.95 1.50 1.25 + 1.30	$2.10 ⎫ 1.95 ⎬ About $1 1.50 1.25 ⎫ About $1 + 1.30 ⎭

The cents make about 2 more dollars. You need to adjust the estimate.

The adjusted estimate is $6 + $2, or $8.

You won't have enough money for all the rides.

GUIDED PRACTICE

THINK ALOUD   Explain how to use front-end estimation to estimate the sum.

1. $2.25
    6.54
    + 9.03

2. $25.75
    + 15.86

3. $7.56
    5.09
    + 2.38

4. $10.01
    13.42
    22.75
    + 15.34

40   LESSON 2-3

PRACTICE

Estimate. Use front-end estimation.

5. $3.90
   + 4.15

6. $6.23
   2.74
   + 3.98

7. $7.89
   4.97
   + 1.10

8. $3.13
   3.87
   + 5.85

9. $4.08 + $3.47 + $1.44

10. $39.85 + $12.23 + $41.20

11. $16.50 + $12.83 + $31.99

12. $24.34 + $33.54 + $15.69 + $24.37

13. $13.94 + $12.61 + $10.88 + $5.00

14. $3.65 + $9.34 + $9.98 + $3.02

NUMBER SENSE Write *true* or *false*. Do not compute.

15. The sum of $5.81, $9.25, and $7.89 is more than $30.

16. A reasonable estimate for $1,029 minus $486 is $500.

17. The sum of $24,598 and $675 is less than $25,000.

18. The difference between $29.07 and $11.85 is closer to $20 than $10.

19. The sum of $49.50, $103.11, $9.99, and $14.92 is not greater than $200.

PROBLEM SOLVING

Use the chart to solve.

20. About how many people went on the 3 most popular rides?

21. Was the total income from all the tickets sold closer to $1,500 or $2,000? Explain.

22. **IN YOUR WORDS** More people rode the Bumper Cars than any of the other rides, but the income was the least. What might be the reason?

23. Of the total income from ticket sales, $1,000 was used for expenses. The rest was profit. How much was profit?

Rides	Tickets Sold	Income
Twister	298	$447.00
Monster Mix	229	$446.55
Crazy Coaster	193	$405.30
Mega Wheel	295	$383.50
Bumper Cars	301	$376.25

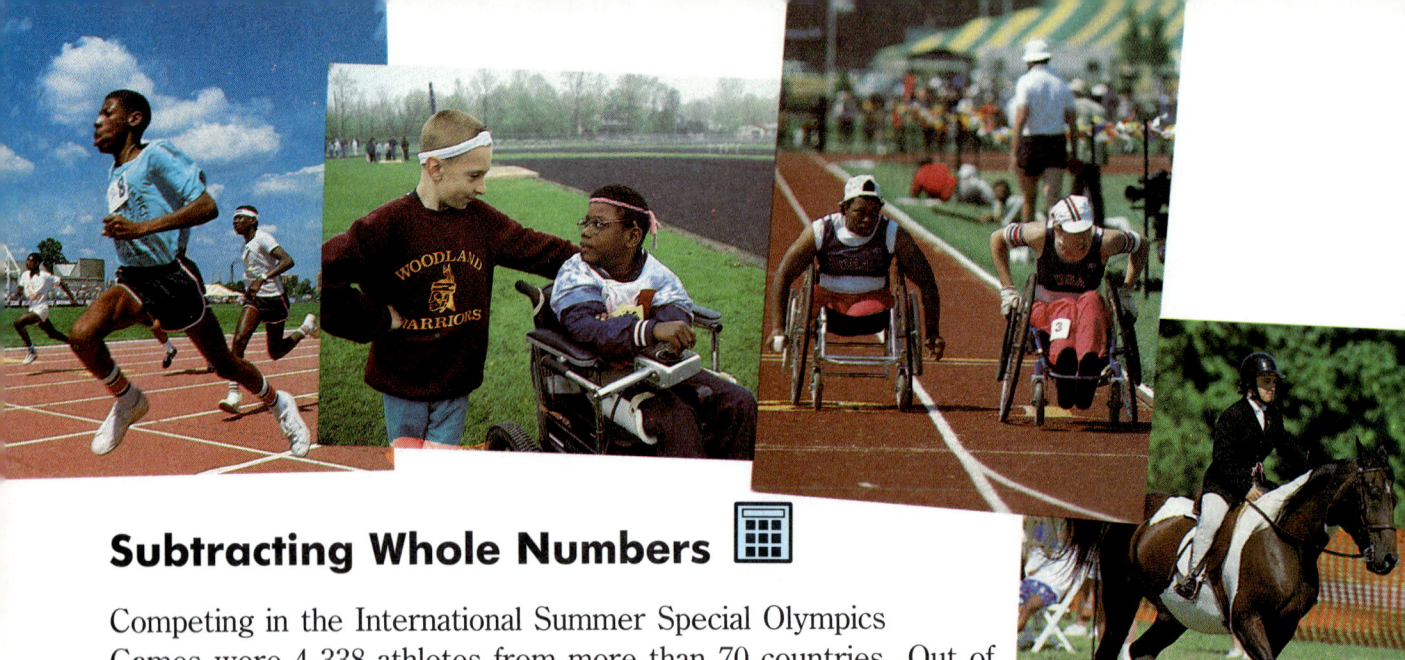

Subtracting Whole Numbers

Competing in the International Summer Special Olympics Games were 4,338 athletes from more than 70 countries. Out of the total number of athletes, 1,419 took part in only track and field events. How many athletes competed in other events?

Subtract to find the difference.

Since 9 > 8, rename the tens and ones. Subtract.	Since 4 > 3, rename the thousands and hundreds. Subtract.	Add to check.
2 18 4,3 3̸ 8̸ − 1,4 1 9 ─────── 1 9	3 13 2 18 4̸,3̸ 3̸ 8̸ − 1, 4 1 9 ─────── 2, 9 1 9	1 1 2,919 + 1,419 ─────── 4,338

Subtraction and addition undo each other. They are **inverse** operations.

Competing in other events were 2,919 athletes.

GUIDED PRACTICE

THINK ALOUD

1. Describe the steps you follow to subtract 991 from 56,789.

2. Explain how you can check that your answer is correct.

PRACTICE

Subtract. Use a calculator, mental math, or paper and pencil.

3. 315
   − 202

4. 763
   − 538

5. 616
   − 129

6. 914
   − 25

7. 1,329
   − 785

8. $4,272
   −   98

9. 1,238
   − 613

10. 21,126
   − 4,458

11. 9,459
   − 888

12. 18,616
   −   700

42 LESSON 2-4

13. 3,347 − 1,798
14. 16,745 − 988
15. $279.83 − 199.42
16. 43,561 − 4,782
17. 276,684 − 87,786

18. 7,652 − 5,621
19. 16,127 − 9,358
20. $126,367 − $79,748

21. 413,629 − 9,488
22. 911,475 − 34,933
23. 469,112 − 368,065

MIXED REVIEW Find the answer.

24. 4,537 − 1,869
25. 59,887 + 13,098
26. 3,735 + 148
27. 4,000 + 1,737
28. 68,165 − 34,497

PROBLEM SOLVING

Use the information given below to solve.

29. How many more athletes took part in track and field events than in gymnastics events?

30. **IN YOUR WORDS** You might conclude that track and field were the favorite events. Explain why this conclusion may not be right.

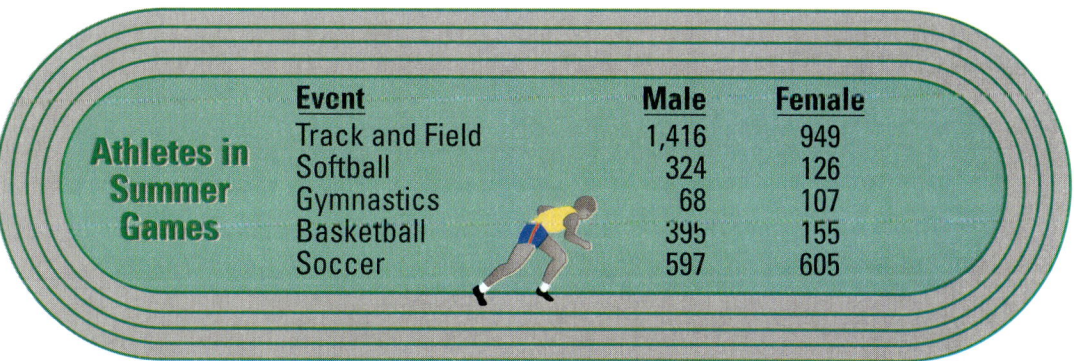

Athletes in Summer Games

Event	Male	Female
Track and Field	1,416	949
Softball	324	126
Gymnastics	68	107
Basketball	395	155
Soccer	597	605

Mental Math

Try subtracting numbers ending in 9's and 8's in this way.

35 − 19 = ▧

Add 1 to 19, and subtract.
35 − 20 = 15

Then, add 1.
15 + 1 = 16

35 − 19 = 16

To subtract numbers ending in 8, add 2 instead of 1.

Subtract mentally by adding on.

1. 33 − 19
2. 45 − 18
3. 56 − 28
4. 87 − 59
5. 42 − 28
6. 68 − 39
7. 125 − 98
8. 191 − 79

Subtracting Across Zeros

The Amazon River begins in the Andes Mountains in Peru and empties into the Atlantic Ocean in Brazil. Of its nearly 4,000-mi length, 1,962 mi are in Brazil. How many miles of the river are in Peru?

To find out, subtract.
4,000 mi − 1,962 mi

Rename 400 tens as 399 tens and 10 ones.		Subtract.
3 9 9 10 4,0 0 0 − 1,9 6 2	Think of 4,000 as 400 tens and 0 ones.	3 9 9 10 4,0 0 0 − 1,9 6 2 2,0 3 8

Is 2,038 a reasonable answer?

You can estimate to find out.

4,000 − 2,000 = 2,000

2,038 mi of the Amazon's length are in Peru.

The Amazon River is longer than the highway route between New York City and San Francisco.

Other examples:

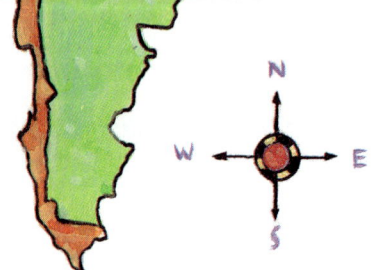

```
      11
  2 9 9 X 10          9 9 13
  3 0,0 2 0         $ 1,0 0 3
 −    1,8 7 5       −    3 8 7
  2 8,1 4 5         $    6 1 6
```

GUIDED PRACTICE

Subtract.

1. 504
    − 179

2. 8,003
    − 4,078

3. $7,060
    − 3,689

4. 36,000
    − 17,591

5. **THINK ALOUD** Explain how to use addition to check Exercise 4.

44   LESSON 2-5

PRACTICE

Subtract. Estimate to see whether your answer is reasonable.

| 6. | 1,060
− 56 | 7. | 2,009
− 119 | 8. | 5,013
− 98 | 9. | 600
− 304 | 10. | 4,000
− 671 |

Subtract. Use a calculator, mental math, or paper and pencil.

| 11. | 5,080
− 34 | 12. | 7,707
− 209 | 13. | 3,008
− 169 | 14. | $4,880.00
− $94.70 | 15. | 602,137
− 104 |

| 16. | $600.00
− 382.20 | 17. | 49,400
− 85 | 18. | 30,080
− 16,754 | 19. | 72,007
− 2,698 | 20. | 150,037
− 3,007 |

21. 10,005 − 2,408   **22.** $39,070 − $195   **23.** 50,000 − 2,798   **24.** 20,709 − 6,909

Complete.

25. 108 − ▨ = 96   **26.** 2,003 − ▨ = 477   **27.** ▨ − 29,101 = 99

MIXED REVIEW   Find the answer.

28. 51,398 + 965

29. 524,870 − 19,505

30. $5,000 − $3,744

31. 9,098 + 1,405

32. NUMBER SENSE Explain how the answer to 1,587 − 100 can help to solve 1,587 − 98.

ESTIMATE   Round to the nearest thousand.

33. 62,008 − 990

34. 18,723 − 1,268

35. 34,007 + 6,704

36. 97,030 − 2,820

PROBLEM SOLVING

37. Giant Brazil-nut trees grow to be 150 ft tall in the Amazon rain forest. How much taller is the Brazil-nut tree than a 28-ft-tall oak tree?

38. The Amazon River begins at 17,200 ft above sea level. It falls about 16,400 ft during the first 600 mi. At that point, how many feet above sea level is the river?

CHAPTER 2   45

Solving Equations Using Mental Math

Magena needs 13 more coupons to get 3 free rolls of film. How many coupons does she have?

You can use an equation to help find the answer.

Free film offer
Send in 25 film carton coupons and get 3 rolls of film, FREE!

> An equation is a mathematical sentence stating that two quantities are equal.

Think: What number plus 13 equals 25?

$$n + 13 = 25$$
$$\square + 13 = 25$$

> A letter, like n, that is used to represent an unknown number is called a **variable**.

> Cover n, the variable, to help you solve the equation.

Check your answer by replacing n with 12.

$12 + 13 = 25$ ✓   12 makes the equation true.

Magena has 12 coupons.

Other examples:

$n - 8 = 19$ $6 \times n = 72$
$\square - 8 = 19$ $6 \times \square = 72$
$n = 27$ $n = 12$

46 LESSON 2-6

GUIDED PRACTICE

Cover the variable to help you solve the equation.

1. $16 + n = 23$
   $n = \rule{1cm}{0.15mm}$
2. $n \times 3 = 36$
   $n = \rule{1cm}{0.15mm}$

Replace n with 5. Now is the equation true?

3. $27 - n = 22$
4. $n + 18 = 24$
5. $12 \times n = 60$

PRACTICE

Solve the equation mentally.

6. $n + 18 = 20$
7. $n - 3 = 20$
8. $29 - n = 2$
9. $23 + n = 34$
10. $2 \times n = 100$
11. $n + 7 = 42$
12. $n - 9 = 4$
13. $n \times 6 = 60$

Replace n with 8. Is the equation true? Write *yes* or *no*.

14. $n + 12 = 20$
15. $14 + n = 21$
16. $n - 8 = 0$
17. $39 + n = 24$
18. $42 - n = 34$
19. $n + 11 = 19$
20. $25 - n = 9$
21. $n \times 3 = 24$

Which number makes the equation true? Choose **2**, **3**, **4**, or **5**.

22. $13 - n = 11$
23. $n + 19 = 24$
24. $31 - n = 29$
25. $49 - n = 44$
26. $n + 7 = 10$
27. $12 \times n = 48$
28. $n + 110 = 113$
29. $37 - n = 33$

PROBLEM SOLVING

Choose the correct equation. Use mental math to solve.

30. When Quickpix has a 2-for-1 sale, 2 prints cost $.40. At that price, how much will 10 prints cost?
    a. $n = 2 \times \$.40$
    b. $n = 10 \times \$.40$
    c. $n = 5 \times \$.40$

31. When Lita's roll of 24 pictures was developed, only 7 of them came out. How many did not come out?
    a. $n + 24 = 7$
    b. $24 - n = 7$
    c. $7 \times n = 24$

32. There are 36 exposures on the large roll of film, 3 times as many as there are on the small roll. How many exposures are on the small roll?
    a. $n \times 36 = 3$
    b. $36 - n = 3$
    c. $n \times 3 = 36$

33. Jack buys film for $4.50 and a camera case for $10.95. His change is $4.55. How much did he give the clerk?
    a. $n = \$4.50 + \$10.95 + \$4.55$
    b. $\$4.50 + \$10.95 + n = \$4.55$
    c. $n = (\$4.50 + \$10.95) - \$4.55$

MIDCHAPTER CHECKUP

LANGUAGE & VOCABULARY

Explain how the following properties can help you add. Give an example of each.
1. Commutative Property of Addition
2. Associative Property of Addition
3. Zero Property of Addition

QUICK QUIZ

Complete. Name the property that is used. *(pages 36–37)*

1. $(5 + 7) + 4 = \blacksquare + (7 + 4)$
2. $0 + \blacksquare = 2{,}453$

Use front-end estimation. Estimate the sum or difference. *(pages 40–41)*

3. $4.74 + $3.22 + $6.54 + $3.46
4. $11.65 + $9.25 + $16.70 + $3.25

Add. *(pages 38–39)*

5. 4,326
   4,954
   + 3,473

6. 746,315
   46
   9,804
   + 25,927

Subtract. *(pages 42–45)*

7. 82,741
   − 58,693

8. $350.00
   − 196.47

Use mental math to solve the equation. *(pages 46–47)*

9. $n - 79 = 52$
10. $100 - n = 25$

Write the answer in your learning log.

1. How are front-end estimating with money and rounding to the nearest dollar similar? How are they different?

2. If your answer to a column addition problem is not reasonable, what kinds of things should you go back to check?

3. Describe a quick way to mentally do the subtraction example 1,682 − 98.

DID YOU KNOW . . . ? Hawaii is our 50th state. It became a state in 1959. How many years ago did your state join the United States?

CALCULATOR What is the sum of all numbers from 1 through 100? There's a quick way to find this sum. See if you can discover it!

Pick and Choose

Problem Solving: Choosing Mental Math, Pencil and Paper, or Calculator

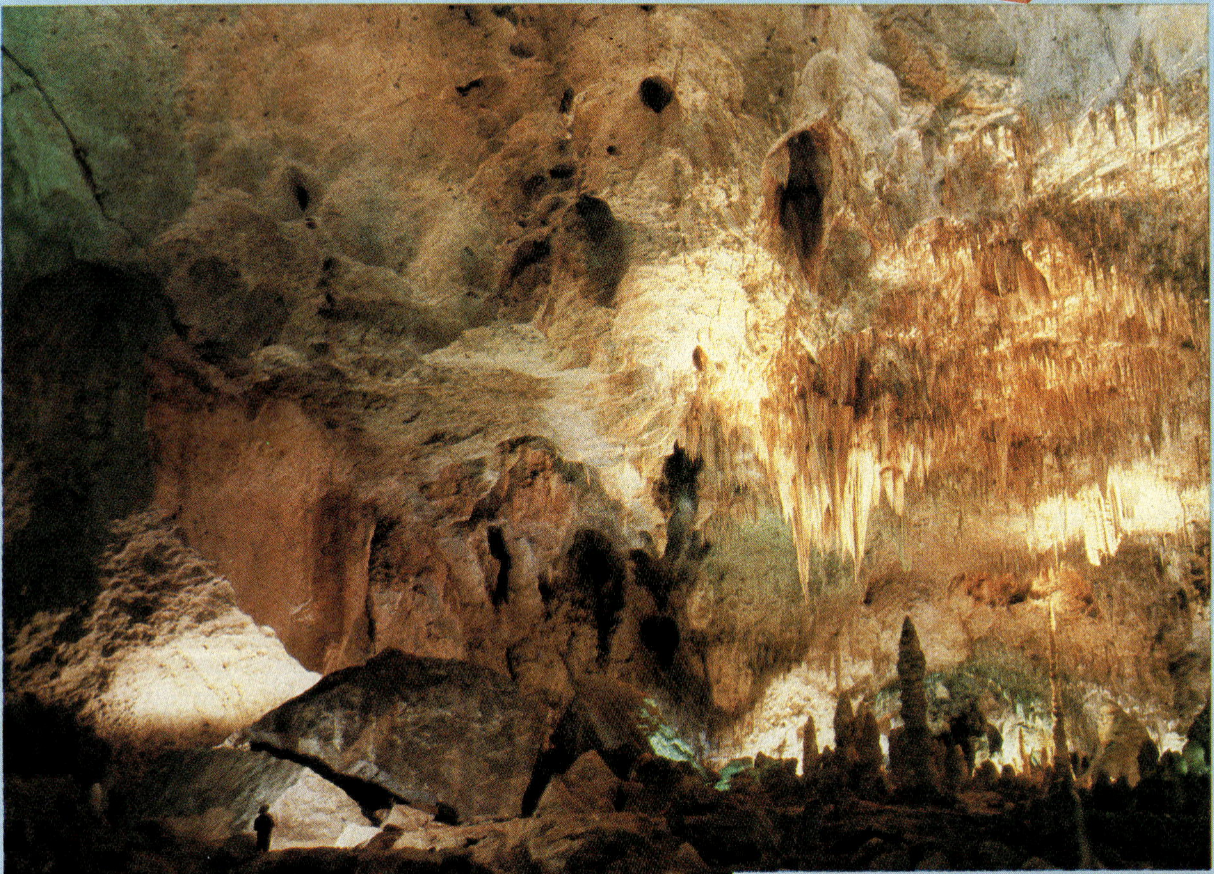

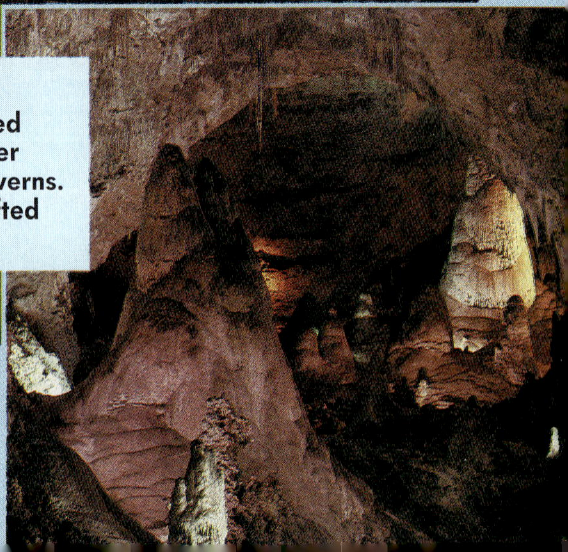

Geologists believe the caverns formed 60 to 70 million years ago when Earth's movements caused cracks to open up in the limestone. Ground water flowed through the cracks, hollowing out the caverns. About 3 million years ago, Earth's movements lifted the region and the ground water drained away.

READ ABOUT IT

Carlsbad Caverns is one of the largest series of caves in the world. It is located in Carlsbad Caverns National Park in New Mexico.

You can choose mental math, pencil and paper or calculator to solve the problems about this natural wonder.

THINK ABOUT USING...

MENTAL MATH when you know an easy way to solve in your head.

CALCULATOR when the numbers are large, or getting the answer is complex.

PAPER AND PENCIL when the work is not hard enough for a calculator, but is too hard to solve mentally.

TALK ABOUT IT

Which method would you choose? Solve.

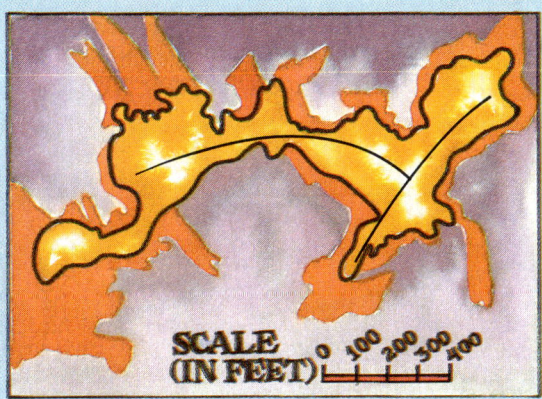

You can take an elevator down 754 ft to the Big Room, an enormous cave shaped roughly like a T. This cave is big enough to hold 14 football fields!

1. If the cave's length is 1,800 ft, how many times longer is the cave than a 120-yd football field?

2. The highest dome in the Big Room is 255 ft high. Each story of a skyscraper is 10 ft high. How many stories would fit in this dome?

The Scenic Rooms, the lowest caves open to the public, are found 829 ft below the surface. Before entering these caves, you pass Iceberg Rock, a 100,000-ton chunk of limestone that fell during the last stage of cave formation.

3. The lowest known point in Carlsbad Caverns is the Lake of the Clouds, 1,037 ft below the surface. How many feet deeper is this than the area that is open to the public?

4. How many pounds does Iceberg Rock weigh? (*Hint*: 1 t = 2,000 lb)

WRITE ABOUT IT

5. Use the information in this lesson to write a word problem about Carlsbad Caverns. Ask a partner to choose a method of solving and then to solve the problem.

Adding Decimals

What was the total amount deposited during the month of January?

To find out, add.

Line up the decimal points.	Add. Place a decimal point in the answer.
$26.45 20.00 + 5.65	¹¹ ¹ $26.45 20.00 + 5.65 $52.10

Deposits for January totaled $52.10.

THINK ALOUD Why is it so important to line up the decimal places?

Estimate to check your answer.

$26 + $20 + $6 = $52

═══ **GUIDED PRACTICE** ═══

THINK ALOUD Explain the steps needed to add the numbers.

1. 29.3 + 145.009 + 3.16
2. $12 + $.98 + $495

═══ **PRACTICE** ═══

Add. Use a calculator, mental math, or paper and pencil.

3. $6.70
  + 5.90

4. 67.23
  + 5.49

5. 12.06
  + 9.40

6. 1.599
  + 70.5

7. $15.99
  + 7.05

8. $21.60
  + 93.52

9. 68.237
  + 17.309

10. 85.076
   + 6.925

11. 5.670
   + 28.792

12. 41.75
   + 9.763

13. 3.7
   4.37
   + 6.483

14. $9.03
   3.92
   + 56.16

15. 4.9
   3.876
   + 13.029

16. 3.776
   42.6
   5.09
   + 13.022

17. 21.85
   6.61
   4.8
   + 63.95

18. 48.5 + 95.7
19. 1.07 + 36.97 + 0.56
20. 0.009 + 19.811

LESSON 2-8

MIXED REVIEW Find the answer.

21. 351.6 + 11.75 **22.** 7,890 + 4,788 **23.** 14,816 − 11,899 **24.** 8,167 − 992

Find the answer if $n = 6.75$.

25. $n + 8.43 =$ ▨ **26.** $999.2 + n =$ ▨ **27.** ▨ $+ n = n$

28. $n + 0.84 =$ ▨ **29.** $n + 0.90 + n =$ ▨ **30.** $15 + n +$ ▨ $= 35$

ESTIMATE Use your number sense and the numbers below.

$.68 \quad \$.23 \quad \$.31 \quad \$.80 \quad \$.55$

31. Choose two numbers whose sum is close to $1.00.

32. Choose three numbers whose sum is close to $2.00.

33. Choose four numbers whose sum is sure to be less than $2.50.

34. Estimate the sum of all five numbers.

PROBLEM SOLVING

Use the bank statement to solve.

35. How much money was in Maria's account by the end of September?

36. During which month did Maria deposit the most money? What is the total for that month?

37. How much interest was earned in September and October?

38. Maria deposited 2 rolls of 40 quarters each on December 12. How much money was in her account then?

39. How much more money did Maria deposit in October than in November?

40. **IN YOUR WORDS** Will the amount of interest continue to increase each month even if Maria doesn't deposit or withdraw any money from her account? Why or why not?

Centrobank

MARIA ARIAS
ACCOUNT # 4528601

DATE	DEPOSIT	BALANCE
SEPT 12	$5.25	$5.25
SEPT 22	3.00	
SEPT 30 INTEREST	.45	
OCT 13	10.55	19.25
OCT 21	9.65	
OCT 27	5.60	
OCT 30 INTEREST	1.90	
NOV 10	2.25	
NOV 21	5.35	44.00
NOV 30 INTEREST	2.42	

Subtracting Decimals

Emeralds have been treasured for centuries. The national bank in Bogotá, Colombia holds a giant emerald crystal weighing 1,759 carats. If a 167.52-carat piece was cut from this crystal, how much would the remaining piece weigh?

To find the difference, subtract 167.52 from 1,759.

Write 1,759 as 1,759.00.	Subtract as with whole numbers.
1,759.00 − 167.52	1,759.00 − 167.52 1,591.48

Don't forget to place the decimal point.

The Muzo Valley in Colombia is a center for mining.

Estimate to check your answer.
  1,800 − 200 = 1,600.

The remaining piece would weigh 1,591.48 carats.

GUIDED PRACTICE

Not all of these examples are correct. Explain what is wrong with each incorrect example and rewrite it correctly.

1. 6,215.9 − 98.021 = 6,117.921
2. 509.802 − 0.5 = 509.302
3. $67 − 0.65 = $.02
4. 16.002 − 8.7 = 24.702

PRACTICE

Subtract. Use a calculator when needed.

5. 4.5 − 1.7
6. 37.8 − 14.9
7. 83.78 − 9.9
8. 70.4 − 35.7
9. 147.6 − 35.47

10. $350.35 − 179.48
11. 20.851 − 0.984
12. 86.005 − 14.32
13. 100.83 − 98.135
14. 773.85 − 187.03

15. $14,006 − $7.88
16. 516.086 − 9.77
17. $333 − 187.46
18. 11.9 − 9.603

54 LESSON 2-9

MIXED REVIEW Find the answer.

19. 98.2 + 25.46   **20.** 4,672 + 1,009   **21.** $7,002 − $1,455   **22.** 6,005 + 98.604

CALCULATOR Make an equation using the numbers in each group. An operation sign, + or −, may be used more than once.

23. 218.3 4.9 223.2   **24.** 0.024 1.08 1.104   **25.** 2.9 8.54 5.64

26. 1.45 8.9 2.48 7.87   **27.** 1.6 98.89 100 2.71   **28.** 3.493 0.007 0.2 3.3

PROBLEM SOLVING

Use the chart to solve.

29. A steel-blue 112-carat stone was mined in India in 1668. Louis XIV of France had it cut into a 67-carat gem. Later, the Hope diamond was cut from this gem. How much smaller is the Hope than the original stone?

30. **IN YOUR WORDS** Explain how you can make a quick estimate to find out whether the Devonshire emerald weighs more than all the other gems on the chart combined.

Some Famous Gems

Name	Weight (carats)
Devonshire emerald	1,343
Mogul emerald	217.8
Hope diamond	44.25
Koh-i-noor diamond	108.93
Star of the East diamond	94.8
Logan sapphire	423
Star of Asia sapphire	330

31. If two paper clips equal the weight of a 10-carat diamond, about how many paper clips would you need to equal the weight of the Star of Asia? the Logan sapphire?

32. The Koh-i-noor diamond was found in 1304. It has been cut several times and now weighs 108.93 carats. How many years ago was this diamond discovered?

Critical Thinking

Copy this arrangement of circles.

1. Write the digits **1-10** in the circles. Make the bottom row add up to 29, the third row to 15, the second to 7 and the top to 4.

2. Make up a different set of rules for arranging the digits **1-10**. Try out your rules.

3. Exchange your rules with a partner and solve.

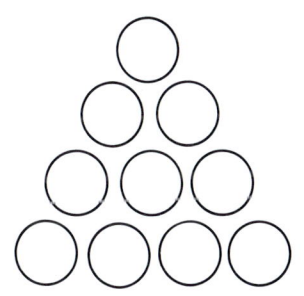

CHAPTER 2 55

Making Change

If you buy a delta kite on sale, how much change should you get back from $20?

You can count on to find out:

$13.59 $13.60 $13.65 $13.75 $14 $15 $20
     +1¢   +5¢   +10¢   +25¢   +$1   +$5

Your change is $6.41.

Carmen bought a dragon kite and a reel of string for $14.78. She gave the cashier $20.03. What was her change?

Cashier said: Carmen received:
  "$15,   One quarter $.25
  $20."   One $5 bill   + $5.00
                                     $5.25

THINK ALOUD Why did Carmen give the cashier $20.03, not $20?

GUIDED PRACTICE

The bill is $11.79. List, in order, the fewest number of coins and bills you could give as change for each amount of money.

1. $12
2. twenty dollars
3. $11.85
4. $15.04

PRACTICE

In your wallet you have three $5 bills, six $1 bills, 3 quarters, 5 dimes, 3 nickels, and 7 pennies. List the fewest number of coins and bills you could use to pay each amount.

5. $.88
6. $6.62
7. $14.36
8. $9.49
9. $19.97

Find the change. List the fewest coins and bills possible.

10. Owe: $4.75   11. Owe: $3.65   12. Owe: $8.84   13. Owe: $.80
    Give: $10   Give: $5   Give: $10   Give: $1.05

14. Spent: $.79   15. Spent: $19.89 16. Spent: $8.03 17. Spent: $13.27
    Give: $5.04   Give: $20   Give: $10.03   Give: $20.02

Explain why the amount of change is wrong and how to correct it.

18. Spent: $8.95
    Give: $10
    Change: $.05

19. Spent: $11.26
    Give: $12.01
    Change: $1.25

20. Spent: $12.80
    Give: $15.05
    Change: $3.25

Toy World Weekly Sale
- Fish Tank $19.98
- Delta Kite $13.59
- String $4.90
- Dragon Kite $9.88
- Puzzles $1.59
- Race Track $27.00
- Race Cars $5.49
- Football Cards $1.29
- Car Models $4.29
- Ship Models $3.97

PROBLEM SOLVING

CHOOSE Choose estimation, mental math, or paper and pencil to solve.

21. Alphonso buys 2 ship models and 2 puzzles. He gives the clerk $12. What is his change?

22. Maria spent $9.17. The cashier is out of quarters. List 2 ways the cashier can give Maria change from $10.

23. What are three items that you can buy with $9?

24. If you bought 1 of each sale item, about how much money would you spend?

25. Sam bought 2 picture puzzles, a pack of football cards, and a dragon kite. José bought 3 car models. Who spent more money? How much more?

26. Karin wants a ship model, a delta kite, and a fish tank. So far, she has saved $10.50. How much more money will she need to save?

27. Leroy cuts lawns for $5 an hour. How many hours will he have to work to buy a racetrack and 2 race cars?

28. Bob received $6.34 in change from a $20 bill. How much money did he spend?

Problem Solving: Four-Part Process

The space shuttle re-enters the Earth's atmosphere at about 16,000 mi/h. By the time it lands, its speed is about 220 mi/h. How much slower is the shuttle's landing speed than its re-entry speed?

Understand the problem.

Know what to find.	What is the question?
Read for information.	At about what speed does the shuttle re-enter the atmosphere?
	At about what speed does it land?
	How much slower is the shuttle's landing speed than its re-entry speed?

Make a plan.

Ask questions.	Will the answer be greater than or less than 16,000 mi/h?
Decide what to do.	You can use an equation to represent the problem. $16{,}000 - 220 = n$

Carry out the plan.

Work through the plan.	$16{,}000 - 220 = n$
	$16{,}000 - 220 = 15{,}780$ mi/h

Look back.

Check that your answer is reasonable.	Make a quick estimate.
	$16{,}000 - 200 = 15{,}800$
	15,800 is close to 15,780.
Check that you answered the question.	Reread the question. Check your labels.

LESSON 2-11

GUIDED PRACTICE

Use the four-part process to solve. Explain how you got your answer.

1. During re-entry a spacecraft's heat shield reached a temperature of 4,200°F. The cabin temperature was 4,120°F cooler. What was the cabin temperature?

2. *Pioneer 10* traveled into outer space at 30,450 mi/h. *Voyager 1*'s speed was 13,120 mi/h faster. What was *Voyager*'s speed?

PRACTICE

Use the four-part process to solve these problems.

3. In 1926 Robert Goddard, an American scientist, launched the first liquid-fuel rocket. How many years ago was this launching?

4. The fastest rocket plane traveled at 4,534 mi/h. *Helios B*, a research spacecraft, reached a speed of 149,125 mi/h. How much faster did *Helios B* travel?

5. In 1963, Cosmonaut Valentina Tereshkova, the first woman in space, made 45 revolutions around Earth in a 71 h flight. Did each revolution take more than 2 h?

6. A person weighs about six times more on Earth than on the Moon. If your Moon weight is 12 lb, what is your Earth weight?

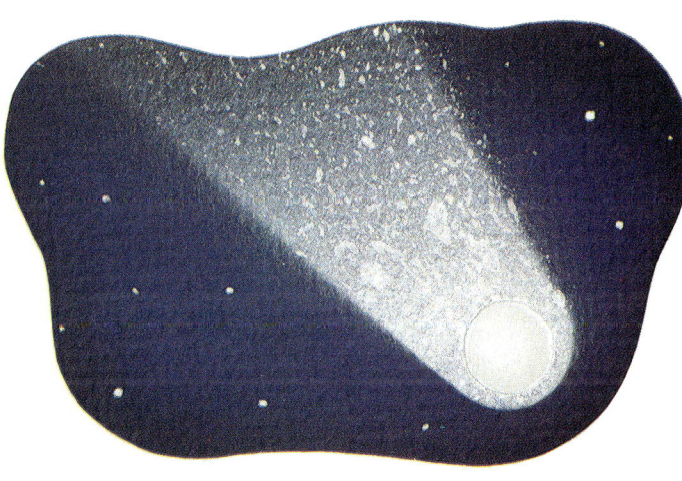

CHOOSE Choose any strategy to solve.

7. An astronomical unit (A.U.) is 93 million miles. The distance from the Sun to Saturn is almost 900 million miles. Is this more than or less than 10 A.U.'s?

8. Encke's Comet can be observed from Earth once about every 3 yr. Can the comet be seen more than a dozen times in a century? Justify your answer.

9. **IN YOUR WORDS** Distances in space are measured in light-years. If light travels at about 186,000 mi/s explain how to find the distance light travels in a year.

10. A 6 day 12 h 9 min Space Shuttle mission, piloted by Frederick Gregory, an African American, was launched at 11:02 P.M. on April 29, 1985. When did it land?

CHAPTER 2 59

CHAPTER CHECKUP

LANGUAGE & VOCABULARY

Choose the word that best completes the sentence.

addend   difference   sum   estimate   equation   variable

1. A __?__ can be used in place of a number in an equation.

2. When you subtract, you find the __?__.

3. Rounding and adding is one way to __?__ the answer.

4. The __?__ is the answer in addition.

TEST

CONCEPTS

Complete. Name the property that is used. *(pages 36–37)*

1. 25 + ▨ = 14 + 25
2. ▨ + 157 = 157

Is the answer *reasonable* or *unreasonable*? Use rounding and estimating to check. *(pages 38–39, 44–45)*

3. 42,386 + 5,702 + 893 = 48,981
4. 58,007 − 4,514 = 54,593

Use front-end estimation. Estimate the sum. *(pages 40–41)*

5. $3.17 + 5.85
6. $9.36 + $2.42 + + $7.38

SKILLS

Add. *(pages 38–39 and 52–53)*

7. 3,547
   256
   + 74,981

8. 5.2
   4.709
   + 3.28

9. 246 + 7,359 + 43 + 29,154

10. 4.52 + 3.8 + 25.427

Subtract. *(pages 42–45 and 54–55)*

11. 21,462
    − 8,985

12. 50,003
    − 24,328

13. 5.204
    − 0.876

14. $16,541 − $978

15. 47 − 3.28

Solve the equation. Use mental math. *(pages 46–47)*

16. $25 + n = 75$ **17.** $90 - n = 45$ **18.** $n \times 6 = 54$

Find the change. List the fewest coins and bills possible. *(pages 56–57)*

19. Owe: $15.75
    Give: $20

20. Owe: $3.36
    Give: $10

21. Owe: $12.28
    Give: $20.03

PROBLEM SOLVING

Solve. *(pages 58–59)*

22. It is 2,451 mi by plane from New York City to Los Angeles. By car, the distance from New York City to Los Angeles is 2,825 mi.

   a. What is the distance between New York City and Los Angeles if you are traveling by plane?

   b. What would you need to do to find how much farther it is from New York City to Los Angeles by car than by plane?

   c. How much farther is it from New York City to Los Angeles by car than by plane?

23. Jared bought a shirt for $20 and a pair of jeans for $30. How much did he spend?

24. The Wiltons' car cost $16,541. They made a $978 down payment. How much did they still owe after that?

LEARNING LOG

Write the answer in your learning log.

1. What do you know about the decimal point in a number when you don't see one?

2. Your mother's grocery bill was $24.27. She gave the cashier $25.02 instead of $25. Why did she do this?

EXTRA PRACTICE

Complete. Name the property that is used. *(pages 36–37)*

1. ▨ + 0 = 326
2. 45 + (36 + 51) = (45 + ▨) + 51
3. $a + b = $ ▨ $+ a$

Round to the underlined place. Estimate the sum or difference. *(pages 38–39 and 44–45)*

4. 4̲51 + 6̲38 + 2̲87
5. 8̲39 − 4̲73
6. $3̲,256 + $4̲,532 + $1̲,097 + $2̲,674
7. 32,040 − 4̲,168

Estimate. Use front-end estimation. *(pages 40–41)*

8. $7.29 + 5.64
9. $4.37 + 6.15 + 9.51
10. $4.49 + $2.06 + $1.45
11. $6.72 + $2.10 + $2.25

Find the sum or difference. *(pages 38–39, 42–45, 52–55)*

12. $3,097 + 2,984
13. 2,341 + 19 + 872
14. 47.28 + 5.46
15. 17.9 + 32.86 + 3.558
16. 18,325 − 4,897
17. 8,000 − 321
18. 94,070 − 4,836
19. 8.7 − 4.9
20. 45.64 − 9.7
21. 15 − 3.6

22. 341 + 296 + 572 + 138
23. $47,398 + $26 + $805 + $9,277
24. 913,455 − 25,786
25. 70,000 − 6,235
26. 32.5 + 48.7 + 53.6
27. 72.46 + 3.1 + 9.478
28. $13,005 − $16.52

Solve the equation. Use mental math. *(pages 46–47)*

29. $99 + n = 125$
30. $40 - n = 20$
31. $n \times 7 = 42$

List the change. Use the fewest coins and bills. *(pages 56–57)*

32. Owe: $6.25 Give: $10
33. Owe: $12.63 Give: $20
34. Owe: $4.32 Give: $10.02

Solve. *(pages 58–59)*

35. A 4-engine jet can travel 625 mi/h. A 3-engine jet can go 7 mi/h faster. What is a 3-engine jet's speed?
36. A plane ticket to Paris is regularly $600. Use mental math to find its price when it sold for $200 less.

ENRICHMENT

PALINDROME IS NOT A PALINDROME

Any number that reads the same when written backward or forward is a **palindrome.**

| 121 | 3,223 |
| 87578 | 14,741 |

Study these examples. They make numbers into palindromes.

```
   123
 + 321    ← 1st step
   444    ← palindrome
```

```
    489
  + 984    ← 1st step
  1,473
 + 3,741   ← 2nd step
  5,214
 + 4,125   ← 3rd step
  9,339   ← palindrome
```

```
    238
  + 832    ← 1st step
  1,070
 + 0,701   ← 2nd step
  1,771   ← palindrome
```

- Explain the steps needed to change a number into a palindrome.
- Use this method to see how many steps it takes to make 68, 489, 518, and 567 into palindromes.
- Try to find a two-digit number that cannot be made into a palindrome. Now try a three-digit number.
- Words such as **Otto** and **did** are palindromes. What other names or words can you think of that are palindromes?

LINKING LOGIC

Names	Favorite School Subject	Favorite Color
Alma	art	pink
Orji	math	red
Matt	gym	yellow
Lisa	reading	green
Jan	science	orange

Work with your group. Match the names, subjects and colors any way you wish. Then write clues like the one below so that another group can find how you matched the favorite color and subject of all five children. Trade with another group and solve.

Sample clue: Alma does not like science best. Her favorite color is not pink, red, or orange.

CUMULATIVE REVIEW

Find the number expressed in standard form.

1. 72 thousand, four
   - a. 72,400
   - b. 72,004
   - c. 72,040
   - d. none of these

2. 3 billion, 4 thousand, 516
   - a. 3,004,516
   - b. 3,000,400,516
   - c. 3,000,004,516
   - d. none of these

3. 52 and 9 thousandths
   - a. 52.009
   - b. 52.0009
   - c. 52.900
   - d. none of these

Identify the value of the underlined digit.

4. 9,2<u>7</u>5,368,014
   - a. 70
   - b. 7
   - c. 70,000,000
   - d. none of these

5. 32.5<u>4</u>7
   - a. 400
   - b. 0.04
   - c. 0.4
   - d. none of these

6. 806.<u>2</u>07
   - a. 0.02
   - b. 2
   - c. 0.002
   - d. none of these

Choose the correct number.

7. ▨ > 25,329
   - a. 25,392
   - b. 25,239
   - c. 25,229
   - d. none of these

8. 4.17 = ▨
   - a. 4.017
   - b. 4.170
   - c. 4.107
   - d. none of these

9. 3.7 < ▨
   - a. 3.57
   - b. 3.70
   - c. 3.72
   - d. none of these

Round to the underlined place.

10. <u>6</u>,845
    - a. 6,900
    - b. 6,000
    - c. 6,800
    - d. none of these

11. 741.8<u>4</u>2
    - a. 700
    - b. 741.84
    - c. 741.85
    - d. none of these

12. <u>9</u>87,012
    - a. 1,000,000
    - b. 900,000
    - c. 100,000
    - d. none of these

13. 4.0<u>3</u>
    - a. 4.3
    - b. 4.1
    - c. 4.0
    - d. none of these

14. 3<u>5</u>.61
    - a. 40
    - b. 35.6
    - c. 35
    - d. none of these

15. 8<u>1</u>5,426
    - a. 820,000
    - b. 810,000
    - c. 800,000
    - d. none of these

PROBLEM SOLVING REVIEW

Remember the strategies and types of problems you have had so far. Solve.

Problem Solving Check List
- Choosing the operation
- Using a pattern
- Using estimation
- Too much information
- Too little information
- Multistep problems
- Writing a problem

1. Of the 1,284 students at Westwood Middle School, 449 participate in school sports.
   a. How many students attend Westwood Middle School?
   b. What would you need to do to find out how many students do not participate in sports?
   c. To the nearest hundred, how many students do not participate in sports?

2. For lunch Marla bought a hamburger for $1.59 and juice for $.68. She gave the cashier a $5 bill. How much change did she get?

3. For Sunday's football game, tickets for 1,248 box seats, 42,306 regular seats, and 25,924 bleacher seats were sold. How many tickets were sold for Sunday's game?

4. A personal computer regularly sells for $1,378. Tanya bought it for $100 off the regular price. How much did she pay for the computer?

5. At the hardware store Mrs. Wong spent $7.89 for a hammer, $2.26 for nails, and $.72 for sandpaper. About how much did she spend?

6. Mr. Clay planted his garden as follows: Row 1: 3 plants; Row 2: 6 plants; Row 3: 10 plants; Row 4: 15 plants. If he continues in this pattern, how many plants will he put in the 5th row?

7. Jason worked 3.25 h on Friday and 6.5 h on Saturday. Marie worked 5.75 h on Saturday. How many more hours did Jason work than Marie on Saturday?

8. Use the information in the chart to write a word problem. Then solve it.

Country	Population
China	1,103,900,000
U.S.S.R.	289,000,000
United States	248,800,000

TECHNOLOGY

FILL IT UP

In the computer game "Master Math," you determine missing digits in addition and subtraction exericises.

Complete this activity to sharpen your strategies for finding missing digits. You do not need a computer. Fill in the missing numbers for each exercise.

1. ▆,38▆
    + 6,▆06
    ─────
    9,1▆1

2. 8,▆0▆
    − ▆,4▆3
    ─────
      724

3. ▆0,56▆
    + 19,▆91
    ──────
    4▆,1▆0

4. 96,▆03
    − ▆8,1▆4
    ──────
    68,39▆

SWITCH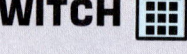

1. Switch one ⊕ to a ⊖ to get an answer of 10.65.

   4.6 ⊕ 5.8 ⊕ 2.9 ⊕ 3.15

2. Switch one ⊖ to a ⊕ to get an answer of 98.45.

   122 ⊖ 45.6 ⊖ 2.45 ⊖ 24.5

3. Switch one ⊕ to a ⊖ to get an answer of 145.4.

   69.71 ⊕ 25.38 ⊕ 46.25 ⊕ 54.82

ONE TOO MANY

Each addition exercise below has one addend too many. Decide which addend does not belong.

0.3	0.36	2.43
0.2	0.41	1.77
0.7	2.23	0.59
1.1	6.03	0.44
+ 0.6	+ 0.95	+ 1.88
1.8	7.75	5.34

Make up exercises like these for a classmate to solve.

Multiplication: Whole Numbers and Decimals   3

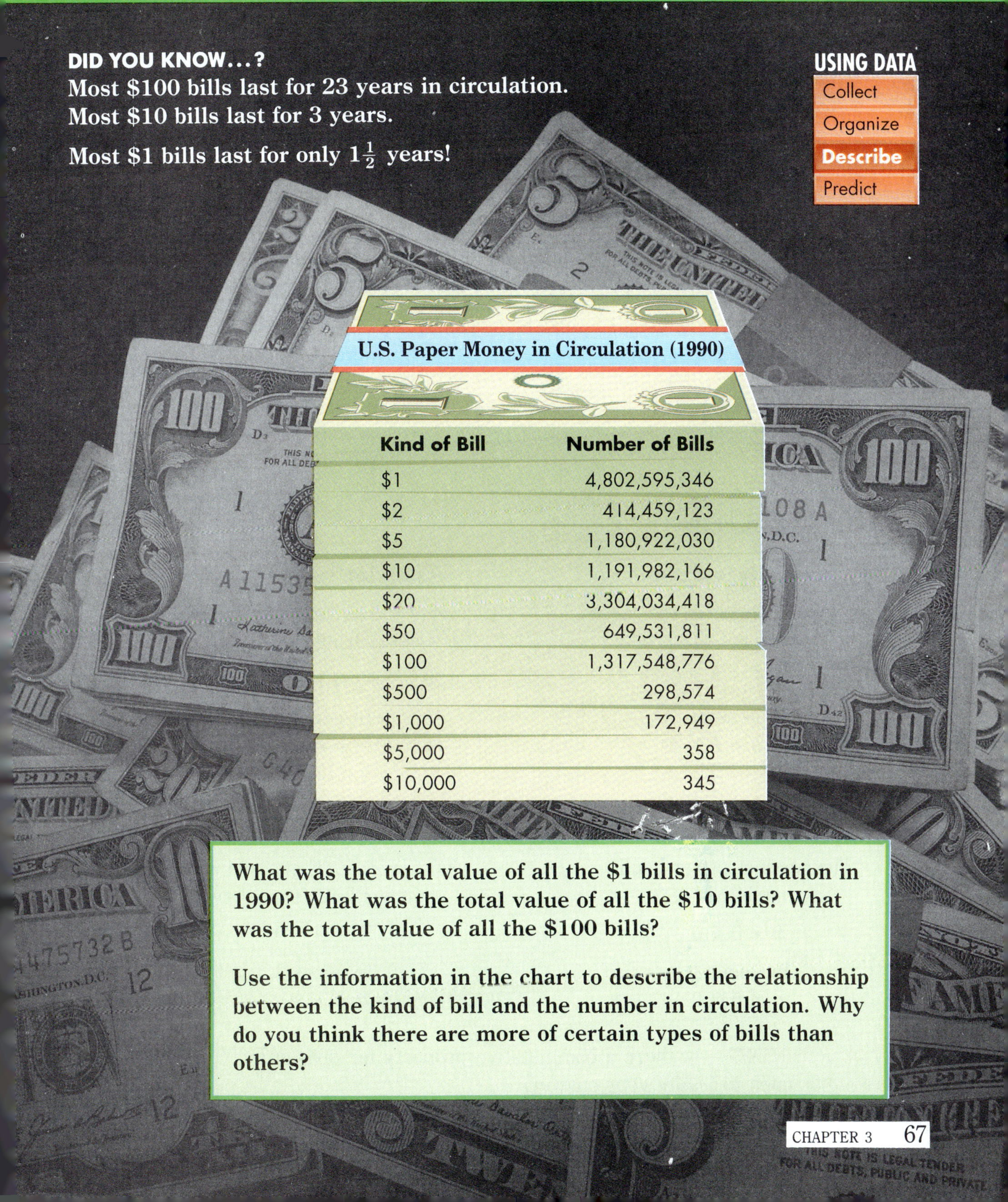

DID YOU KNOW...?
Most $100 bills last for 23 years in circulation.
Most $10 bills last for 3 years.
Most $1 bills last for only $1\frac{1}{2}$ years!

USING DATA
Collect
Organize
Describe
Predict

U.S. Paper Money in Circulation (1990)

Kind of Bill	Number of Bills
$1	4,802,595,346
$2	414,459,123
$5	1,180,922,030
$10	1,191,982,166
$20	3,304,034,418
$50	649,531,811
$100	1,317,548,776
$500	298,574
$1,000	172,949
$5,000	358
$10,000	345

What was the total value of all the $1 bills in circulation in 1990? What was the total value of all the $10 bills? What was the total value of all the $100 bills?

Use the information in the chart to describe the relationship between the kind of bill and the number in circulation. Why do you think there are more of certain types of bills than others?

CHAPTER 3   67

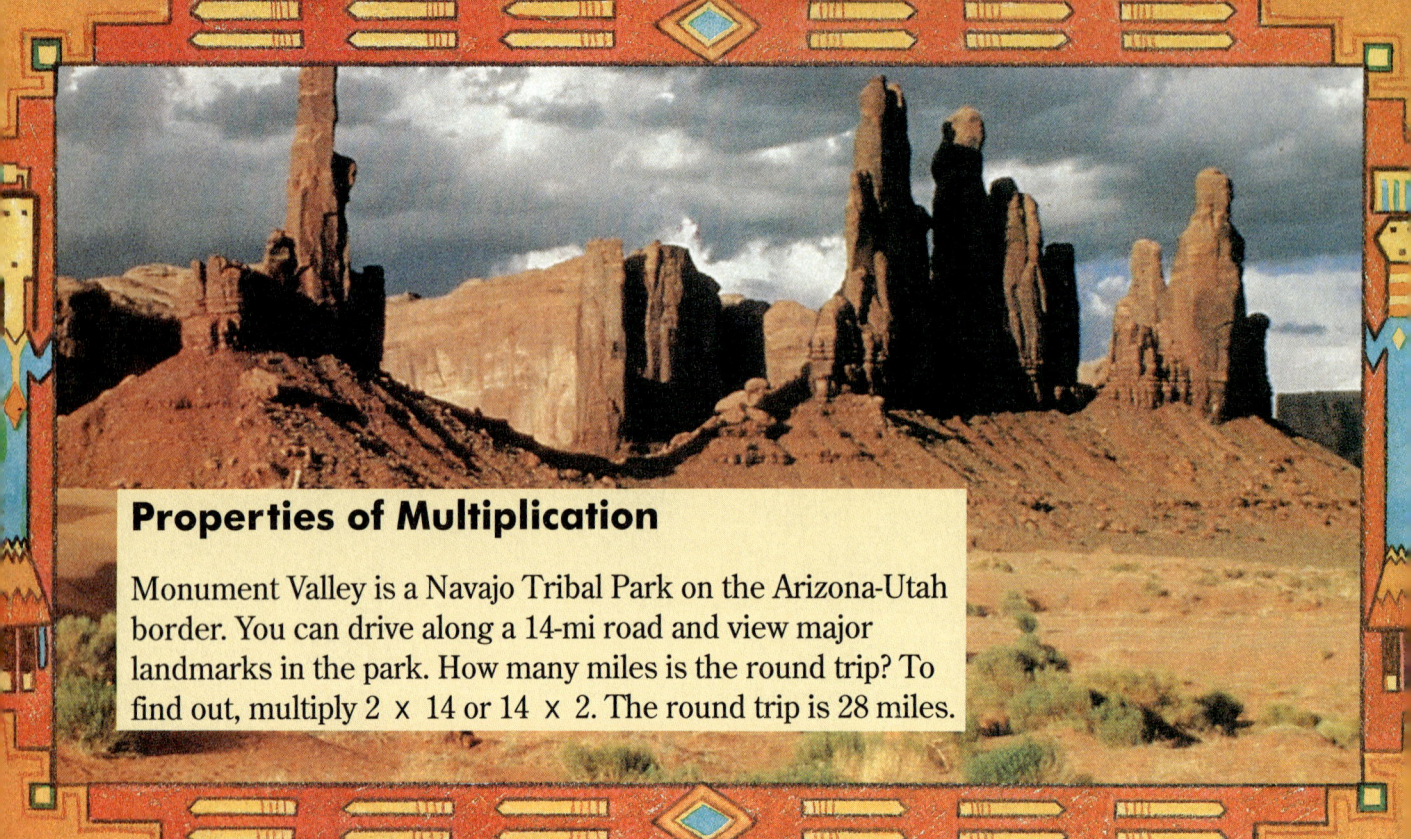

Properties of Multiplication

Monument Valley is a Navajo Tribal Park on the Arizona-Utah border. You can drive along a 14-mi road and view major landmarks in the park. How many miles is the round trip? To find out, multiply 2 × 14 or 14 × 2. The round trip is 28 miles.

The properties of multiplication can help you multiply mentally.

Commutative Property

Changing the order of the factors does not change the product.

$9 \times 8 = 8 \times 9$

In general:
$a \times b = b \times a$

Associative Property

Changing the grouping of the factors does not change the product.

$(9 \times 8) \times 6 = 9 \times (8 \times 6)$

In general:
$(a \times b) \times c = a \times (b \times c)$

Zero Property

The product of zero and any number is zero.

$9 \times 0 = 0$

In general:
$a \times 0 = 0$

Property of One

The product of one and any number is that number.

$1 \times 9 = 9$

In general:
$a \times 1 = a$

CRITICAL THINKING Is there a commutative property for division? Give two examples to justify your answer.

GUIDED PRACTICE

Write an example of the property.

1. zero
2. property of one
3. associative
4. commutative

PRACTICE

Which equation shows the property? Choose *a, b,* or *c.*

5. commutative
   a. $3 \times 4 = 6 \times 2$
   b. $(7 \times 1) \times 3 = 7 \times (1 \times 3)$
   c. $146 \times 9 = 9 \times 146$

6. associative
   a. $23 \times 3 + 1 = 70$
   b. $(3 \times 4) \times 5 = 3 \times (4 \times 5)$
   c. $98 \times 0 = 0 \times 98$

Complete. Name the property that is shown.

7. $(5 \times 3) \times 4 = 5 \times (\blacksquare \times 4)$
8. $126 \times 49 = 49 \times \blacksquare$
9. $13 \times 0 = \blacksquare$
10. $8 \times (23 \times 11) = (8 \times 23) \times \blacksquare$
11. $(41 \times 6) \times 5 = 41 \times (6 \times \blacksquare)$
12. $8 \times \blacksquare = 12 \times 8$
13. $(b \times \blacksquare) = a \times b$
14. $a \times \blacksquare = a$
15. $\blacksquare \times (b \times c) = (a \times b) \times c$
16. $(8 \times 2) \times 4 = 8 \times (\blacksquare \times 4)$

NUMBER SENSE Use the properties to answer *true* or *false* without multiplying.

17. $7 \times 29 < 28 \times 7$
18. $12 \times 265 = 265 \times 1 \times 12$
19. $(13 \times 20) \times 4 > 13 \times (20 \times 4)$
20. $16 \times (53 \times 0) = (17 \times 53) \times 0$

MENTAL MATH Use the properties to help you find the product mentally.

21. $3 \times 3 \times 7$
22. $0 \times 8 \times 6$
23. $4 \times 1 \times 25$
24. $8 \times 3 \times 2 \times 0$
25. $9 \times 2 \times 4 \times 5$
26. $5 \times 4 \times 5 \times 2$

PROBLEM SOLVING

27. A guided overnight camping trip in Monument Valley costs $45 per person, including meals, for a group of four or more. Three or fewer people are charged $60 per person. How much would a group of 5 pay?

28. A $1\frac{1}{2}$-h horseback ride through the park with a Native American guide costs $15 per person. How many such rides would be covered by a $50 gift certificate?

CHAPTER 3 69

One-Digit Multipliers

Each person in the United States produces an average of about 1,380 lb of trash each year. Imagine how much trash is produced by a family of 4 people!

- Ways You Can Help
- Recycle aluminum and glass containers
- Recycle newspapers
- Reuse grocery store bags
- Use fewer throwaways
- Recycle plastics

You can multiply to find that number.

Multiply the ones and tens. Rename as needed.	Multiply the hundreds and thousands. Rename as needed.
3 1,380 1,380 × 4 × 4 20	13 1,380 × 4 5,520

Remember to add on the numbers from renaming.

A four-person family produces about 5,520 lb of trash per year.

Estimate to check that your answer is reasonable.

1,380 → 1,400
× 4 → × 4
             5,600

Other examples:

   7   5 3
 708   2,905
 × 9   × 6
6,372   17,430

5,520 is close to 5,600

LESSON 3–2

MATH AND ECOLOGY

GUIDED PRACTICE

Multiply. Estimate to check.

1. 89 × 6
2. 296 × 7
3. 407 × 8
4. 5,395 × 3
5. 3,706 × 5

PRACTICE

Multiply.

6. 376 × 8
7. 54 × 4
8. 2,579 × 2
9. 428 × 7
10. 790 × 5
11. 3,708 × 8
12. 3,468 × 5
13. 407 × 9
14. 78 × 3
15. 8,672 × 6

16. 9 × 384
17. 6 × 46,003
18. 2 × 537
19. 6 × 72,398
20. 6 × 5,096
21. 7 × 7,560
22. 9 × 685
23. 7 × 5,736
24. 8 × 942
25. 3 × 8,697

MIXED REVIEW   Find the answer.

26. 65 + 3.49
27. 42.1 − 6.34
28. 25 − 6.8
29. 6.48 + 3.7 + 2.539

NUMBER SENSE   Without computing, decide whether or not the answer is reasonable. Explain how you know.

30. 8 × 705 = 375
31. 2 × 349 = 698
32. 7 × 3,492 = 2,444
33. 6 × 2,045 = 12,270
34. 9 × 2,370 = 12,330
35. 5 × 393 = 1,565

PROBLEM SOLVING

36. **IN YOUR WORDS**   On the average, for every dollar you spend on food, about 13¢ is for packaging costs. About how much is for other costs? Explain what you think some of these might be.

37. In the U.S. in 1990, 1 out of every 10 lb of trash — about 45,000 tons — was recycled per day. How many tons of trash per day was not recycled?

38. If recycling all copies of one Sunday's *New York Times* would save 75,000 trees, how many trees would be saved by recycling all copies of the Sunday New York Times this month?

Mental Math: Multiplying by Multiples of 10, 100, and 1,000

An adult's heart normally beats about 70 times per minute. At that rate, how many times does it beat in 1 h?

Multiply 60 × 70 to find out.

If you know that 7 × 6 = 42 and you know how to find a pattern, you can make a **MATH CONNECTION** to find the product of 70 × 60.

7 × 6 = 42
7 × 60 = 420
70 × 60 = 4,200

CRITICAL THINKING Will the number of zeros in the product always be the same as the total number of zeros in the factors? Explain.

GUIDED PRACTICE

Complete.

1. 7 × 1 = 7
   7 × 10 = ▨
   70 × 100 = ▨
   700 × 1,000 = ▨

2. 6 × 9 = 54
   6 × 90 = ▨
   60 × 900 = ▨
   60 × 9,000 = ▨

3. 6 × 5 = ▨
   6 × 50 = ▨
   60 × 500 = ▨
   60 × 50,000 = ▨

PRACTICE

Multiply mentally.

4. 6 × 20
5. 9 × 50
6. 4 × 600
7. 30 × 50
8. 4 × 80
9. 30 × 8
10. 9 × 30
11. 400 × 20
12. 3 × 300
13. 6 × 3,000
14. 600 × 10
15. 20 × 700
16. 2,000 × 30
17. 3,000 × 500
18. 44,000 × 20
19. 4 × 37 × 25

Compare. Write >, <, or =.

20. 20 × 80 ▨ 40 × 40
21. 120 × 40 ▨ 250 × 30
22. 200 × 1,100 ▨ 600 × 9,000
23. 980 × 1,000 ▨ 98 × 10,000
24. 600 × 5,000 ▨ 1,500 × 3,000
25. 6,000 × 600 ▨ 20 × 600 × 100

LESSON 3–3

Stroboscopic photography shows the motion of the runner at different intervals.

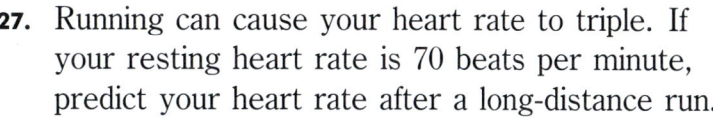

PROBLEM SOLVING

CHOOSE Choose mental math or paper and pencil to solve.

26. An elephant's heart beats about 30 times per minute. At that rate, how many times does it beat in 1 h? In 10 h?

27. Running can cause your heart rate to triple. If your resting heart rate is 70 beats per minute, predict your heart rate after a long-distance run.

28. A horse's heart beats about 3,400 times per hour. An elephant's heart beats about 1,800 times per hour. What is the difference between their heartbeat rates?

29. **CREATE YOUR OWN** Find the number of times your heart normally beats in 1 min. Then find how many times your heart beats after these activities:

    • walking for 1 min   • running for 1 min

    Make a display of your data.

Critical Thinking

An **analogy** shows a relationship between two pairs of words or numbers.

Example: Subtract : difference :: add : ▒
Read   *Subtract* is to *difference* as *add* is to *what?*

When you subtract, you find the difference. Find a word that makes the second pair of words relate in the same way.

Sum is the answer because when you *add* you get the *sum*.

Complete.

1. circle : sphere :: square : ▒
2. tens : tenths :: hundreds : ▒
3. 100 : one hundred :: 3,000 : ▒
4. 50 : 50 thousand :: 25 : ▒
5. $3 \times 50 : 3 \times 500 :: 5 \times 40 :$ ▒
6. minute : second :: hour : ▒

Estimating Products

A wild African elephant eats more than 770 lb of plants each day. At this rate, about how many pounds of plants does it eat in a week?

You can estimate the product to find out.

Round the factors.	Multiply.
770 → 800 × 7 → 7	800 × 7 5,600

The wild African elephant eats about 5,600 lb of plants in a week.

THINK ALOUD Why didn't we round 7 to 10?

Another example: 776 × 389
                   ↓   ↓
                   800 × 400 = 320,000

GUIDED PRACTICE

Estimate the product by rounding to the underlined place.

1. 6 × <u>8</u>8
2. 4 × <u>3</u>19
3. <u>3</u>7 × <u>4</u>3
4. <u>2</u>3 × <u>6</u>38
5. <u>6</u>89 × <u>9</u>43

PRACTICE

Estimate the product by rounding to the underlined place.

6. 9 × <u>3</u>7
7. 7 × <u>6</u>3
8. 9 × <u>7</u>34
9. <u>7</u>2 × <u>5</u>2
10. <u>7</u>3 × <u>1</u>5
11. <u>9</u>8 × <u>5</u>09
12. <u>1</u>87 × <u>6</u>3
13. <u>5</u>9 × <u>2</u>85
14. <u>1</u>23 × <u>8</u>74
15. <u>4</u>76 × <u>4</u>07
16. 6 × <u>3</u>78
17. 3 × <u>5</u>,729
18. <u>3</u>7 × <u>6</u>9
19. <u>3</u>4 × <u>7</u>,692
20. <u>5</u>15 × <u>3</u>09
21. <u>4</u>,989 × <u>4</u>10
22. <u>9</u>,257 × <u>3</u>12
23. <u>6</u>,890 × <u>7</u>,204
24. <u>9</u>,985 × <u>7</u>98

LESSON 3-4

MIXED REVIEW Use rounding to estimate the answer.

25. 485 − 276   **26.** 78 − 23   **27.** 826 + 907   **28.** 3,765 + 2,895

29. CRITICAL THINKING You can round numbers different ways to estimate. Predict which estimate will be closer to the actual product. Then use a calculator to check.

  a. 73 × 456:   70 × 500 or 100 × 456?
  b. 8 × 723:   10 × 723 or 8 × 700?
  c. How did changing the factors different ways affect the closeness of the estimate to the actual product? Write a sentence about what you noticed.

30. NUMBER SENSE If one factor is 98 and the estimated product is 2,000, can the other factor be 23?

CALCULATOR Find the missing digits.

31. 5▨ × 9▨ = 5,225   **32.** ▨4 × 9▨ = 5,130   **33.** ▨9 × 4▨ = 2,058
34. 5▨ × ▨6 = 4,992   **35.** 5▨ × 5▨ = 2,856   **36.** 8▨ × 2▨ = 1,936

PROBLEM SOLVING

Solve.

37. A wild elephant drinks almost 40 gal of water per day. At this rate, about how much water does it consume in 1 month?

38. Modoc, an Asian elephant, lived to be 78 yr old. She died in 1975. In what year was Modoc born?

Estimation

To find the **range** of a product, find an estimate above and below the actual product.

Round factors up.	Round factors down.
$90 × 60 = $5,400 ← 87 × 53 → $80 × 50 = $4,000	

The product is between $4,000 and $5,400.

Find the range.

1. 63 × 37   **2.** $8.29 × 45   **3.** 691 × 732   **4.** 4,328 × 634

Two- and Three-Digit Multipliers

A railroad tank car can hold 1,250 barrels of oil. How many barrels of oil can be transported on a train of 25 tank cars?

To find out, multiply 25 × 1,250.

If you know how to multiply two-digit numbers, how can you make a **MATH CONNECTION** to multiply three- or four-digit numbers?

Multiply by 5 ones.	Multiply by 2 tens.	Add.
1,250 × 25 6,250	1,250 × 25 6250 25000	1,250 × 25 6250 + 25000 31,250

Oil travels from the drilling site to a refinery. It then goes to market by trucks, railroad tank cars, ships, or pipelines.

On a train of 25 tank cars, 31,250 barrels of oil can be transported.

Estimate to check that your answer is reasonable.

$$\begin{array}{rcr} 1{,}250 & \rightarrow & 1{,}000 \\ \times 25 & \rightarrow & \times 30 \\ \hline & & 30{,}000 \end{array}$$

31,250 is close to 30,000.

GUIDED PRACTICE

Explain the errors in the example.
Rewrite the example correctly.

1. 87
    × 27
    ───
    569
    + 1640
    ─────
    2,209

2. 708
    × 32
    ────
    1426
    + 21640
    ──────
    23,066

3. $385
    × 559
    ────
    3474
    1925
    + 1925
    ─────
    $5,399

PRACTICE

Multiply. Use a calculator or pencil and paper.

4. 68
    × 38

5. 79
    × 59

6. $.77
    × 84

7. 385
    × 18

8. $666
    × 67

| 9. 578
×38 | 10. $8.67
×22 | 11. 509
×78 | 12. 604
×419 | 13. $9,658
×353 |

14. 636 × 641 15. $1.69 × 339 16. 7,080 × 555 17. 3,009 × 156

Multiply. Estimate to check.

| 18. 95
×46 | 19. $427
×23 | 20. 778
×329 | 21. $843
×796 | 22. 457
×287 |

If $n = 19$, is the number sentence true? Answer *yes* or *no*.

23. $37 \times n = 703$ 24. $n \times n \times n > 7{,}000$ 25. $n \times 196 < 3{,}000$

MIXED REVIEW Find the answer.

26. $18 \times 5 \times 0$ 27. $\$5{,}766 + \$86.08$ 28. $189.009 - 45$ 29. 36×38

CRITICAL THINKING Write *true* or *false*. Justify your answer.

30. The product of a 2-digit factor and a 2-digit factor never has 5 digits.

31. The product of a 2-digit factor and a 1-digit factor always has 2 digits.

PROBLEM SOLVING

CHOOSE Choose mental math, paper and pencil, or calculator.

32. In a recent year, home heating oil cost $.89/gal for 100 gal or $.80/gal for 300 gal. How much money could you have saved by buying 300 gal at once?

33. A large refinery can process 600,000 barrels of crude oil per day. How many barges with a 15,000-barrel capacity are needed to transport this amount of oil?

Mental Math

Use the **distributive property** to help you multiply.

$9 \times 52 = 9 \times (50 + 2)$
$ = (9 \times 50) + (9 \times 2)$
$ = 450 + 18$
$ = 468$

Break 52 apart to make multiplying easier.

Use mental math to solve.

1. 7×36 2. 8×47 3. 9×84 4. 71×30 5. 20×46

Zeros in the Multiplier

An advertising robot can be rented from ShowAmerica, Inc. for $905 per day. How much income will the company take in from the 1-day rental of 127 robots?

Multiply to find the answer.

Multiply by 5 ones.	There are no tens. Write a zero to show this.	Multiply by 9 hundreds.	Add.
127 × $905 635	127 × $905 635 00	127 × $905 635 114300	127 × $905 635 + 114300 $114,935

ShowAmerica, Inc. will take in $114,935 in rental income.

Estimate to check that your answer is reasonable.

    127 → 100
 × $905 → × $900
                    $90,000

THINK ALOUD Is $114,935 a reasonable answer? Explain why the estimate is lower than the actual answer.

GUIDED PRACTICE

Multiply. Estimate to check.

1. 389 × 407
2. 389 × 470
3. 756 × 805
4. 706 × 805
5. 593 × 8,000

PRACTICE

Multiply. Use a calculator or pencil and paper.

6. $4.17 × 304
7. $764 × 780
8. 308 × 205
9. 806 × 310
10. $9.31 × 707

11. 590 × 208
12. 318 × 20
13. 608 × 900
14. 978 × 750
15. 324 × 902
16. 1,493 × 305
17. 830 × 500
18. 766 × 503
19. $90.05 × 60
20. 7,007 × 707

21. 84 × 300
22. 804 × 60 × 50
23. 305 × 600 × 32
24. 6 × 3,068

Computers control this robot. It is used in the building of automobiles.

MIXED REVIEW Find the answer.

25. 1,936 + 18,990
26. 8,926 − 8,727
27. 5,492 × 99
28. 84.72 − 38
29. 7 × $104.37
30. $225.98 − $37.09

Complete the table.

31.
n	n × 30
8	
10	
12	

32.
n	n × 120
6	
19	
35	

CALCULATOR Use the numbers **19, 102, 274, 65, 26,** or **349**. Choose two numbers whose product is within the range.

33. 400–600
34. 1,000–2,000
35. 2,000–3,000
36. 15,000–20,000

PROBLEM SOLVING

CHOOSE Choose mental math, paper and pencil, or calculator to solve.

37. The Rhino Charger, an industrial robot, loads thirty-five 50-lb boxes of cans per hour. How many boxes are loaded in 8 h?

38. A fully assembled robot costs $2,499.85. You can save $1,200.95 if you assemble it yourself. How much does the robot kit cost?

39. When you order a custom-made talking robot, you must pay a deposit of half the price. If the price is $7,500, how much is the deposit?

40. Robopaint Company sold 1,200 spray-painting robots for $98,000 each. How much did the company take in on the sale of these robots?

CHAPTER 3   79

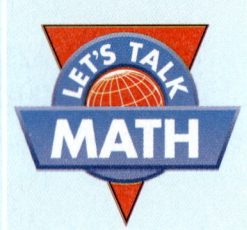

It's Show Time!

Problem Solving: Representing a Problem

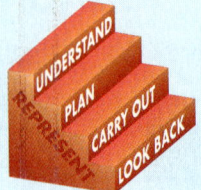

READ ABOUT IT

There are many ways to represent a problem. Sometimes making a model or drawing a picture can help you solve a problem.

> You are building a fence to enclose an area that is 9 ft by 12 ft. If the fence posts are to be 3 ft apart, how many posts do you need?

Sometimes writing an equation can help you solve a problem.

> With 6 players per team, how many teams can you make from a class of 24 students?

$6 \times n = 24$ or $24 \div 6 = n$

Sometimes finding a pattern can help you solve a problem.

> You put $.01 in your bank the first day and double the previous day's deposit each following day. After depositing $5.12 on the 10th day, how much money do you have in your bank?

Day	Deposit	Money in Bank
1	$ .01	$ .01
2	$ .02	$ .03
3	$ .04	$ .07
4	$ .08	$ .15
—	—	—
—	—	—
10	$5.12	$
11	$10.24	—

Sometimes organizing the information in a table can help you solve a problem.

> Frozen yogurt comes in vanilla, strawberry, and chocolate. The three toppings are coconut, bananas, and nuts. What combinations of one flavor and one topping can you make?

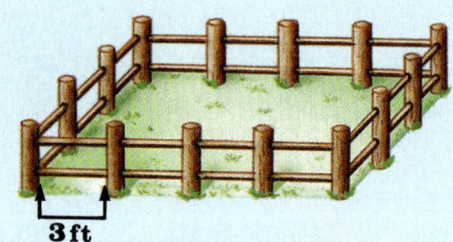

	Coconut	Banana	nuts
Vanilla			
Strawberry			
Chocolate			

LESSON 3–7

TALK ABOUT IT

How can you represent the problem?
Talk about the following things:
- Why did you represent the problem that way?
- Is there more than one way to represent the problem?

1. How can you arrange 12 square-foot tiles in a rectangle so that the perimeter (distance around the outside) is the least? The greatest?

2. You have some half dollars, quarters, dimes, nickels, and pennies. Use 10 coins to make $2.71. To make $1.50.

3. Yuma doesn't work on Mondays and Saturdays. If today is Monday, October 6, what are the dates of Yuma's next six days off?

4. Kesia lent Oni $60, spent $10, and still had $40 left. How much money did she start with?

5. A table seats 4 people on each side and 2 at each end. How many can sit at 4 of these tables placed end-to-end?

6. How many triangles do you get if you draw lines from 1 corner of a 10-sided figure to all the other corners? (*Hint*: Look for a pattern.)

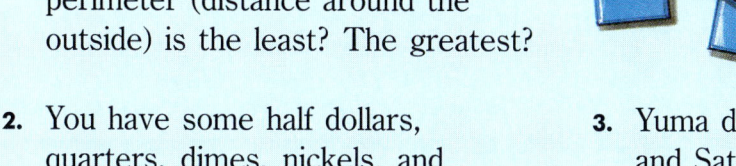

3-sided   4-sided   5-sided   6-sided
1 triangle 2 triangles ▇ triangles ▇ triangles

WRITE ABOUT IT

Work with a partner to write a problem that can be solved with the representation.

7.
Lemons	3	6	9	12
Lemonade (qt)	2	4	6	8

8. ($13.95 − $10) + $56.75 = ▇

9.
3	$6	$18
4	$4	$16
5	$3	$15

10.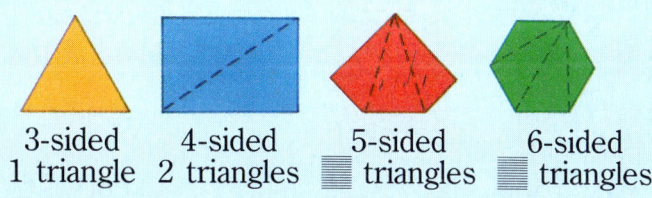

CHAPTER 3 81

MIDCHAPTER CHECKUP

LANGUAGE & VOCABULARY

Describe how the following properties are similar or different for addition and multiplication.

1. Commutative Property
2. Associative Property
3. Zero Property

QUICK QUIZ

Complete. Name the property that is used. *(pages 68–69)*

1. (48 × 52) × 37 = 48 × (■ × 37)
2. 524 × ■ = 524

Choose the best estimate. *(pages 74–75)*

3. 364 × 52   a. 15,000   b. 20,000   c. 2,400

Multiply. *(pages 70–73 and 76–79)*

4. 426
   × 8

5. 396
   × 24

6. 7,915
   × 603

7. 6,000 × 700

8. 8,103 × 649

Solve. *(pages 72–73)*

9. It took the Ortega family 20 h to drive from their home to Yellowstone National Park. They drove at an average speed of 50 mi/h.

   a. On the average, how many miles did the Ortegas drive each hour?
   b. What should you do to find the distance from the Ortegas' home to Yellowstone National Park?
   c. Use mental math to find the distance from the Ortegas' home to Yellowstone National Park.

LEARNING LOG

Write the answers in your learning log.

Explain to a friend how you remember what the Associative Property means.

MATH AMERICA

Did you know that the White House is part of the National Park system? It is the second smallest national park. It covers only 18.07 acres. Find the size of the largest park in your state. How much larger or smaller is it than the park which includes the White House?

CALCULATOR Find these products.

222,222 × 81

333,333 × 81

444,444 × 81

Use the pattern to find the product of 999,999 × 81.

Exploring Multiplication of Decimals

Just as you can use arrays to show multiplication of whole numbers, you can use arrays to show multiplication of decimals.

Let's build an array for 0.7×0.6.

Start with a single square.

Divide each side into 10 parts.

1. **NUMBER SENSE** How do you know that the array for 0.7×0.6 will take up less than one whole square?

2. How many sections are in the square after you divide each side into 10 parts?

3. One section makes up what part of the square?

LESSON 3-8

Now shade up 7 tenths (0.7) of the square and across 6 tenths (0.6) of the square.

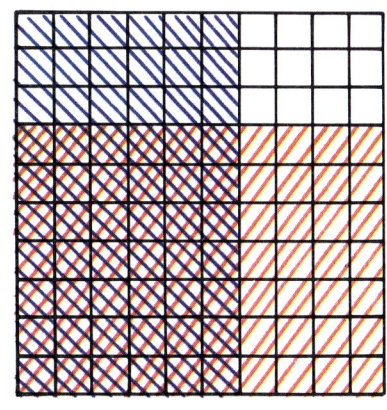

4. How many parts were shaded for both 0.7 and 0.6? (*Hint*: Look at the squares where the shading overlaps.)

5. What part of the whole square does the 0.7 × 0.6 array cover?

6. What is 0.7 × 0.6?

7. When you multiply two numbers that are tenths, what do you notice about the product?

Now let's think about making an array to multiply 0.7 and 0.25.

8. Why will the array for 0.7 × 0.25 take up less than one whole square?

9. **IN YOUR WORDS** Describe how you would need to divide the array in order to shade both 0.7 and 0.25?

10. Into how many parts will the whole array be divided?

11. How many of these thousandths will be shaded to show 0.7 × 0.25?

12. **IN YOUR WORDS** Explain why multiplying tenths by hundredths gives a product in the thousandths.

> **The number of decimal places in the product is equal to the total number of decimal places in the factors.**

Think about 0.575 × 2.3.

13. How many decimal places does the first factor, 0.575, have?

14. How many decimal places does the second factor, 2.3, have?

15. How many decimal places will the product have? How do you know?

16. Find the product.

Sometimes you need to write place-holding zeros so that the product has the correct number of decimal places.

17. **IN YOUR WORDS** The product of 0.2 × 0.03 has 3 decimal places. Is it 0.006, 0.060, or 0.600? Explain how you know. How could using an array help you decide?

Multiplying Decimals

Alexander Graham Bell's hydrofoil, *HD-4*, set a world record in 1918 by traveling at 61.6 knots. How many miles per hour was *HD-4* traveling when it set a world record?

A ship's speed is measured in knots. One knot = 1.15 mi/h. A ship moving at one knot travels 1.15 miles in an hour.

To change *HD-4*'s speed from knots to miles per hour, multiply 61.6 × 1.15.

Pictured here is the HD-4. Shown below is a modern hydrofoil from Hong Kong.

First, estimate the product.	Multiply as with whole numbers.	Use your estimate to place the decimal point.
61.6 → 62 ×1.15 → ×1 ——— 62	61.6 ×1.15 ——— 3080 6160 + 61600 ——— 70840	61.6 ×1.15 ——— 3080 6160 + 61600 ——— 70.840

Your estimate was 62. Place the decimal point to give a two-digit whole number.

HD-4 traveled at 70.84 mi/h.

Another example:

There are four decimal places in the factors. Write a zero to make four decimal places in the product.

$$\begin{array}{r} 5.1 \\ \times\,0.006 \\ \hline 0.0306 \end{array}$$

=========== GUIDED PRACTICE ===========

Place the decimal point in the product.

1. 3.2
   ×0.2
   ———
   64

2. 1.66
   ×0.04
   ———
   664

3. 5,431
   ×0.36
   ———
   195516

4. 0.002
   ×0.9
   ———
   18

Multiply. Explain the steps you followed.

5. 0.46 × 6.8
6. 0.006 × 0.029
7. 5.7 × 0.039
8. 0.38 × 0.045

LESSON 3-9

PRACTICE

Find the product. Use a calculator or pencil and paper.

9. 1.3
    ×2.4

10. 0.34
    ×0.07

11. 0.305
    ×1.6

12. 2.39
    ×0.5

13. 20.8
    ×3.5

14. 38.5
    ×4.9

15. 0.605
    ×0.231

16. 5.17
    ×0.75

17. 0.436
    ×0.08

18. 9.32
    ×4.87

19. 3.6 × 5.16
20. 1.05 × 0.009
21. 0.006 × 0.2
22. 1.9 × 2.091

23. 0.4 × 0.78
24. 2.4 × 0.036
25. 0.38 × 25.19
26. 4.35 × 0.261

MIXED REVIEW Find the answer.

27. $36.89 × 8
28. 0.935 − 0.048
29. 12.013 + 3.01
30. 191 × 6,000

31. 3.24 + 6 + 9.8
32. 0.63 − 0.348
33. 6,000,004 − 589,762

CALCULATOR Use a calculator to find at least one answer.

34. three decimals with a sum of 16.9
35. four decimals with a sum of 58
36. two decimals with a product of 6
37. two decimals with a product of 60

PROBLEM SOLVING

38. The top speed of the fastest sailboard is 32.35 knots. The top speed of the United States Navy's *SES-100B* hovercraft is 91.9 knots. How much faster is the hovercraft?

39. Passenger hydrofoils travel at speeds ranging from 30 to 55 knots. What are those speeds in miles per hour?

Critical Thinking

1. Look at the table to the right. What do you notice about the product when one factor is increased 10 times? Decreased 10 times?

2. What will be the product if 3,456 is multiplied by 20,000? By 0.00002?

3. Explain how to mentally find 0.006 × 789 if you know 0.6 × 789.

Factor	× Factor	= Product
3,456	2,000	6,912,000
3,456	200	691,200
3,456	20	69,120
3,456	2	6,912
3,456	0.2	691.2
3,456	0.02	69.12
3,456	0.002	6.912
3,456	0.0002	0.6912

Multiplying Whole Numbers and Decimals

Each $100 bill costs 2.5 cents to make. How many cents does it cost to make a sheet of 32 bills?

To find out, multiply the number of cents it costs to make 1 bill by the total number of bills.

Multiply as with whole numbers.	Place the decimal point in the product.
32 × 2.5 ─── 160 + 640 ─── 800	32 × 2.5 ─── 160 + 640 ─── 80.0

Think of the whole number as a decimal with zero decimal places.

It costs 80¢ to produce a sheet of $100 bills.

GUIDED PRACTICE

Find the product.

1. 67.7 × 12
2. 129 × 2.008
3. 475 × 7.05
4. 0.58 × 35

PRACTICE

Find the product. Use a calculator when needed.

5. 3.6 × 45
6. 0.003 × 21
7. 569 × 5.4
8. $5.22 × 73
9. 600 × 8.91

10. 6.11 × 35
11. 3.008 × 27
12. 48 × 2.009
13. 1,296 × 0.12
14. $9.50 × 905

15. $2.01 × 59
16. 396 × 0.007
17. 0.062 × 555
18. 98.6 × 300

19. 492 × $.53
20. 62.7 × 530
21. $.08 × 107
22. 0.428 × 536

Compare. Write >, <, or =.

23. 28 × 0.009 ▮ 12 × 0.021
24. 0.043 × 15 ▮ 14 × 0.34 + 0.1
25. 0.7 × 5 ▮ 50 × 0.007
26. 1.9 × 0.9 × 10 ▮ 19 × 0.09

MIXED REVIEW Find the answer.

27. 34,008 + 89 + 8,633   **28.** 98 × 30 × 400   **29.** $56,806 − $185.35

NUMBER SENSE Write *true* or *false*.
Do not compute. Explain your answer.

30. The product of 5 and 0.75 is less than 5.

31. The product 6 × 7 is greater than the product 0.9 × 100.

32. The product of 2 and 0.476 is less than 0.8.

33. The product of two decimals less than 1 is always less than 1.

At the Philadelphia mint, $100 bills are bundled into stacks of 10,000 sheets before cutting.

PROBLEM SOLVING

CHOOSE Choose mental math, calculator, or paper and pencil to solve.

34. If it costs the Philadelphia Mint $.05 to make 1 fifty-cent piece, how much does it cost to produce 1,000 fifty-cent pieces?

35. There are about 150 pennies in 1 lb. If your penny collection weighs 20 lb, about how much money do you have?

36. An 1879 silver dollar with a Carson City mint mark (CC) is worth $2,500. Without the CC, it's worth only $17. What is the difference between their values?

37. A nickel weighs 5.0 grams and a penny weighs 3.1 grams. How much money do you have in a pocketful of nickels and pennies weighing 41.0 grams?

CHAPTER 3   89

Problem Solving: Too Much, Too Little Information

Some problems give you more facts than you need. Others don't give enough facts.

In 1907, close to a million immigrants passed through Ellis Island's Main Building, 11,747 of them on one day. The number of immigrants that year was how many times the number the building was planned for.

- What are you supposed to find?
- Do you have all the facts you need?
- Are there extra facts given?

About 2 out of every 5 Americans have an ancestor who entered the country through Ellis Island.

GUIDED PRACTICE

Read the passage. Answer the questions.

Ellis Island reopened in 1990 as an immigration museum. The Main Building cost $1.5 million to build in 1900. The restoration project cost about 100 times as much. More than 20 million Americans contributed to this project.

1. **a.** The restoration cost was how many times the cost of building the Main Building?

   b. Which facts are *not* needed to find the cost of the restoration?

   c. How much did the restoration cost?

2. What additional fact is needed to find the total amount of money contributed by individual Americans?

PRACTICE

If there are too many facts, list the extra fact and solve.
If a fact is missing, write what other information is needed.

3. Ellis Island grew from a single building with 3 acres to a complex of 42 structures on 27 acres of landfill. By how many times did the size of the land increase?

4. About two out of every hundred immigrants were sent back home, many because of health problems. About how many people were sent home from Ellis Island?

Tell whether you can answer each question from the graph alone. If not, tell which of the additional facts is needed. Then estimate the answer.

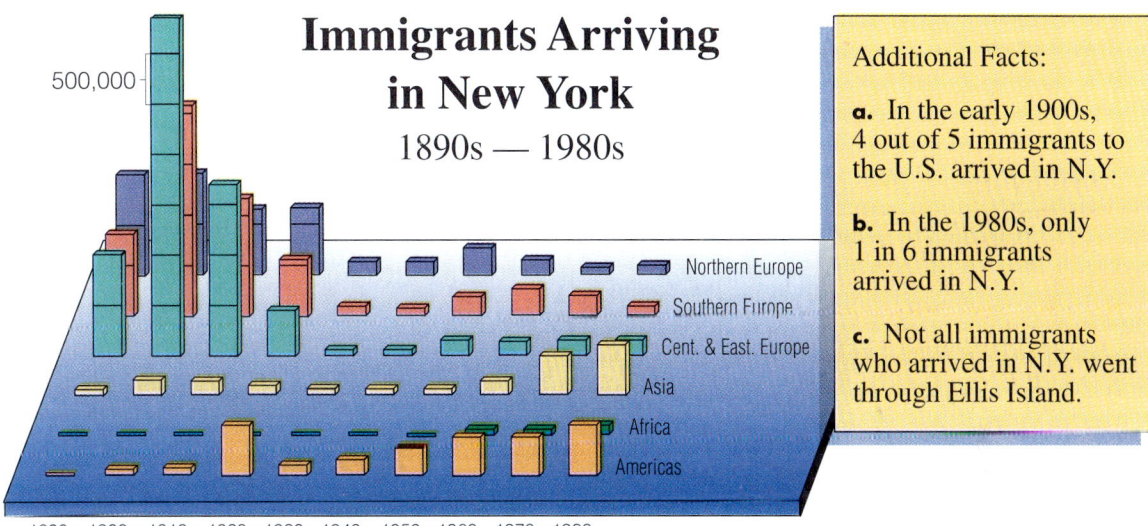

5. About how many immigrants came to New York from Europe in the decade 1900–1910?

6. About how many immigrants arrived in the country in the 1980s?

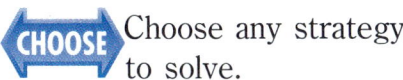 Choose any strategy to solve.

7. Most families traveled steerage class. To pay, many had saved for months or even years. A first-class ticket cost four times as much as a steerage ticket, and $40 more than the $40 cabin-class ticket. How much did a steerage ticket cost?

8. About twice as many of the immigrants left N.Y. as stayed there. On a typical busy day, 3,400 immigrants could have headed for places outside N.Y. Estimate how many immigrants arrived at Ellis Island on such a day.

CHAPTER CHECKUP

LANGUAGE & VOCABULARY

Explain the steps you would need to follow to multiply 0.07×0.003.

TEST ✓

CONCEPTS

Complete. Name the multiplication property that is used. *(pages 68–69)*

1. $27 \times \blacksquare = 46 \times 27$
2. $\blacksquare \times 1 = 439$
3. $a \times (b \times c) = (a \times b) \times \blacksquare$
4. $b \times 0 = \blacksquare$

Choose the best estimate. *(pages 74–75)*

5. 28×495 **a.** 15,000   **b.** greater than 15,000   **c.** less than 15,000

SKILLS

Multiply. *(pages 70–73, 76–79, 84–89)*

6. $4 \times 9{,}000$
7. $8{,}000 \times 300$
8. 7×93

9. 458×6
10. 64×59
11. 709×28
12. 825×374
13. 973×508

14. 0.2×0.6
15. 5.28×4.7
16. 0.34×0.09
17. 76×3.8
18. $\$4.97 \times 54$

PROBLEM SOLVING

Solve. If there is not enough information to solve the problem, name the information that is still needed. *(pages 90–91)*

19. Erika bought 8 rolls of film to take on vacation. Each roll of film has 36 exposures and costs $3.28.

   a. How many pictures can Erika take with each roll of film?

   b. What information do you need to find the total number of pictures Erika can take with the film she bought?

   c. What is the total number of pictures Erika can take with the film she bought?

20. Fifty-nine rows of chairs were set up for the concert. Each row contains 64 chairs. How many chairs were set up for the concert?

21. The library charges $.05 for each day a book is overdue. How much did Jenny have to pay for her overdue book?

LEARNING LOG

Write the answers in your learning log.

1. How is setting up an addition example with decimals different from setting up a multiplication example with decimals?

2. Your friend thinks that when you multiply decimals the answer will always be greater than the factors. Explain what is wrong with this thinking.

3. Explain what is meant by decimal places.

EXTRA PRACTICE

Complete. Name the multiplication property that is used. *(pages 68–69)*

1. $(4 \times 7) \times 3 = 4 \times (7 \times \blacksquare)$
2. $248 = 248 \times \blacksquare$
3. $512 \times 0 = \blacksquare$

Estimate the product by rounding to the place of the underlined digit. *(pages 74–75)*

4. $\underline{2}9 \times 6$
5. $\underline{3}17 \times 4$
6. $9\underline{2} \times 5\underline{8}$
7. $\underline{4}53 \times \underline{3}2$
8. $\underline{8},756 \times \underline{3}28$

Multiply. *(pages 70–73, 76–79, and 84–89)*

9. 500×6
10. 70×80
11. 300×900
12. $8,000 \times 600$
13. 4×38
14. 627×7
15. 6×409
16. $9 \times 5,923$
17. 46×52
18. $\$97 \times 33$
19. 185×64
20. 302×89
21. 743×276
22. 26×40
23. 873×600
24. $\$544 \times 308$
25. 396×780
26. $6,047 \times 509$
27. 3.4×7.6
28. 7.9×0.8
29. 0.25×0.043
30. 18.68×0.92
31. 0.607×5.8
32. 64×3.7
33. 2.8×71
34. $\$8.97 \times 54$
35. $1,234 \times 0.28$
36. 400×9.46

Solve. *(pages 80–81 and 90–91)*

37. A sandwich shop has four kinds of bread: white, rye, wheat, and pita. It has three kinds of meat: turkey, chicken, and ham. How many combinations of bread and meat can be made?

38. Adult tickets cost $12.75. Student tickets cost $8.50. One night, 392 adult tickets and 124 student tickets were sold. How much money was received for student tickets?

ENRICHMENT

What is Your Number?

Work with a partner. Use these steps to find a person's secret number. Try to find out how it works.

Tell your partner to mentally:

- Choose a number from 1–10.
- Multiply that number by 3.
- Add 1.
- Multiply the result by 3.
- Add the number you started with.
- Say the final answer aloud.

What was the secret number?

I Predict...

Multiply the largest two-digit number by itself. Then multiply the largest three-digit number by itself. Use your results to find the product of the largest four-digit number multiplied by itself.

Use patterns to think of the products. Multiply 11 by itself. Then try 111 multiplied by itself. Predict the product of 1,111 × 1,111.

Check with a calculator.

HOW MUCH DO YOU BURN UP?

Keep a record of all that you eat for one day. Record both the type and amount of food. Then find the number of calories for that type and amount of food. Find the total number of calories in the food you ate.

Keep a record of your activities that same day. Record the number of minutes you spend walking, jogging, reading, eating, sleeping, playing, and so on. Find the total number of calories you used during the day.

Now answer these questions.

- Did you use more calories or fewer calories than you ate? how many more or less?
- What happens if you use more calories one day than you eat that day? if you use fewer calories?

CUMULATIVE REVIEW

Find the number expressed in standard form.

1. eighty million, fifty
   a. 8,000,050
   b. 18,000,050
   c. 80,000,000,050
   d. none of these

2. thirty thousandths
   a. 0.0030
   b. 0.030
   c. 0.003
   d. none of these

3. 15 and 7 hundredths
   a. 0.157
   b. 15.700
   c. 15.07
   d. none of these

Choose the property represented in the equation.

4. $12 + 0 = 12$
   a. Associative
   b. Commutative
   c. Zero
   d. none of these

5. $2 \times 1 = 1 \times 2$
   a. Associative
   b. Commutative
   c. Zero
   d. none of these

6. $3 + (4 + 5) = (3 + 4) + 5$
   a. Associative
   b. Commutative
   c. Zero
   d. none of these

Solve.

7. $25{,}843 + 95 + 676 = n$
   a. 26,514
   b. 26,614
   c. 27,324
   d. none of these

8. $97{,}326 - 8{,}547 = n$
   a. 91,221
   b. 105,873
   c. 11,856
   d. none of these

9. $45{,}000 - 17{,}378 = n$
   a. 27,622
   b. 27,621
   c. 28,378
   d. none of these

10. $n + 85 = 100$
    a. $n = 15$
    b. $n = 85$
    c. $n = 185$
    d. none of these

11. $n \times 8 = 48$
    a. $n = 4$
    b. $n = 4$
    c. $n = 6$
    d. none of these

12. $78 - n = 75$
    a. $n = 3$
    b. $n = 5$
    c. $n = 8$
    d. none of these

Choose the correct change.

13. Owe $16.77; give $20.
    a. $4.25
    b. $3.23
    c. $3.75
    d. none of these

14. Owe $13.90; give $15.
    a. $.10
    b. $2.10
    c. $1.10
    d. none of these

15. Owe $7.30; give $10.05.
    a. $2.75
    b. $2.25
    c. $3.35
    d. none of these

PROBLEM SOLVING REVIEW

Remember the strategies and types of problems you have done so far. Solve.

Problem Solving Check List
- Choosing the operation
- Using equations
- Using a pattern
- Making a table
- Using estimation
- Multistep problems

1. Marlo's school starts on August 25. His first football game is 2 weeks later.
   a. On what date does Marlo's school start?
   b. What do you need to do to find the date 2 weeks later than August 25?
   c. What is the date of Marlo's first football game?

2. Michael practices his saxophone 40 min each day. If he starts practicing at 4:05 P.M., when will he be finished?

3. An afghan has this pattern of stripes: 3 red, 2 blue, 3 white, 2 blue, 3 red. What color stripe should be next?

4. Joe bought 23 packages of vegetable and flower seeds. Each package cost $.89. About how much did Joe spend for seeds?

5. Kara gives her cat 0.15 kg of food each day. How much food does Kara need to feed her cat for a week?

6. Sweatshirts come in 4 sizes: S, M, L, and XL, and in 4 colors: blue, white, black, and red. How many different combinations are available?

7. The Quans drove 92 km in 3 h before lunch. After lunch they drove 157 km. How many kilometers did they drive in all?

8. Sarah worked 3.5 h Friday and 6.25 h Saturday at $4.50 per hour. How much did she earn? (Round to the nearest cent.)

Write a problem that could be solved using the equation. Then solve.

9. $3 \times n = 15$

10. $18 - n = 8$

TECHNOLOGY

DECIMAL SENSE

In the computer game "Multi-Maze," you multiply decimals to find your way through a forest maze. Use this paper and pencil activity to practice.

Estimate the product. Then rearrange the digits in the box to predict the exact product. Check your product with a calculator.

1. 4.25 × 2.1 = ▓ | 2 5 8 9 |
2. 7.3 × 0.42 = ▓ | 0 3 6 6 |
3. 0.34 × 0.9 = ▓ | 0 0 3 6 |
4. 0.59 × 7.8 = ▓ | 0 4 2 6 |
5. 57 × 0.47 = ▓ | 2 6 7 9 |
6. 0.35 × 8.6 = ▓ | 0 1 3 |

NUMBER JUGGLING

Copy both sets of exercises. Find the least and the greatest possible products for each set using the numbers 6, 7, 8, and 9.

1. ▓ × ▓▓.▓
   ▓ × ▓▓.▓

2. ▓▓ × ▓.▓
   ▓▓ × ▓.▓

Now make up your own set of exercises using a different set of numbers. Challenge a classmate to complete them.

PREDICT THE ANSWER

Study the completed number sentences. Do you recognize the pattern? Predict each of the remaining answers *before* you calculate.

(*Hint:* Multiply before you add.)

1. 1 × 8 + 1 = 9
2. 12 × 8 + 2 = 98
3. 123 × 8 + 3 = ▓
4. 1,234 × 8 + 4 = ▓
5. 12,345 × 8 + 5 = ▓
6. 123,456 × 8 + 6 = ▓
7. 1,234,567 × 8 + 7 = ▓

Division: Whole Numbers and Decimals 4

DID YOU KNOW . . . ?

In one day the U.S. Postal Service prints 93,000,000 stamps and uses 3200 lb of glue on them. This many stamps would cover more than 22 acres of land!

YEAR OF PRICE CHANGE FIRST CLASS POSTAGE STAMP

Year	Price	Year	Price	Year	Price
1863	4¢	1958	4¢	1975	13¢
1883	6¢	1963	5¢	1978	15¢
1885	2¢	1968	6¢	1981	18¢
1917	3¢	1971	8¢	1981	20¢
1919	2¢	1974	10¢	1985	22¢
1932	3¢			1988	25¢

USING DATA
- Collect
- Organize
- Describe
- Predict

Use the information in the chart to make a bar graph of the price of a stamp. Start with 1863 and show the price every 20 yr to 1983.

Compare the chart and the graph. What do you notice about the rise in prices on each?

Why do you think someone would want to show the information in a graph instead of a chart?

Exploring Inverse Operations

Did you open the door on your way out of your home this morning? Did you then shut it? If you did, you used an **inverse** activity. Inverse activities undo each other.

Look at this flow chart.

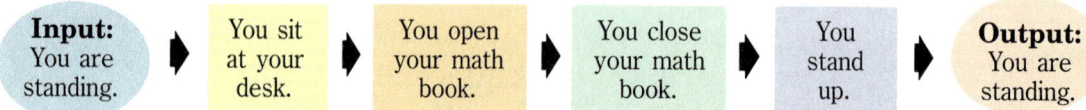

1. What inverse activity undoes "You sit at your desk"?

2. What other inverse activity is shown in the flow chart?

3. If "You pick up your pencil" were added to the flow chart, what inverse activity would undo it?

4. What do you notice about the input and the output when you use one or more inverse activities?

5. **IN YOUR WORDS** Is the statement, "You do not sit down." the inverse of "You sit down."? Explain.

6. **CRITICAL THINKING** Not all activities can be undone. For example, you cannot undo speaking. Name some others.

Complete the flow chart using inverse operations.

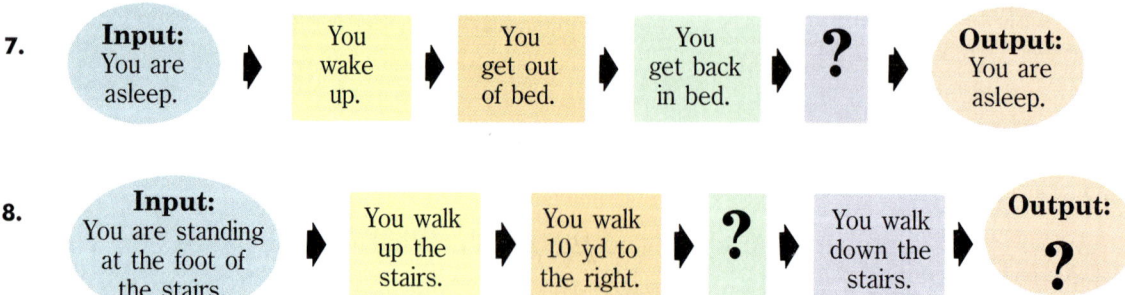

100   LESSON 4-1

In mathematics, when two operations undo each other, they are called **inverse operations**.

Input: 10 ▶ + 6 ▶ − 6 ▶ Output: 10

9. Since 6 was added to 10, what had to be done to the sum to make the output 10?

10. If the input number in this flow chart were changed to 8, what would the output number be?

11. If 4 were subtracted from the input 10, what inverse operation would you use to make the output 10?

12. If the input 10 were divided by 2, what inverse operation would you use to make the output 10?

Complete the flow chart using inverse operations.

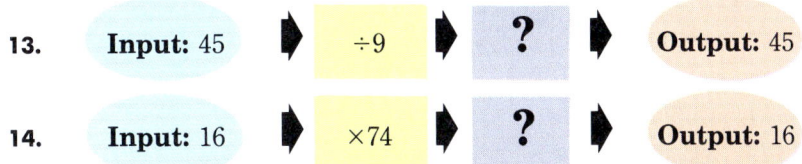

13. Input: 45 ▶ ÷9 ▶ ? ▶ Output: 45

14. Input: 16 ▶ ×74 ▶ ? ▶ Output: 16

Addition and subtraction are inverse operations. They undo each other.	Multiplication and division are inverse operations. They undo each other.

15. **CRITICAL THINKING** Explain how addition and multiplication are related. Are they inverses?

Complete the flow chart using inverse operations.

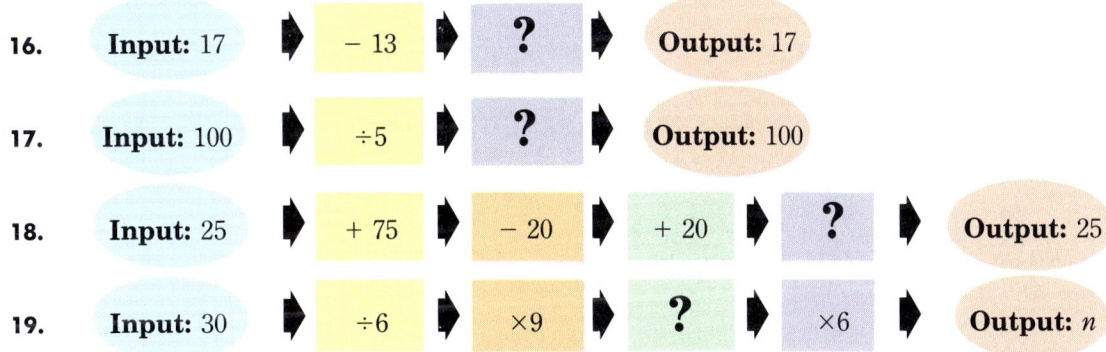

16. Input: 17 ▶ − 13 ▶ ? ▶ Output: 17

17. Input: 100 ▶ ÷5 ▶ ? ▶ Output: 100

18. Input: 25 ▶ + 75 ▶ − 20 ▶ + 20 ▶ ? ▶ Output: 25

19. Input: 30 ▶ ÷6 ▶ ×9 ▶ ? ▶ ×6 ▶ Output: n

20. **CREATE YOUR OWN** Write a flow chart that uses inverse operations. The input number should be 36.

One-Digit Divisors

Suppose you volunteer to help set up for Minnesota Ethnic Days, a twelve-day celebration held each year in Chisholm, Minnesota. You are given 109 flags and told to set them up in groups of 8. How many groups can you set up?

You can divide to find the answer.

$$\text{divisor} \rightarrow 8\overline{)109} \begin{array}{l} n \leftarrow \text{quotient} \\ \leftarrow \text{dividend} \end{array}$$

THINK ALOUD How many places will the quotient have? How do you know?

Divide 10 tens by 8. Multiply and subtract.	Divide 29 ones by 8. Multiply and subtract.	Multiply to check your answer.
1 $8\overline{)109}$ $\underline{-8}$ 2	$13\text{ R}5$ $8\overline{)109}$ $\underline{-8\downarrow}$ 29 $\underline{-24}$ $5 \leftarrow \text{Remainder}$	13 $\underline{\times8}$ 104 $\underline{+5}$ $109\checkmark$

You can set up 13 groups of 8 flags each.

THINK ALOUD Will every flag be used? Explain.

GUIDED PRACTICE

Tell whether the quotient is correct. Fix any error.

1.
$$\begin{array}{r} 86\text{ R}3 \\ 6\overline{)519} \\ -48 \\ \hline 39 \\ -36 \\ \hline 3 \end{array}$$

2.
$$\begin{array}{r} 118 \\ 3\overline{)355} \\ -3 \\ \hline 05 \\ -3 \\ \hline 25 \\ -25 \\ \hline 0 \end{array}$$

3.
$$\begin{array}{r} 8\text{ R}1 \\ 7\overline{)571} \\ -56 \\ \hline 1 \end{array}$$

4.
$$\begin{array}{r} \$183 \\ 2\overline{)\$336} \\ -2 \\ \hline 16 \\ -16 \\ \hline 06 \\ -6 \\ \hline 0 \end{array}$$

PRACTICE

Divide. Check your answer.

5. $2\overline{)562}$ 6. $3\overline{)47}$ 7. $8\overline{)94}$

8. $5\overline{)78}$ 9. $4\overline{)63}$ 10. $7\overline{)775}$

11. $3\overline{)550}$ 12. $9\overline{)295}$ 13. $6\overline{)838}$

14. $4\overline{)386}$ 15. $6\overline{)352}$ 16. $9\overline{)\$819}$

17. $2\overline{)179}$ 18. $7\overline{)5,975}$ 19. $8\overline{)5,836}$

20. $5\overline{)315}$ 21. $8\overline{)2,875}$ 22. $6\overline{)1,450}$

23. $3\overline{)5,920}$ 24. $7\overline{)\$25.76}$ 25. $4\overline{)9,685}$

26. $6\overline{)\$74.04}$ 27. $5\overline{)43,769}$ 28. $9\overline{)516,537}$

29. $\$252 \div 7$ 30. $2,625 \div 3$ 31. $3,834 \div 9$

Write $+$, $-$, $\times$, or $\div$ to make the equation true.

32. $2,254 \; \blacksquare \; 14 = 161$ 33. $1,575 \; \blacksquare \; 25 = 63$ 34. $\$19.98 \; \blacksquare \; 23 = \$459.54$

MIXED REVIEW   Find the answer.

35. $2.58 + 4.6$ 36. $789 - 398$ 37. $3,025 \times 1.8$ 38. $15.01 - 0.9$

39. 177.4×2.5 40. $557 \div 9$ 41. $n - 28.4 = 4.8$ 42. $62.4 + n = 100$

PROBLEM SOLVING

IN YOUR WORDS Use the chart to solve.
Try not to use paper and pencil.
Explain how you got your answer.

43. The same number of people enter the Bocce Ball Tournament and the International Games Derby. Which event has more teams?

44. The Polka Competition and the Bocce Ball Tournament have the same number of teams. Which event has more players?

45. If 149 people signed up for the Bocce Ball Tournament, could they all play exactly one game in the first round?

Minnesota Ethnic Days Field Events

Events	People per Team
Highland Games Tournament	9
Polka Competition	2
Bocce Ball Tournament	4
International Games Derby	6

CHAPTER 4   103

Dividing by Multiples of Ten

The front section of the Airphibian traveled as an ordinary car.

The Airphibian, built in 1946, was a combination car and plane. It flew as far as 390 mi on 30 gal of gas. How many miles per gallon did the Airphibian get in the air?

First, find the number of digits in the quotient. Then divide.

$$\frac{?}{30 \overline{)390}}$$

30 > 3
The quotient will not have hundreds.

$$\frac{?}{30 \overline{)390}}$$

30 < 39
The quotient will have tens.

Divide 39 tens by 30.	Multiply the divisor by the quotient. Subtract.	Divide 90 ones by 30. Multiply and subtract.
$\begin{array}{r}1\\30\overline{)390}\end{array}$	$\begin{array}{r}1\\30\overline{)390}\\-30\\\hline 9\end{array}$	$\begin{array}{r}13\\30\overline{)390}\\-30\\\hline 90\\-90\\\hline 0\end{array}$

The Airphibian got up to 13 mi/gal in the air.

GUIDED PRACTICE

Divide. Check by multiplying.

1. $20\overline{)570}$
2. $70\overline{)350}$
3. $80\overline{)9,779}$
4. $54,587 \div 400$

PRACTICE

Divide.

5. $40\overline{)89}$
6. $70\overline{)770}$
7. $50\overline{)750}$
8. $30\overline{)9,860}$
9. $60\overline{)2,152}$
10. $20\overline{)1,750}$
11. $30\overline{)197}$
12. $70\overline{)5,720}$
13. $700\overline{)41,890}$
14. $800\overline{)49,781}$
15. $600\overline{)4,756}$
16. $200\overline{)5,540}$

LESSON 4-3

17. 600⟌1,000   **18.** 900⟌3,101
19. 43,216 ÷ 200   **20.** 5,692 ÷ 700
21. 35,416 ÷ 800   **22.** 4,925 ÷ 200
23. 543 ÷ 60   **24.** 324 ÷ n = 8 R4
25. 100,000 ÷ 500   **26.** n ÷ 60 = 83 R20

MIXED REVIEW Find the answer.

27. 578
   − 87.6

28. 473.6
   + 15.08

29. 0.154
   − 0.08

30. 15.85
   × 9

31. 65.6
   × 128

32. 8,564 ÷ 60

IN YOUR WORDS Describe the rule.

33.
In	300	600	900	1,200	1,500
Out	10	20	30	40	50

34.
In	2	3	4	5
Out	101	151	201	251

CRITICAL THINKING Explain what happens to the quotient.

35. The dividend increases and the divisor remains the same.

36. The divisor increases and the dividend remains the same.

PROBLEM SOLVING

37. On December 17, 1903, Orville Wright made his first flight—a distance of 120 ft. He flew 10 ft/s. How long did the flight last?

38. In 1896 Samuel Langley's Aerodrome flew about 2,700 ft in 90 s. How many feet per second did he fly?

Mental Math

If both the divisor and dividend end in zeros, cross out the same number of zeros in each. Then divide mentally.

1,600 ÷ 200 → 16 ÷ 2 = 8

Divide mentally.

1. 100 ÷ 50   **2.** 8,000 ÷ 400   **3.** 90,000 ÷ 90,000   **4.** 90,000 ÷ 200

5. 30⟌210   **6.** 90⟌18,000   **7.** 5,000⟌75,000   **8.** 30⟌21,300

Two- and Three-Digit Divisors

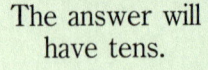

Greg LeMond won his third Tour de France bicycle race in 1990. He raced 2,112 mi in 21 stages. His total pedaling time was about 91 h. Find his average speed per hour.

The answer will have tens.	Divide 211 tens by 91. There are 29 tens left.	Divide 292 ones by 91. There are 19 ones left.
$\;\;\;\;n$ $91\overline{)2{,}112}$ $\begin{cases} 91>2 \\ 91>21 \\ 91<211 \end{cases}$	$\;\;\;2$ $91\overline{)2{,}112}$ $\underline{-\;182}$ 29	$\;\;23\;\;\text{R}19$ $91\overline{)2{,}112}$ $\underline{-\;182}$ 292 $\underline{-\;273}$ 19

His average speed was about 23 mi/h.

CRITICAL THINKING Is the speed closer to 23 mi/h or 24 mi/h? Explain.

Other examples:

$$\begin{array}{r} 42\;\text{R}143 \\ 200\overline{)8{,}543} \\ -\;800 \\ \hline 543 \\ -\;400 \\ \hline 143 \end{array} \qquad \begin{array}{r} 115\;\text{R}23 \\ 25\overline{)2{,}898} \\ -\;25 \\ \hline 39 \\ -\;25 \\ \hline 148 \\ -\;125 \\ \hline 23 \end{array} \qquad \begin{array}{r} 431\;\text{R}106 \\ 145\overline{)62{,}601} \\ -\;580 \\ \hline 460 \\ -\;435 \\ \hline 251 \\ -\;145 \\ \hline 106 \end{array}$$

GUIDED PRACTICE

THINK ALOUD Explain the steps to complete the example.

1. $\begin{array}{r} 5 \\ 16\overline{)912} \\ -\;80 \\ \hline 112 \end{array}$
2. $\begin{array}{r} 1 \\ 37\overline{)4{,}295} \\ -\;37 \\ \hline 5 \end{array}$
3. $521\overline{)1{,}357}$

PRACTICE

Divide. Use a calculator or pencil and paper.

4. $31\overline{)78}$
5. $42\overline{)\$168}$
6. $25\overline{)185}$
7. $70\overline{)1,477}$
8. $81\overline{)18,883}$
9. $32\overline{)460}$
10. $71\overline{)9,308}$
11. $57\overline{)\$8,778}$
12. $29\overline{)\$98,948}$
13. $66\overline{)13,928}$
14. $121\overline{)248}$
15. $32\overline{)999}$
16. $57\overline{)13,856}$
17. $600\overline{)5,912}$
18. $617\overline{)42,110}$
19. $14\overline{)10,094}$
20. $612\overline{)895}$
21. $89\overline{)\$3,916}$
22. $425\overline{)50,960}$
23. $589\overline{)67,158}$
24. $546 \div 12$
25. $7,429 \div 125$
26. $592 \div n = 16$
27. $576 \div n = n$

MIXED REVIEW Find the answer.

28. $7.068 + 5.92$
29. $185 - 75.5$
30. 65.8×45
31. $1,700 \div 60$

NUMBER SENSE Choose the best estimate — **3**, **30**, or **300**.

32. $56\overline{)167}$
33. $27\overline{)8,075}$
34. $324\overline{)93,847}$
35. $924\overline{)30,238}$

PROBLEM SOLVING

36. The Tour de France course is usually about 2,500 mi long, and the race lasts about 24 days. About how many miles are raced per day?

37. Greg LeMond's 1990 time of 90 h 43 min 20 s was 2 min 16 s faster than the time of the second-place winner. What was the second-place time?

Mental Math

If you can multiply by 100 and divide by a one-digit number, you can make a **MATH CONNECTION** to multiply by 25.

	Multiply by 100.	Divide by 4.	
$23 \times 25 = n$	$23 \times 100 = 2,300$	$4\overline{)2,300}$ = 575	$23 \times 25 = 575$

How do you use this method to multiply by 50?

Multiply mentally.

1. 40×25
2. 64×25
3. 42×50
4. 66×50
5. 82×25

CHAPTER 4

Zeros in the Quotient

To break away from the pull of the planet Mercury's gravity, a rocket would need to be traveling at least 9,619 mi/h. At that rate, how far would it travel in 1 min?

Divide the hundreds.	Divide the tens.	Divide the ones.
1 $60\overline{)9,619}$ $\underline{-60}$ 36	160 $60\overline{)9,619}$ $\underline{-60}$ 361 $\underline{-360}$ 1	160 R19 $60\overline{)9,619}$ $\underline{-60}$ 361 $\underline{-360}$ 19 $\underline{-0}$ 19

60 > 19
No ones are shared. Record a zero in the ones' place.

The rocket would travel about 160 mi in 1 min.

THINK ALOUD In the example above, why is it important to write a zero in the quotient?

Another example:

```
      108 R5          Multiply to check.
  45)4,865
    - 45                    108
      36                   × 45
    -  0                   540
      365               + 4,320
    - 360                4,860
        5              +     5   ooo Remainder
                        4,865  ✓
```

GUIDED PRACTICE

Divide. Multiply to check.

1. $7\overline{)1,442}$
2. $3\overline{)2,881}$
3. $28\overline{)5,852}$
4. $19\overline{)1,343}$

PRACTICE

Divide. Use a calculator, mental math, or paper and pencil.

5. $3\overline{)27{,}186}$
6. $5\overline{)5{,}062}$
7. $4\overline{)1{,}204}$
8. $11\overline{)10{,}785}$
9. $33\overline{)10{,}070}$
10. $54\overline{)33{,}804}$
11. $75\overline{)9{,}016}$
12. $92\overline{)9{,}880}$
13. $18\overline{)8{,}111}$
14. $36\overline{)18{,}150}$
15. $29\overline{)8{,}875}$
16. $71\overline{)50{,}300}$
17. $122\overline{)3{,}672}$
18. $15\overline{)6{,}138}$
19. $28\overline{)36{,}600}$
20. $345\overline{)71{,}775}$
21. $37{,}505 \div 5$
22. $1{,}778 \div 22$
23. $16{,}892 \div 28$
24. $39{,}091 \div 13$
25. $6{,}042 \div 12$
26. $2{,}496 \div n = 24$
27. $1{,}639 \div n = 52$ R27

MIXED REVIEW Find the answer.

28. $56.8 - 7.51$
29. $45\overline{)4{,}605}$
30. 15.5×0.8
31. $\$2.40 \div 4$

PROBLEM SOLVING

CHOOSE Choose estimation, paper and pencil, or calculator to solve.

32. In 1972 astronauts Cernan and Schmitt explored the moon for about 75 h. How many days and hours is that?

33. The space shuttle enters the atmosphere at 283 mi/min. What is its speed in miles per hour?

Mental Math

Many problems can be solved mentally using short division.

Divide the hundreds. Solve mentally.	Divide the tens. Solve mentally.	Divide the ones. Solve mentally.
1 $6\overline{)7^155}$ Trade the remainder for tens.	$1\,2$ $6\overline{)7^15^35}$ Trade the remainder for ones.	$1\,2\,5$ R5 $6\overline{)7^15^35}$

Use short division to solve.

1. $4\overline{)24{,}232}$
2. $8\overline{)16{,}042}$
3. $6\overline{)7{,}745}$
4. $30\overline{)270{,}090}$

CALL NOW

To order your
BOOMING PERSONAL STEREO

Just 1 payment of
$19.95

or

12 easy payments of
$2.25 each

Limited-Time Offer!

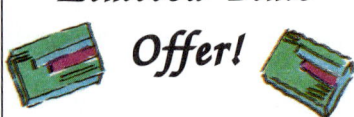

Blank Cassette Tapes
2 for **$6**   5 for **$10**

Problem Solving: Multistep Problems

Have you ever seen ads like this one for the personal stereo? What is the difference in the total price between a single payment and 12 payments?

Some problems need two or more steps to solve. It may help to ask yourself questions before solving.

What is the problem about?	Finding the difference between the single-payment price and the 12-payment price.
What information do you already know?	The stereo's price as a single payment is $19.95. The installment plan requires 12 payments of $2.25 each.
What do you need to do to solve?	**STEP 1:** Figure the total installment-plan price. **STEP 2:** Find the difference between the single-payment price and the installment-payment price.

Look back and check your answer.

GUIDED PRACTICE

Read the ad and answer the question.

1. a. How many tapes can you buy for $10?

   b. How can you find the cost of a single cassette at each price?

   c. Which offer is the better buy? Explain.

PRACTICE

Use the advertisements. Solve.

2. Tami bought 2 tapes on sale. One tape was regularly priced at $4.25, the other at $8. How much money did she save?

3. Manuel and Jade each ordered *Rocking in the 90s*. Manuel spent $5 more than Jade. What did Manuel buy if Jade's payment was $11.49?

4. Jason said that he could buy 2 cassettes, including the postage, for less than the price of the CD alone. Is he right?

Save! Save! Save!
All tapes
$1.79
Regularly
$3.50 to $8.00

Buy now!
Order your very own
Rocking in the 90s

Available only through this offer

Send **$9.99** for album, **$6.99** for cassette, or **$14.99** for CD.

Add $1.50 for postage and handling.

CHOOSE Choose any strategy to solve. Use the advertisement below.

5. Make a table that shows the cost of buying 1 CD, 2 CDs, and so on, up to 10 CDs.

6. How much less is the single-payment sale price than the original price of the stereo?

7. Niabi has 5 CDs at home. She bought 7 CDs at the clearance sale. How much money did she spend?

8. Kalia has $225. If she buys a CD-cassette player for $99, how many CDs can she buy to play on it?

Year-End Clearance Sale

Bargains! Bargains!
Prices slashed!

Super Surround Sound Stereo
Was $268   Now $225
Or just 10 payments of $25 each!

CD-Cassette Player
Was $139   Now $99

Special-purchase CDs
4 for $34   or $9 each

MIDCHAPTER CHECKUP

LANGUAGE & VOCABULARY

Write a sentence to explain how the following are related to each other.

dividend
divisor
quotient

QUICK QUIZ

Write *4*, *40*, or *400* as the best estimate. *(pages 106–107)*

1. 632)27,493
2. 37)13,049

Divide. *(pages 102–109)*

3. 457 ÷ 3
4. 5,760 ÷ 60
5. 7,689 ÷ 43
6. 7)$392
7. 92)9,718
8. 842)49,613
9. 57)63,889

Use the advertisement. Solve. *(pages 110–111)*

10. **PORTABLE CD PLAYERS**
Only $65.99
Or just 8 installment payments of $8.75 each

a. What is the amount of each installment payment?
b. What operation(s) would you use to find the difference in price between a single payment and 8 installment payments?
c. What is the difference in price between a single payment and 8 installment payments?

Write the answer in your learning log.

Sometimes you cross out zeros in the divisor and dividend to make dividing easier. Explain what you are really doing when you do this.

The Erie Canal, completed in 1825, is composed of 83 locks. Find out what locks are and how they are used in waterways. Is there a waterway in your state where locks are used? Find the number of locks used in the waterway, the average and total length of the locks.

Three music store clerks can unpack, price, and shelve 720 cassette tapes in 4 h. How many clerks are needed to get a shipment of 1,710 tapes unpacked, priced, and shelved in 6 h?

CHAPTER 4   113

Dividing a Decimal by a Whole Number

Using echolocation, a dolphin can find a small steel ball placed 72.2 m away. Traveling at its normal speed of 2 m/s, how long does it take a dolphin to reach a ball that far away?

Write the decimal point in the quotient.	Divide as with whole numbers.
$2 \overline{)72.2}$ with decimal point above	$\begin{array}{r} 36.1 \\ 2\overline{)72.2} \\ -6 \\ \hline 12 \\ -12 \\ \hline 02 \\ -2 \\ \hline 0 \end{array}$

It takes 36.1 s.

Another example:

$\begin{array}{r} 0.087 \\ 16\overline{)1.392} \\ -128 \\ \hline 112 \\ -112 \\ \hline 0 \end{array}$

You need to write a zero in the tenths' place to show that there are no tenths in the quotient.

Echolocation is similar to the sonar system ships use to find water depths.

114   LESSON 4–7

GUIDED PRACTICE

Divide. Multiply to check.

1. $6\overline{)127.8}$
2. $8\overline{)4.976}$
3. $24\overline{)\$238.80}$
4. $5\overline{)0.425}$

PRACTICE

Divide. Use a calculator, mental math, or paper and pencil.

5. $3\overline{)3.42}$
6. $5\overline{)367.5}$
7. $9\overline{)2.934}$
8. $8\overline{)7.52}$
9. $7\overline{)\$584.85}$
10. $6\overline{)0.48}$
11. $12\overline{)1,660.8}$
12. $23\overline{)1,664.28}$
13. $72\overline{)0.216}$
14. $95\overline{)\$11.40}$
15. $122\overline{)198.86}$
16. $248\overline{)972.16}$
17. $16.68 \div 2$
18. $\$12.32 \div 4$
19. $401.1 \div 7$
20. $113.88 \div 13$
21. $\$21.60 \div 30$
22. $382.823 \div 119$
23. $3,520 \div 40$
24. $48.02 \div 49$

Complete.

25. $17.55 \div n = 5$
26. $n \div 4 = 17.92$
27. $n \div 2.85 = 6$

MIXED REVIEW   Find the answer.

28. $645.8 + 29.38$
29. $74 - 8.16$
30. $848 \div 9$
31. 30.7×49
32. 16.4×23

ESTIMATE   Give a whole-number estimate for the quotient.

33. $37.02 \div 4$
34. $\$79.45 \div 7$
35. $124.92 \div 12$
36. $2,000.36 \div 110$

NUMBER SENSE   Look at the examples in order. Do not solve.

a. $335 \div 5$ b. $33.5 \div 5$ c. $3.35 \div 5$

37. Will the quotients increase or decrease? Explain.
38. By how much will the quotient increase or decrease each time? Explain.

PROBLEM SOLVING

39. Each dolphin at the New England Aquarium is fed about 11.3 kg of fish per day. How much fish is that per year?

40. A bottlenose dolphin is 4.5 m long. A killer whale — also a dolphin — is twice as long as the bottlenose. How long is the killer whale?

41. At an average speed of 5 km/h, how long does it take a dusky dolphin to travel 12.5 km?

42. At top speed, a dolphin can swim at 40.25 km/h. How many meters can it swim in $\frac{1}{2}$ h?

CHAPTER 4

Exploring Multiplying and Dividing by Powers of Ten

When you multiply 10 by itself any number of times you get a **power of 10**.

$10 \times 10 = 100$
$10 \times 10 \times 10 = 1{,}000$

CALCULATOR Multiply to find the power of 10.

1. $10 \times 10 \times 10 \times 10$
2. $10 \times 10 \times 10 \times 10 \times 10$
3. $10 \times 10 \times 10 \times 10 \times 10 \times 10$
4. $10 \times 10 \times 10 \times 10 \times 10 \times 10 \times 10$

5. What pattern do you notice about the number of tens that is multiplied and the number of zeros in the product?

Look for a pattern in the products.

$5 \times 10 = 50$	$45 \times 10 = 450$	$749 \times 10 = 7{,}490$
$5 \times 100 = 500$	$45 \times 100 = 4{,}500$	$749 \times 100 = 74{,}900$
$5 \times 1{,}000 = 5{,}000$	$45 \times 1{,}000 = 45{,}000$	$749 \times 1{,}000 = 749{,}000$

6. What pattern do you notice? (*Hint*: Look at how the pattern changes when you multiply by 10, by 100, and by 1,000.)

7. **IN YOUR WORDS** Write a rule for mentally multiplying a whole number by any power of ten.

We can also multiply a decimal by a power of 10.

$8.4 \times 10 = 84$	$0.25 \times 10 = 2.5$	$0.0048 \times 10 = 0.048$
$8.4 \times 100 = 840$	$0.25 \times 100 = 25$	$0.0048 \times 100 = 0.48$
$8.4 \times 1{,}000 = 8{,}400$	$0.25 \times 1{,}000 = 250$	$0.0048 \times 1{,}000 = 4.8$

8. When you multiply a decimal by 10, how can you find the product mentally?

9. How would you mentally find the product of a decimal and 100? A decimal and 1,000?

10. **IN YOUR WORDS** Explain what happens to the decimal point when multiplying a decimal by any power of 10.

Multiply mentally.

11. 16×10
12. 64×100
13. $789 \times 1{,}000$
14. 432×100
15. 7.45×10
16. 8.24×100
17. $0.076 \times 1{,}000$
18. 0.0974×10

116   LESSON 4-8

We can also divide by powers of 10.

6,000 ÷ 10 = 600	75,000 ÷ 10 = 7,500	325,000 ÷ 10 = 32,500
6,000 ÷ 100 = 60	75,000 ÷ 100 = 750	325,000 ÷ 100 = 3,250
6,000 ÷ 1,000 = 6	75,000 ÷ 1,000 = 75	325,000 ÷ 1,000 = 325

19. What pattern of zeros did you notice when a whole number was divided by a power of 10?

20. IN YOUR WORDS Write a rule for mentally dividing a whole number by a power of 10.

Divide.

21. 650 ÷ 10 **22.** 23,900 ÷ 100 **23.** 50,000 ÷ 10 **24.** 600,000 ÷ 1,000

We can also divide decimals by powers of 10.

642.5 ÷ 10 = 64.25	110.12 ÷ 10 = 11.012	32 ÷ 10 = 3.2
642.5 ÷ 100 = 6.425	110.12 ÷ 100 = 1.1012	32 ÷ 100 = 0.32
642.5 ÷ 1,000 = 0.6425	110.12 ÷ 1,000 = 0.11012	32 ÷ 1,000 = 0.032

25. What happened to the decimal point when a number was divided by 10?

26. How many places does the decimal point move when dividing by 100? by 1,000?

27. In which direction does the decimal point move when you divide by a power of 10?

Divide.

28. 4.5 ÷ 10 **29.** 234.74 ÷ 10 **30.** 0.4 ÷ 10

31. 0.033 ÷ 10 **32.** 971.4 ÷ 100 **33.** 6,200.9 ÷ 100

34. 84.52 ÷ 100 **35.** 0.1 ÷ 100 **36.** 4,456.23 ÷ 1,000

37. 479.2 ÷ 1,000 **38.** 4,320.28 ÷ 1,000 **39.** 4.625 ÷ 1,000

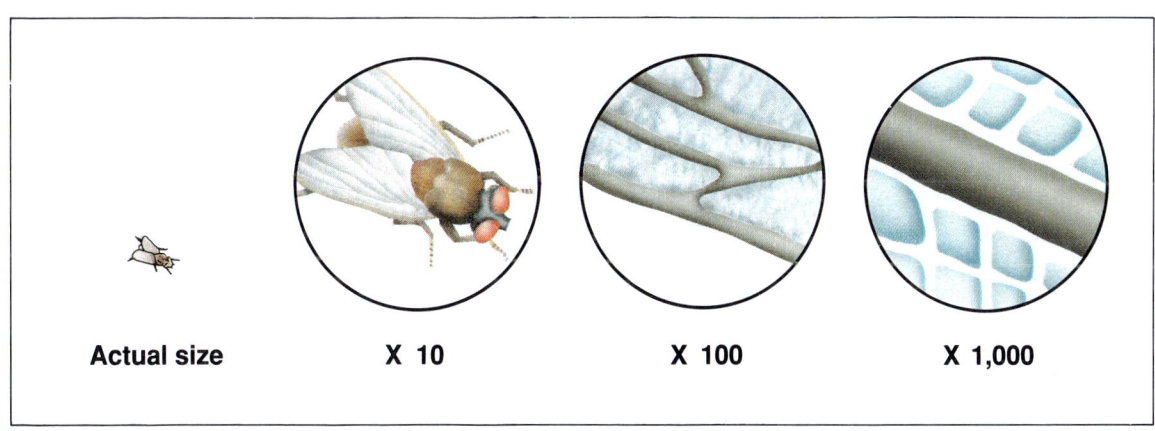

Actual size   X 10   X 100   X 1,000

Exploring Dividing by Decimals

Thinking about dividing with whole numbers can help you understand how to divide with decimals.

$5\overline{)10}$ $10\overline{)20}$

$20\overline{)40}$ $40\overline{)80}$

1. In the above examples, what is the pattern of the divisors? of the dividends?

2. How are the two patterns in Exercise 1 related?

3. What are the next two divisions in the pattern?

4. Work the examples mentally. What do you notice about the quotients?

Do you think the pattern will work with other numbers?

$6\overline{)120}$ $30\overline{)600}$

5. Do you multiply by the same number to get from 6 to 30 as you do to get from 120 to 600?

6. If $120 \div 6 = 20$, what do you think $600 \div 30$ equals? Explain how you know.

Multiplying both the dividend and the divisor by the same number does not change the quotient.

LESSON 4-9

Dividing with decimals works the same way.

Look at the example: $0.5 \overline{\smash{)}3.5}$.

7. By what number can you multiply the divisor to make it a whole number?

8. Why is it easier to multiply the divisor by 10 instead of 20?

9. What must you do to the dividend to keep the quotient the same?

10. Would the quotient have been the same if you had multiplied the dividend and the divisor by 100?

Look at the example: $0.37 \overline{\smash{)}126.17}$

11. By what number do you multiply 0.37 to make it a whole number?

12. What else must you do to keep the quotient the same?

Look at the pattern.

Decimal Divisor	Multiply by	Whole-number Divisor
675.4	10	6754
67.54	100	6754
6.754	?	?

13. How do you know by which number to multiply the divisor, 6.754, to get a whole number?

14. **IN YOUR WORDS** Write a rule for choosing the number by which a decimal divisor and its dividend should be multiplied.

Sometimes the divisor is a decimal but the dividend is a whole number: $2.5 \overline{\smash{)}3,525}$

15. How can you make the divisor a whole number?

16. Is the quotient of $3,525 \div 2.5$ the same as the quotient of $3,525 \div 25$? Explain.

17. What do you need to do to the dividend so that the quotient does not change?

18. **CRITICAL THINKING** What do you know that allows you to make a **MATH CONNECTION** to divide with decimals?

CHAPTER 4   119

Dividing by Decimals

Zina is a speed walker. The length of her pace is half a yard. How many paces does she take in 1 mi?

10 kilometer walk, World Track and Field Championships, Rome, 1987

Since you need to find how many 0.5-yd paces there are in 1 mi, or 1,760 yd, divide 1,760 by 0.5.

Multiply the divisor by 10 to make it a whole number.	Multiply the dividend by 10.	Divide as with whole numbers.
$0.5.\overline{)1,760}$ $\begin{array}{r} 0.5 \\ \times 10 \\ \hline 5 \end{array}$	$5\overline{)17,60\,0}$ $\begin{array}{r} 1,760 \\ \times 10 \\ \hline 17,600 \end{array}$	$\begin{array}{r} 3,520 \\ 5\overline{)17,600} \\ -15 \\ \hline 26 \\ -25 \\ \hline 10 \\ -10 \\ \hline 00 \\ -00 \\ \hline 0 \end{array}$

Zina takes about 3,520 paces per mile.

Another example:

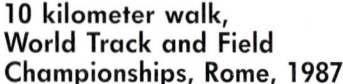

Multiply both the divisor and the dividend by 100.

$$\begin{array}{r} 0.043 \\ 75.\overline{)3.225} \\ -300 \\ \hline 225 \\ -225 \\ \hline 0 \end{array}$$

You need to write a zero in the tenths' place to show there are no tenths in the quotient.

120   LESSON 4-10

GUIDED PRACTICE

THINK ALOUD Use the example $0.82\overline{)24.026}$ to answer.

1. By what number do you multiply 0.82 to make it a whole number?
2. By what number do you multiply the dividend?
3. Does multiplying 0.82 and 24.026 by 100 change the quotient? Why?
4. Is 300 a reasonable estimate for the quotient? If not, why not?

Divide.

5. $0.36\overline{)12.6}$
6. $1.5\overline{)6}$
7. $0.6\overline{)0.0192}$
8. $0.008\overline{)4.032}$

PRACTICE

Divide.

9. $0.28\overline{)2.24}$
10. $0.7\overline{)0.021}$
11. $0.05\overline{)4}$
12. $0.008\overline{)0.10056}$
13. $0.35\overline{)2.1}$
14. $0.4\overline{)0.0008}$
15. $9\overline{)3.33}$
16. $0.034\overline{)14.586}$
17. $1.1\overline{)22.66}$
18. $0.2\overline{)43}$
19. $3.2\overline{)0.0672}$
20. $0.22\overline{)0.19866}$
21. $0.9\overline{)40.5}$
22. $0.25\overline{)75}$
23. $0.004\overline{)1.72}$
24. $0.238\overline{)0.50932}$
25. $42 \div 1.2$
26. $21.875 \div 6.25$
27. $n \div 0.22 = 0.0396$

MENTAL MATH Divide mentally.

28. $5\overline{)0.25}$
29. $8\overline{)4.24}$
30. $0.4\overline{)1.608}$
31. $0.2\overline{)0.006}$

MIXED REVIEW Find the answer.

32. 6.428×2.2
33. $50 - 6.45$
34. $39.888 \div 0.72$
35. 5.067×0.4

PROBLEM SOLVING

36. Tor's walking pace is 1.5 ft. How many paces does he walk per mile? (*Hint:* 1 mi = 5,280 ft)
37. Joni walks 0.6 mi to school each day in 12 min. At that rate, how far can she walk in 1 hr?

Critical Thinking

Is the quotient greater than or less than the dividend when you divide a whole number or a decimal by

1. a whole number greater than one?
2. a decimal less than one?

CHAPTER 4

Remainders in Decimal Division

At an exchange rate of 8 Hong Kong dollars to 1 United States dollar, how much is 1 Hong Kong dollar worth in U.S. currency?

Sometimes when you divide with decimals you need to write zeros in the dividend.

Write a decimal point in the dividend. Write a zero in the tenths' place. Divide.	Write a zero in the hundredths' place in the dividend. Divide.	Write a zero in the thousandths' place. Divide.
$$\begin{array}{r} 0.1 \\ 8\overline{)1.0} \\ -8 \\ \hline 2 \end{array}$$	$$\begin{array}{r} 0.12 \\ 8\overline{)1.00} \\ -8 \\ \hline 20 \\ -16 \\ \hline 4 \end{array}$$	$$\begin{array}{r} 0.125 \\ 8\overline{)1.000} \\ -8 \\ \hline 20 \\ -16 \\ \hline 40 \\ -40 \\ \hline 0 \end{array}$$

One Hong Kong dollar is worth $.125 or about $.13 in United States currency.

Sometimes the division does not come out evenly no matter how many zeros you write.

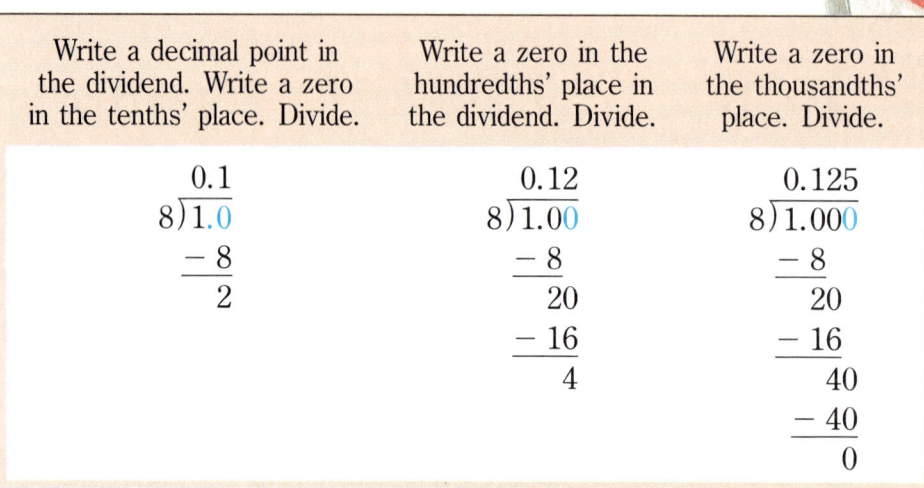

$0.6\overline{)0.01}$

$$\begin{array}{r} 0.016 \\ 6.\overline{)0.100} \\ -6 \\ \hline 40 \\ -36 \\ \hline 4 \end{array}$$

You will keep getting a remainder.

To find the quotient to the nearest hundredth, divide to the thousandths' place and then round.

0.016 rounds to 0.02

THINK ALOUD To round a quotient to the nearest tenth, to which decimal place must you divide? Explain.

───── **GUIDED PRACTICE** ─────

Divide. Write zeros until the division comes out evenly.

1. $4.5\overline{)1.8}$
2. $0.75\overline{)0.006}$
3. $6\overline{)15}$
4. $0.8\overline{)0.09}$

LESSON 4-11

Find the quotient to the nearest tenth or dime.

5. $7 \div 0.3$
6. $\$45.38 \div 7$
7. $0.254 \div 0.11$
8. $50 \div 12$

PRACTICE

Divide. Write zeros until the division comes out evenly.

9. $5\overline{)12}$
10. $0.25\overline{)0.73}$
11. $1.6\overline{)2}$
12. $0.8\overline{)0.5}$
13. $6\overline{)0.003}$
14. $4\overline{)11}$
15. $0.5\overline{)0.74}$
16. $40\overline{)3.1}$

Find the quotient to the nearest hundredth or cent.

17. $0.35\overline{)0.849}$
18. $6\overline{)\$15.62}$
19. $7.6\overline{)5.58}$
20. $0.012\overline{)0.4028}$

Find the quotient to the nearest tenth or dime.

21. $3.17 \div 0.7$
22. $491 \div 0.98$
23. $\$104 \div 56$
24. $0.736 \div 0.009$

Divide. Round to the nearest hundredth if necessary.

25. $0.4\overline{)0.11}$
26. $0.9\overline{)8.56}$
27. $0.5\overline{)0.875}$
28. $0.15\overline{)0.091}$
29. $3.5\overline{)0.014}$
30. $0.08\overline{)0.37}$
31. $0.003\overline{)0.00829}$
32. $1.1\overline{)43.4}$

MIXED REVIEW Find the answer.

33. $1.006 + 0.39$
34. 2.68×0.04
35. $n \times 5 = 2.5$
36. $4.07 + n = 10$

PROBLEM SOLVING

Use the table to solve. Round decimal answers to an appropriate place.

37. About how many Brazilian cruzados equal 1 dime?

38. What is the United States equivalent of 5 British pounds?

39. Which currencies are about half the value of another?

40. About how many yen is 1 Canadian dollar worth?

Foreign Exchange Rates

Currency	Worth in U.S. dollars
Brazilian cruzado	0.0141
British pound	1.8715
Canadian dollar	0.8697
Japanese yen	0.006658

CHAPTER 4

Averages

Diameters of 5 Saguaro Cactuses (in centimeters)

31 35 44 46 48

The flower of the saguaro cactus is the state flower of Arizona. Of the cactuses given, what is the **average** diameter?

> **To find the average, or mean, of any group of numbers:**
> 1. **Find the sum of the numbers.**
> 2. **Divide by the number of addends.**

To find the average diameter:

Add the diameters.	Divide by the number of cactuses.
31 46 44 48 + 35 ─── 204	40.8 5)204.0 − 20 ─── 04 − 00 ─── 40 − 40 ─── 0

The average diameter is 40.8 cm.

CRITICAL THINKING If **35** is changed to **75** in the above example, what happens to the average?

124   LESSON 4-12

GUIDED PRACTICE

Find the average.

1. 2, 4, 6, 8, 10
2. 3, 4, 4, 8, 9, 11
3. $87, $35, $94
4. 1.98, 2.8, 1.26, 2.04

5. **THINK ALOUD** If your test scores are 85, 85, 87, and 83, how can you know without calculating that your average is 85?

6. When finding an average, what can you do if the division does not come out evenly?

PRACTICE

Find the average. Use a calculator or pencil and paper.

7. 32, 46, 57
8. 3, 6, 9, 10, 1, 5, 8, 3, 0
9. 4.216, 0.6194
10. 58.3, 58.5, 58.2, 58.2
11. $12.15, $25, $9.87, $37.50
12. 0.02, 0.002, 0.2
13. 7.918, 6.08, 0.498, 8.18, 2.408
14. 1.5, 2.03, 4.6, 0.96, 2.7

Find the average. Round as directed.

15. 7, 16, 9, 4, 11, 5 (nearest tenth)
16. $5.75, $4, $3.28 (nearest cent)

NUMBER SENSE Find the missing addend mentally.

17. addends: 30, n; average: 40
18. addends: 50, 55, 70, n; average: 60
19. addends: 35, 45, n; average: 30
20. addends: n, 60, 50, n; average: 50

21. **CALCULATOR** Find the average height in inches of the students in your class.

PROBLEM SOLVING

CHOOSE Choose estimation, pencil and paper, or calculator to solve.

22. An average wingspan for an Arizona Red Spotted Purple butterfly is 70 mm. The range is from 56 mm to 84 mm. Write five sizes that will average 70 mm.

23. The populations of Arizona's three largest cities are 789,704; 330,537; and 152,453. Find the average population. Why should it be rounded to a whole number?

24. The average high temperatures in Phoenix for six months were 64°F, 70°F, 75°F, 84°F, 93°F, and 102°F. Make a bar graph. Use it to estimate the average over these months.

25. About 6 out of every 100 Arizona residents are Native Americans. An almanac lists the Arizona population as 2,718,425. About how many are Native Americans?

CHAPTER 4   125

Estimating Quotients

The Pony Express's fastest full-route delivery took 175 h. The trip was made to deliver a copy of President Abraham Lincoln's first address to Congress. About how many days did the trip take?

You can use **compatible numbers** to estimate quotients. Compatible numbers are numbers that are easy to divide.

$$24\overline{)175}^{\,n}$$ *24 is close to 25.*

The Coming and Going of the Pony Express, by Frederic Remington. Mail deliveries from St. Joseph, Missouri to Sacramento, California began in 1860.

$$25\overline{)175}^{\,7}$$ The trip took about 7 days.

CRITICAL THINKING   What happens to the estimate if you use 20 as the divisor?

George Monroe, an African American, was among the famous Pony Express riders. He later became an important stagecoach driver—ranked as the best in California and perhaps the entire West.

Other examples:

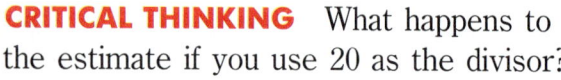

GUIDED PRACTICE

Use compatible numbers to estimate the quotient.

1. $7\overline{)477} \rightarrow 7\overline{)490}$
2. $4\overline{)35.5} \rightarrow 4\overline{)36}$
3. $2.2\overline{)8.900} \rightarrow 2\overline{)8}$
4. $2{,}800 \div 90$
5. $14.8 \div 3.1$
6. $198.7 \div 24.3$

7. **THINK ALOUD**   Could you use other compatible numbers? If so, give an example.

PRACTICE

Use compatible numbers to estimate the quotient.

8. $6\overline{)351}$
9. $9\overline{)62.1}$
10. $12\overline{)4{,}739}$
11. $1.5\overline{)498}$
12. $29\overline{)156}$
13. $4\overline{)2.308}$
14. $45\overline{)0.468}$
15. $2\overline{)0.58}$

MATH AND HISTORY

LESSON 4-13

16. $\$180.35 \div 6$
17. $97.6 \div 24$
18. $2 \div 0.9$
19. $5 \div 0.9$
20. $106 \div 2.1$
21. $745 \div 0.8$
22. $397 \div 9.5$
23. $4{,}910 \div 15.8$
24. $2.196 \div 4$
25. $993 \div 10.5$
26. $239 \div 8.3$
27. $79 \div 0.12$

The actual quotients for Exercises 28–33 are shown below. Estimate with compatible numbers to choose *a*, *b*, *c*, or *d*.

a. 39   b. 62   c. 23   d. 71

28. $81\overline{)3{,}159}$
29. $59\overline{)4{,}189}$
30. $82\overline{)1{,}886}$
31. $39\overline{)897}$
32. $2.1\overline{)149.1}$
33. $3.2\overline{)198.4}$

PROBLEM SOLVING

34. It took 10 24-h days for Pony Express riders to travel 1,966 mi. Estimate how many miles per day the riders averaged.

35. To send a $\frac{1}{2}$-ounce letter by Pony Express cost $5 in 1860. To mail the same letter in 1985 would have cost 25¢. How much more was the Pony Express price?

Estimate

You can use compatible numbers to find a range of quotients.

Multiply the divisor and the dividend by 10 to get a whole number.	A compatible number can be less than the dividend.	A compatible number can be greater than the dividend.
$0.7\overline{)5.7}$ $7.\overline{)57.}$	$7\overline{)56}$ gives 8	$7\overline{)63}$ gives 9

The quotient is between 8 and 9.

Write the range of quotients.

1. $1.1 \div 0.4$
2. $0.76 \div 0.8$
3. $0.063 \div 6$
4. $0.91 \div 0.05$
5. $0.28 \div 3$

Think Again!

Problem Solving: Looking Back

READ ABOUT IT

Looking back at your work after solving a problem can be helpful. Check whether your answer is reasonable. A checklist like the one shown can help you.

Reasonable Answer Checklist
- ✔ Did you answer the question?
- ✔ Did you calculate correctly?
- ✔ Is your answer labeled with the right units?
- ✔ Does your answer need to be rounded to a whole number to make sense?

Louise Bennett, a Jamaican storyteller, has performed often on television, radio, and the stage. Bobby Norfolk's stories are mostly American, African, and European tales.

TALK ABOUT IT

Work with a group. Is the given answer *reasonable* or *unreasonable*? If *unreasonable*, explain why.

1. A shuttle bus can carry 18 people at once to and from the parking lot. How many trips does it take to bring 75 people to the festival?
   Answer: 4 R3 trips

2. Four people paid equal amounts for lunches that totalled $12.15. How much did each person pay?
   Answer: $3.0375

3. A weekend rate of $150 was set for parents with children. Will a family of 2 adults and 2 children save money with this rate?
   Answer: no

4. Each story-telling session lasts 30 min. There is a $\frac{1}{2}$ h break between the sessions. How many sessions can there be on Saturday?
   Answer: 24 sessions

5. How much does an adult save by buying 1 Saturday-Sunday ticket instead of 2 separate tickets?
   Answer: $8

6. How many hours is the festival open on Sunday?
   Answer: 4:30 P.M.

7. Tickets for Saturday's ghost-story contest cost $5 for adults and $4 for children. How much will an adult and 2 children pay for both the festival and the contest?
   Answer: $.78

NATIONAL STORYTELLING FESTIVAL

Jonesborough, Tennessee

Registration Fees

Days of Attendance	Adult	Over 65	Child
3-day pass	$52	$39	$26
Friday only	$28	$21	$14
Saturday only	$33	$25	$16
Sunday only	$13	$10	$6
Saturday and Sunday	$43	$32	$21

Festival Hours

Friday............................1 PM to 10 PM
Saturday......................10 AM to 10 PM
Sunday........................10 AM to 4:30 PM

WRITE ABOUT IT

8. How can you check whether your answer is reasonable?

CHAPTER 4   129

Investigating Gasoline Prices

When you examine the prices for gasoline at service stations, you will notice that there are different prices posted for a gallon of gasoline. Also, different gas stations charge different prices for the same kind of gasoline.

Work with a partner or in a small group. Use a calculator to help find the answers.

- Make three charts like this.

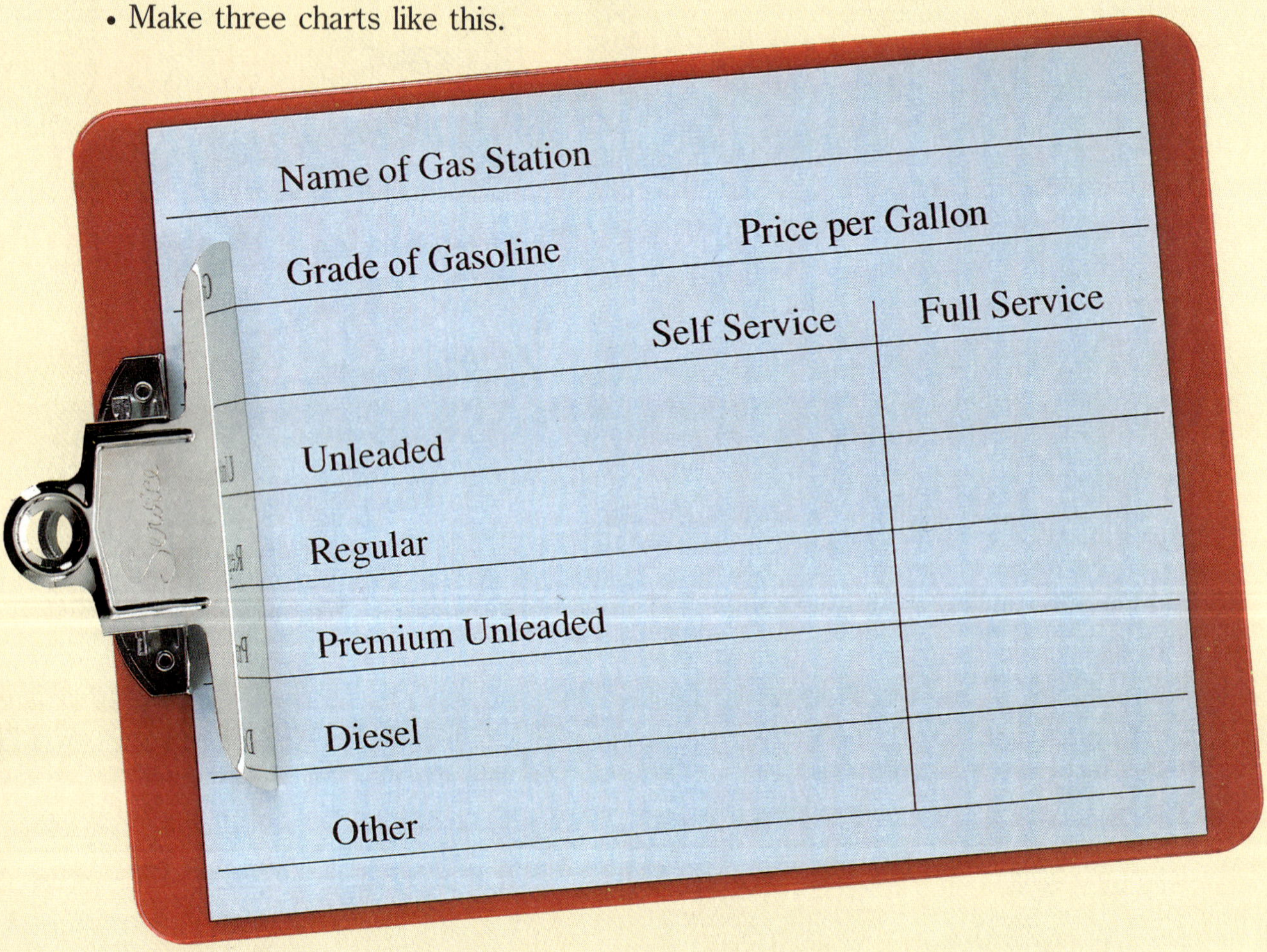

Name of Gas Station		
Grade of Gasoline	Price per Gallon	
	Self Service	Full Service
Unleaded		
Regular		
Premium Unleaded		
Diesel		
Other		

Visit or call three different gas stations. At each station record the *grade*s of gasoline sold, such as "regular", "unleaded", or "premium unleaded."

Next record the price per gallon for each grade sold, such as $1.50^9 for regular. Be sure to record whether there is a different price for self-service and for full service.

130   LESSON 4–15

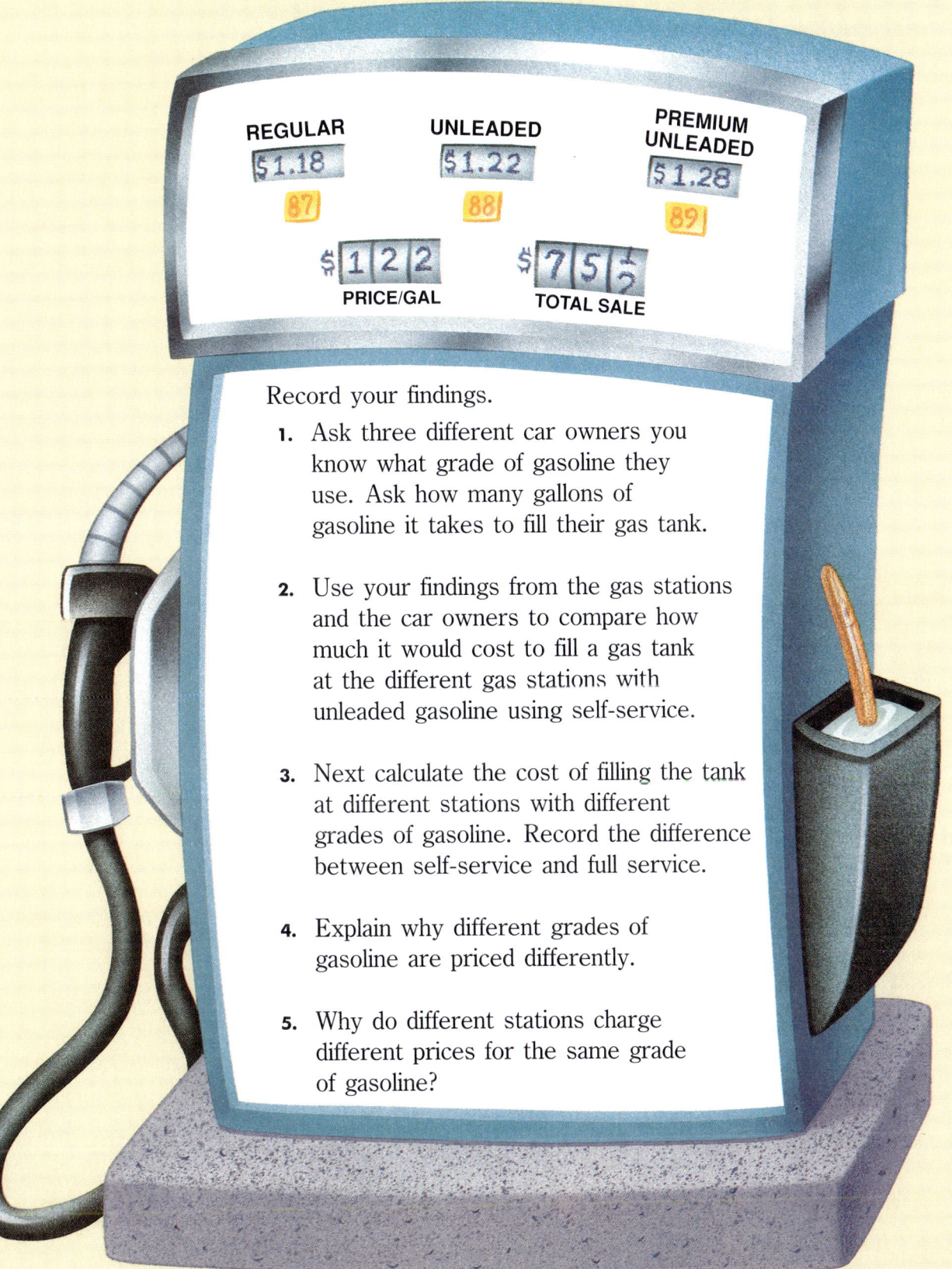

REGULAR	UNLEADED	PREMIUM UNLEADED
$1.18	$1.22	$1.28
87	88	89

PRICE/GAL: $1 2 2

TOTAL SALE: $7 5 ½

Record your findings.

1. Ask three different car owners you know what grade of gasoline they use. Ask how many gallons of gasoline it takes to fill their gas tank.

2. Use your findings from the gas stations and the car owners to compare how much it would cost to fill a gas tank at the different gas stations with unleaded gasoline using self-service.

3. Next calculate the cost of filling the tank at different stations with different grades of gasoline. Record the difference between self-service and full service.

4. Explain why different grades of gasoline are priced differently.

5. Why do different stations charge different prices for the same grade of gasoline?

CHAPTER 4   131

CHAPTER CHECKUP

LANGUAGE & VOCABULARY

1. Explain how to use short division to solve 5)348.

2. Explain why estimates may differ when people use compatible numbers to divide. Give an example.

TEST

CONCEPTS

By what number do you multiply each divisor to make it a whole number? *(pages 120–121)*

1. 0.8)76
2. 3.76 ÷ 0.027
3. 9.24)426.8

Rewrite each problem using compatible numbers. Then estimate the quotient. *(pages 126–127)*

4. 5.3)26.7
5. 4,256 ÷ 78
6. 0.9)15.75

SKILLS

Multiply or divide. *(pages 116–117)*

7. 3.756 ÷ 1,000
8. 5.24 × 10
9. 0.085 ÷ 100

Divide. Give your answer as a whole number with a remainder. *(pages 102–109)*

10. 6)796
11. 40)34,715
12. 8)45,928
13. 563)57,194
14. 77)3,542
15. 92)8,025
16. $474 ÷ 3
17. 96,356 ÷ 381
18. 5,782 ÷ 24

Divide. Give your answer as a decimal. *(pages 114–115)*

19. 3)6.21
20. 63)1.26
21. 17)$7.31
22. 8.43 ÷ 3
23. $15.25 ÷ 25
24. 1,508.7 ÷ 321

Divide. Write zeros until the division comes out evenly. (pages 122–123)

25. $0.08 \overline{)4.5}$
26. $1.2 \overline{)6}$
27. $6 \overline{)75}$

Divide. Round the quotient to the nearest hundredth or cent if necessary. (pages 122–123)

28. $3.8 \overline{)17.138}$

29. $54 \overline{)\$37.95}$

30. $0.063 \overline{)0.0798}$

Find the average. Round the quotient as directed if necessary. (pages 124–125)

31. 6, 8, 18, 5, 12, 15 (nearest tenth)

32. $4.38, $17, $3.20, $7.85 (nearest cent)

PROBLEM SOLVING

Solve. (pages 110–111)

33. For lunch 3 friends ordered a pizza for $12.85 and a pitcher of lemonade for $2.95. They shared the costs evenly.
    a. How many people shared the cost of lunch?
    b. What would you need to do first to find each person's share of the cost of lunch?
    c. What was each person's share of the cost of lunch? Round your answer to the nearest cent.

34. A stadium seats 45,928 people and has 8 gates. On the average, how many people could be expected to use each gate when the stadium is sold out?

LEARNING LOG

Write the answers in your learning log.

1. Explain how the rules for multiplying by powers of ten help with division of decimals.

2. Explain why this is wrong: When you divide by a decimal the quotient is always smaller than the dividend.

EXTRA PRACTICE

Divide. Give your answer as a whole number with remainder. *(pages 102–109)*

1. $5\overline{)92}$
2. $3\overline{)\$267}$
3. $7\overline{)38,481}$
4. $20\overline{)478}$
5. $800\overline{)91,952}$
6. $32\overline{)584}$
7. $6\overline{)3,612}$
8. $573\overline{)4,907}$
9. $8\overline{)2,460}$
10. $17\overline{)6,957}$
11. $9,073 \div 6$
12. $360 \div 90$
13. $4,295 \div 30$
14. $\$66,144 \div 48$
15. $98,765 \div 234$
16. $8,753 \div 23$

Divide. Give your answer as a decimal. *(pages 114–115, 120–121)*

17. $1.696 \div 8$
18. $32 \div 1.6$
19. $12.85 \div 5$
20. $3\overline{)0.264}$
21. $0.9\overline{)3.33}$
22. $36\overline{)\$208.08}$
23. $0.087\overline{)0.04176}$
24. $5.22\overline{)39.672}$

Multiply or divide. *(pages 116–117)*

25. $3.14 \div 10$
26. $5.3 \times 1,000$
27. $0.0097 \div 100$

Divide. Write zeros until the division comes out evenly. *(pages 122–123)*

28. $8\overline{)15}$
29. $0.25\overline{)0.8}$
30. $2.7 \div 3.6$
31. $17 \div 4$

Divide. Round as directed if necessary. *(pages 122–123)*

32. $9 \div 7.5$ (tenth)
33. $0.48\overline{)3.29}$ (hundredth)
34. $52\overline{)64}$ (tenth)

Find the average. Round to the nearest hundredth or cent if necessary. *(pages 124–125)*

35. 2, 5, 7, 9, 10, 8
36. $47, $32, $58
37. 3.2, 4.05, 1.27, 5.6

Rewrite each problem using compatible numbers. Then estimate the quotient. *(pages 126–127)*

38. $8\overline{)622}$
39. $44.9 \div 7$
40. $17,351 \div 28$
41. $3.7\overline{)1.54}$

Solve. *(pages 110–111 and 124–125)*

42. Ayo bought 2 caps on sale for $16 each. The regular price was $24.99 each. How much did he save?

43. Marc scored 85, 92, 91, and 88 on his last 4 math tests. What is the average of his test scores?

ENRICHMENT

That's a Lot of Horses!

Have you ever heard the power of an engine referred to as horsepower? Well, you have horsepower too!

It takes horsepower to climb a flight of stairs. To find how much horsepower it takes you to climb, work with a partner and follow the steps to the right.

- Use a watch with a second hand. Time the number of seconds it takes to get to the top of the stairs.
- Measure or estimate the vertical height of the stairs in feet.
- Divide your weight by 500.
- Multiply the quotient by the height of the stairs
- Divide the product by the number of seconds you climbed.

Did you use a lot of horsepower? Check your work with a calculator

CONDUCT A SURVEY

- Work in a small group. Develop a list of 10 songs, recording groups, or TV shows.
- Then have each group member ask family and friends to rate each item on the list using a scale of 0 (don't like at all) to 9 (enjoy the most).
- Now find an average rating for each item. Divide the total of the ratings by the total number of people surveyed. Decide how you will round each average.
- By placing the average ratings in order, compile a Top 10 list.

CUMULATIVE REVIEW

1. Which statement is true?
    a. 2.536 > 2.356
    b. 0.30 > 0.3
    c. 4.008 = 4.08
    d. none of these

2. Which statement is false?
    a. 22,152 > 22,125
    b. 5,497 < 5,947
    c. 436,007 < 436,700
    d. none of these

3. Which set is in order from least to greatest?
    a. 7, 7.03, 7.003, 77.003
    b. 7, 7.03, 7.003, 77.003
    c. 77.003, 7.03, 7.003, 7
    d. none of these

Round 728,352.596 as directed.

4. to the nearest ten thousand
    a. 730,000
    b. 728,000
    c. 720,000
    d. none of these

5. to the nearest whole number
    a. 700,000
    b. 728,352
    c. 728,353
    d. none of these

6. to the nearest hundredth
    a. 728,400
    b. 728,352.60
    c. 728,352.59
    d. none of these

Choose the property shown by each equation.

7. $(7 \times 5) \times 3 = 7 \times (5 \times 3)$
    a. Commutative
    b. Distributive
    c. Associative
    d. none of these

8. $17 \times 14 = 14 \times 17$
    a. Property of One
    b. Associative
    c. Distributive
    d. none of these

9. $2 \times (4 + 5) = (2 \times 4) + (2 \times 5)$
    a. Associative
    b. Distributive
    c. Commutative
    d. none of these

Find the answer.

10. 35,010 − 4,927
    a. 30,083
    b. 31,917
    c. 30,082
    d. none of these

11. 47.36 − 29.58
    a. 17.78
    b. 76.94
    c. 22.22
    d. none of these

12. 56 − 3.24
    a. 53.24
    b. 52.76
    c. 23.6
    d. none of these

13. 5,000 × 600
    a. 300,000
    b. 3,500,000
    c. 3,000,000
    d. none of these

14. 457 × 38
    a. 5,027
    b. 17,364
    c. 16,366
    d. none of these

15. 596 × 902
    a. 54,832
    b. 537,592
    c. 537,492
    d. none of these

PROBLEM SOLVING REVIEW

Remember the strategies and types of problems you have done so far. Solve.

Problem Solving Check List
- Choosing the operation
- Using a pattern
- Drawing a diagram
- Using estimation
- Too much information
- Too little information
- Multistep problems
- Writing problems

1. Jordan lives 3.8 km further from school than Marie.
   a. Who lives further from school, Jordan or Marie?
   b. What else would you need to know to find how far Jordan lives from school?
   c. Make up appropriate data for the missing information. Then find how far Jordan lives from school.

2. A jet flew 6.25 h from New York to San Jose at an average speed of 625.7 km/h. To the nearest kilometer, how many kilometers did it fly?

3. For the bookcase Mrs. Lee is building, she needs 5 boards 1.65 m long and 2 boards 1.9 m long. How many boards does she need?

4. Each of the King School's 493 students had to sell 20 school fair tickets. If all students sell all their tickets, how many will be sold?

5. Tawanna had softball practice on April 2, 5, 8, and 11. Based on this information, when would you expect her next practice to be?

6. Ana rented a video game for $1.76. She gave the clerk a $10-bill. Name the fewest bills and coins Ana could have received as change.

7. During one week a recycling center collected 6,852 kg of paper, 2,807 kg of glass, and 276 kg of cans. About how many kilograms did the center collect that week?

8. A tape and CD player are priced at $259.95 and speakers are priced at $325.49. They can be purchased as a set for $549.99. How much is saved by buying these items as a set?

Write a problem that can be solved using the model. Then solve.

9.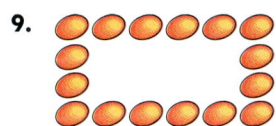

10.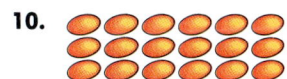

TECHNOLOGY

DIVISION TARGETS

In the computer game "Tug of War," opponents estimate quotients. Sharpen your division estimation skills with this paper-and-pencil activity.

Work with a partner. Follow these rules for each game shown below:

1. Choose a divisor you think will result in the quotient shown.
2. Compute your quotient with a calculator.
3. Compare your quotient with the given quotient.

The player whose quotient is closer to the target quotient wins the game.

GAME A	GAME B	GAME C
1,950 ÷ ▩ = 400	8,330 ÷ ▩ = 120	3,923 ÷ ▩ = 5

TRY, TRY AGAIN

Estimate the missing factor in each exercise below. Multiply to see how close your estimate is.

Then find the exact factor with 5 or fewer tries.

1. 16 × ▩ = 864
2. 43 × ▩ = 2,408
3. 84 × ▩ = 47,208
4. ▩ × 325 = 56,225
5. 8.6 × ▩ = 31.82

INFLATION

The Dazzlers are a team in the International Basketball League. Annual salaries for the twelve team members are shown here.

$3,400,000 $2,800,000 $1,900,000
$1,650,000 $1,320,500 $1,000,000
$975,000 $825,000 $810,000
$775,000 $400,000 $344,500

What does the team's coach earn if his annual salary is one-fourth of the average player's salary?

Measurement 5

DID YOU KNOW...?
In one 24-hour period, 76 in. of snow fell in Silver Lake, Colorado. That's over 6 ft of snow!

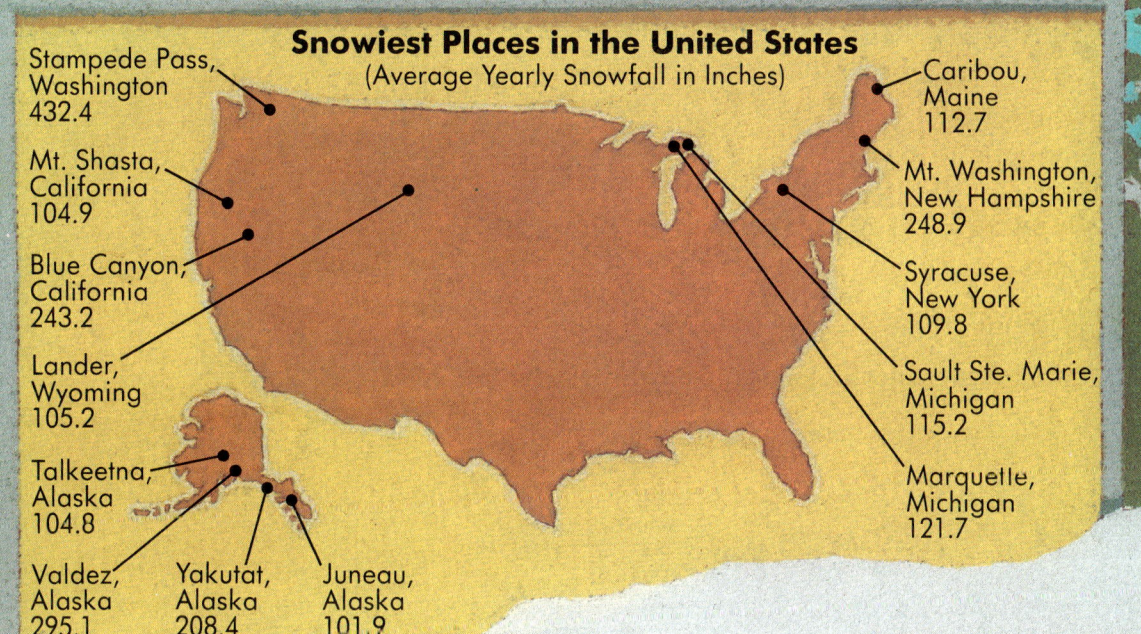

Snowiest Places in the United States
(Average Yearly Snowfall in Inches)

- Stampede Pass, Washington 432.4
- Mt. Shasta, California 104.9
- Blue Canyon, California 243.2
- Lander, Wyoming 105.2
- Talkeetna, Alaska 104.8
- Valdez, Alaska 295.1
- Yakutat, Alaska 208.4
- Juneau, Alaska 101.9
- Caribou, Maine 112.7
- Mt. Washington, New Hampshire 248.9
- Syracuse, New York 109.8
- Sault Ste. Marie, Michigan 115.2
- Marquette, Michigan 121.7

Organize the information about the snowiest places in the United States into a bar graph.

Find out the average yearly snowfall for the place where you live. How does this amount compare to those shown on the graph?

USING DATA
Collect
Organize
Describe
Predict

Exploring Measurement

Each pencil is 3 paper clips long.

1. Are both pencils the same length? Explain your answer.

2. Measure the width of your desk with small paper clips and then with large paper clips. Did you use the same number of paper clips both times? Explain why or why not.

A paper clip is a nonstandard unit of length because paper clips vary in length. The inch is a standard unit of length.

3. Why is the inch a standard unit of length?

4. Name two other standard units of length.

5. Name two other nonstandard units of length.

6. Why, do you think, were standard units invented?

To find the weight of this bunch of grapes, you can use metal cubes and a balance scale.

7. How do you know when you have used the right number of metal cubes?

8. Suppose you used cubes of two different weights. If you could balance the grapes with 8 cubes, would it be accurate to say that the grapes weigh 8 cubes? Why or why not?

9. Describe how the weight of the cube affects the number of cubes you need to balance the grapes.

10. The two different-weight cubes are nonstandard units of weight. What is a standard unit of weight that you know?

Now let's think about how much the pitcher can hold, or its **capacity**.

11. How can you use the glasses to find out how much the pitcher can hold?

12. Which glass would you need to fill more times in order to find out how much the pitcher can hold?

13. The two glasses are nonstandard units of capacity. What is a standard unit of capacity that you know?

When you find the length, the weight, or the capacity of something, you are finding a measurement.

IN YOUR WORDS Explain how measurement is used in the following situations.

14. at a baseball game

15. at a gas station

16. at home

17. at a supermarket

Customary Units of Length

A true tropical rain forest can have as much as 400 in. of rain per year. How many feet of rain is that?

The **inch (in.)**, **foot (ft)**, **yard (yd)** and **mile (mi)** are standard units of length.

  1 ft = 12 in.
  1 yd = 3 ft = 36 in.
  1 mi = 5,280 ft = 1,760 yd

You can divide 400 by 12 to find the number of feet of rain.

feet → 33 R4 ← extra inches
```
    33 R4
12)400
    36
    40
    36
     4
```
400 in. of rain is 33 ft 4 in. of rain.

THINK ALOUD In your classroom, what measures about 1 yd? Name something that is about 1 mi from your school.

Sometimes you need to add or subtract units of length.   1 ft = 12 in.

```
   7 yd 2 ft  6 in.                                    11 ft  19 in.
 +      2 ft 10 in.   (12 in. = 1 ft)  (3 ft = 1 yd)   12 ft   7 in.
   7 yd 4 ft 16 in. = 7 yd 5 ft 4 in. = 8 yd 2 ft 4 in. − 4 ft  9 in.
                                                        7 ft 10 in.
```

GUIDED PRACTICE

Complete.

1. 5 ft = 1 yd ▨ ft
2. 2 mi = ▨ ft
3. $1\frac{1}{2}$ yd = ▨ in.

Subtract.

4.   5 yd 1 ft 7 in.
   − 2 yd 2 ft 11 in.

Use feet and inches to measure.

5. the height of your desk
6. the width of one window

PRACTICE

Choose the best unit of measurement (*in.*, *ft*, *yd*, or *mi*).

7. An apple is about 3 ▨ wide.
8. A bus is about 30 ▨ long.

MATH AND SCIENCE

142   LESSON 5–2

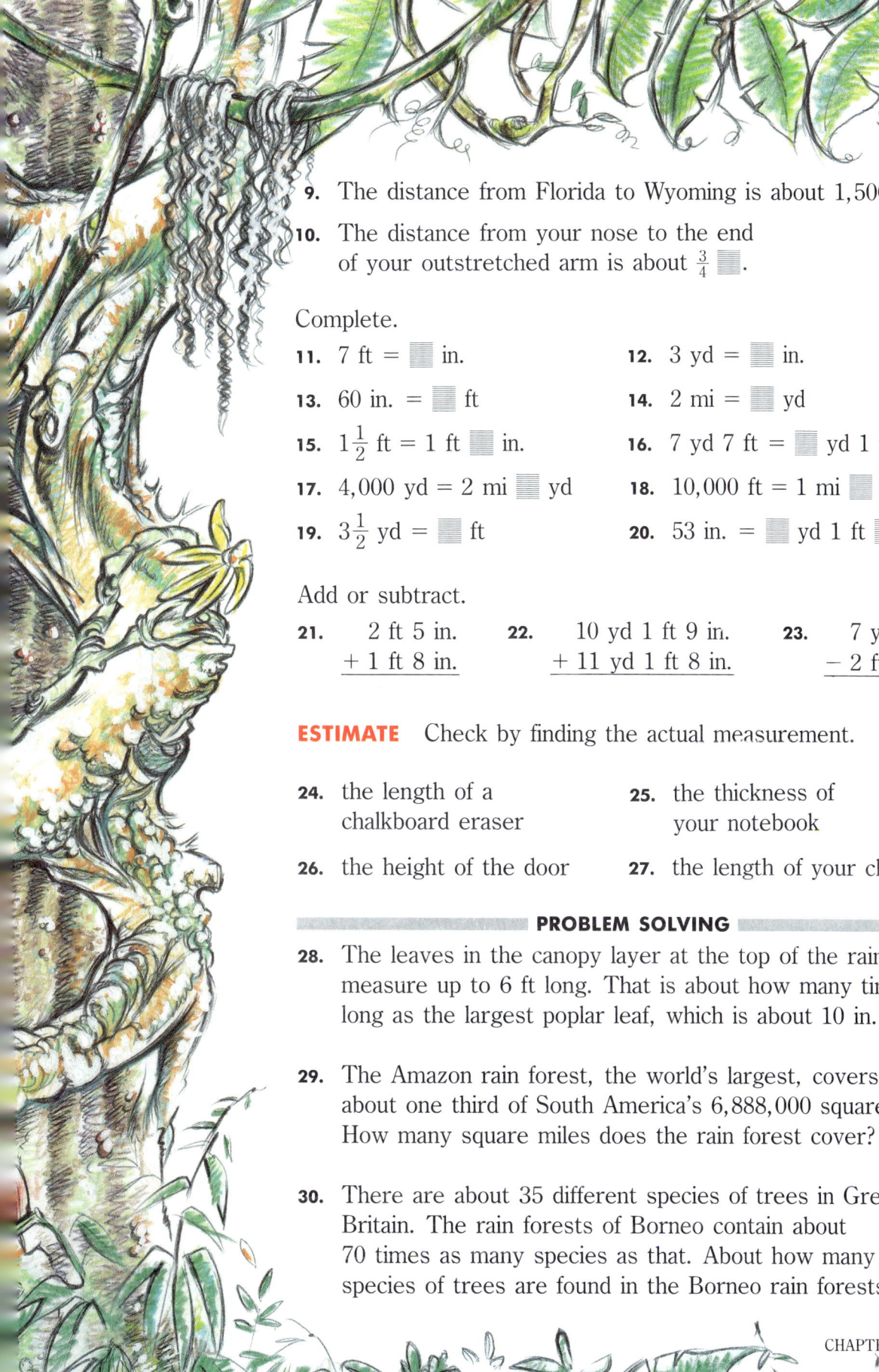

9. The distance from Florida to Wyoming is about 1,500 ▨.

10. The distance from your nose to the end of your outstretched arm is about $\frac{3}{4}$ ▨.

Complete.

11. 7 ft = ▨ in.
12. 3 yd = ▨ in.
13. 60 in. = ▨ ft
14. 2 mi = ▨ yd
15. $1\frac{1}{2}$ ft = 1 ft ▨ in.
16. 7 yd 7 ft = ▨ yd 1 ft
17. 4,000 yd = 2 mi ▨ yd
18. 10,000 ft = 1 mi ▨ ft
19. $3\frac{1}{2}$ yd = ▨ ft
20. 53 in. = ▨ yd 1 ft ▨ in.

Add or subtract.

21. 2 ft 5 in.
    + 1 ft 8 in.

22. 10 yd 1 ft 9 in.
    + 11 yd 1 ft 8 in.

23. 7 yd 1 in.
    − 2 ft 7 in.

ESTIMATE Check by finding the actual measurement.

24. the length of a chalkboard eraser
25. the thickness of your notebook
26. the height of the door
27. the length of your classroom

PROBLEM SOLVING

28. The leaves in the canopy layer at the top of the rain forest measure up to 6 ft long. That is about how many times as long as the largest poplar leaf, which is about 10 in. long?

29. The Amazon rain forest, the world's largest, covers about one third of South America's 6,888,000 square mi. How many square miles does the rain forest cover?

30. There are about 35 different species of trees in Great Britain. The rain forests of Borneo contain about 70 times as many species as that. About how many species of trees are found in the Borneo rain forests?

Customary Units of Capacity and Weight

Families from Vietnam celebrate the mid-autumn festival in September. Bách is making soup with lemon grass for her family's celebration. She has 5 pints of chicken broth. Her recipe calls for 2 quarts. Does she have enough?

Standard units of capacity are **fluid ounces (fl oz)**, **cups (c)**, **pints (pt)**, **quarts (qt)**, and **gallons (gal)**.

$5 \div 2 = 2$ R1, so 5 pt equals 2 qt plus 1 pt. Bách has enough broth for her recipe.

1 c = 8 fl oz
1 pt = 2 c
1 qt = 2 pt
1 gal = 4 qt

THINK ALOUD Why did we divide by 2 to change pints to quarts? How would you change gallons to quarts?

Standard units of weight are the **ounce (oz)**, **pound (lb)**, and **ton (t)**.

1 lb = 16 oz
1 t = 2,000 lb

GUIDED PRACTICE

Choose the more likely measurement, *a* or *b*.

1. the capacity of a kitchen sink   **a.** 12 fl oz   **b.** 12 gal
2. the weight of a loaf of bread   **a.** 2 lb   **b.** 20 lb

Complete.

3. 3 lb = ▨ oz
4. 8,000 lb = ▨ t
5. 4 c = ▨ pt
6. $1\frac{1}{2}$ c = ▨ fl oz

PRACTICE

Choose the more likely measurement. Write *a* or *b*.

7. the capacity of a teapot   **a.** 3 pt   **b.** 3 gal
8. the capacity of a pond   **a.** 25 gal   **b.** 25,000 gal
9. the capacity of a soup spoon   **a.** $\frac{1}{2}$ fl oz   **b.** 5 fl oz
10. the weight of a sixth grader   **a.** 80 oz   **b.** 80 lb
11. the weight of a car   **a.** 1 t   **b.** 100 lb
12. the weight of a slice of bread   **a.** 9 oz   **b.** 1 oz

Complete.

13. 96 oz = ▨ lb
14. 5 t = ▨ lb
15. 34 oz = ▨ lb ▨ oz

16. 1.5 t = ■ lb **17.** $6\frac{1}{4}$ lb = ■ oz **18.** 7 pt = ■ c

19. 16 pt = ■ qt **20.** $5\frac{1}{2}$ gal = ■ qt **21.** 36 pt = ■ gal ■ qt

MENTAL MATH Find these relationships.

22. How many ounces are in 1 t?   **23.** How many fluid ounces are in 1 pt?

MIXED REVIEW Complete.

24. 27 ft = ■ yd   **25.** $\frac{1}{2}$ mi = ■ ft   **26.** 386 in. = ■ yd ■ in.

PROBLEM SOLVING

Use the recipe. Write *true* or *false*.

27. Each serving contains less than $\frac{1}{2}$ oz of chicken or shrimp.

28. If you doubled the recipe, you would need 1 lb of chicken or shrimp.

29. If you want to make 30 servings, you need to double the recipe.

30. You have 1 fl oz of *nuoc mam*. You can make enough *Com Chien* to serve 6 people.

CHAPTER 5   145

Fahrenheit Temperature

The average January temperature is 13°F in Anchorage, Alaska, and −13°F in Fairbanks, Alaska.

Temperature is measured in degrees. The thermometer shows **degrees Fahrenheit (°F)**.

Temperatures that are less than zero are called "below zero." You read −13°F as *thirteen degrees below zero*.

Water boils at 212°F. At what temperature does water freeze?

CRITICAL THINKING How many degrees are between 13°F and −13°F?

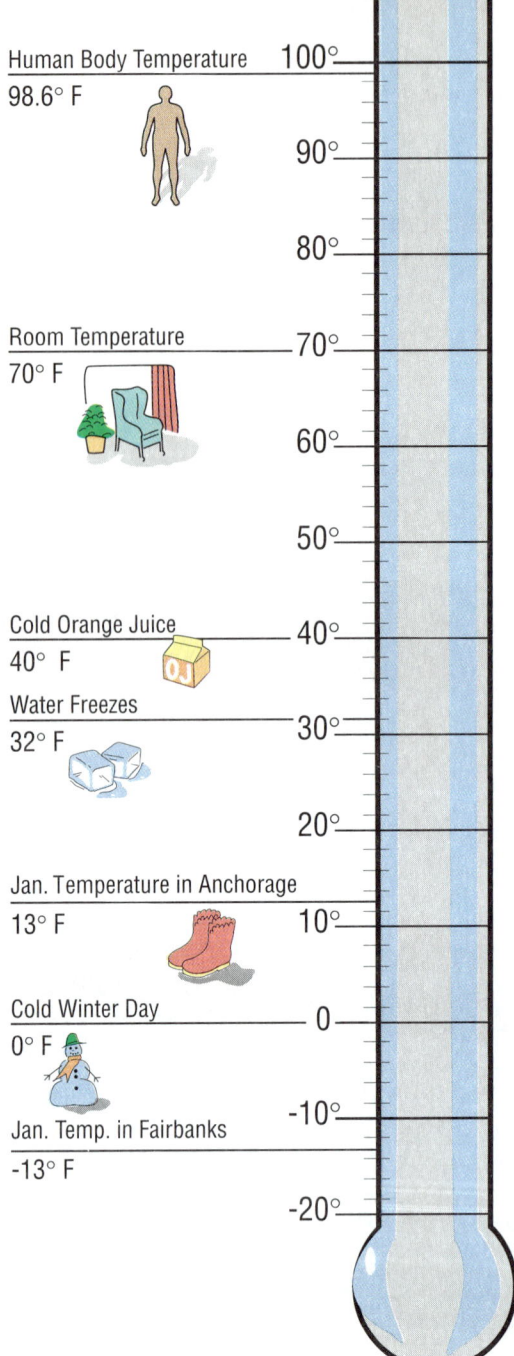

GUIDED PRACTICE

Choose the most likely temperature.

1. for swimming
   a. 55°F b. 92°F c. −10°F

2. for sledding
   a. 27°F b. 50°F c. 42°F

Use the thermometer to answer.

3. What temperature is 10° below room temperature?

4. What temperature is 20° below the freezing point of water?

PRACTICE

Choose the most likely temperature. Write *a*, *b*, or *c*.

5. in an air-conditioned room   a. 10°F   b. 65°F   c. 100°F
6. of bath water   a. 40°F   b. 100°F c. 20°F
7. in a freezer   a. 20°F   b. 35°F   c. 50°F
8. of a bowl of hot soup   a. 75°F   b. 150°F c. 215°F
9. for baking bread   a. 120°F b. 600°F c. 350°F

NUMBER SENSE Use the thermometer on page 146 to help you answer the questions. Explain how you decided.

10. Do you need to wear a sweater when the temperature is 50°F?

11. Is it safe to go ice-skating outdoors when the temperature is 32°F?

ESTIMATE Give an estimate of the temperature.

Explain how you arrived at your estimate.

12. the inside of a refrigerator

13. your classroom

14. a hot summer day

15. a hot day under a tree

PROBLEM SOLVING

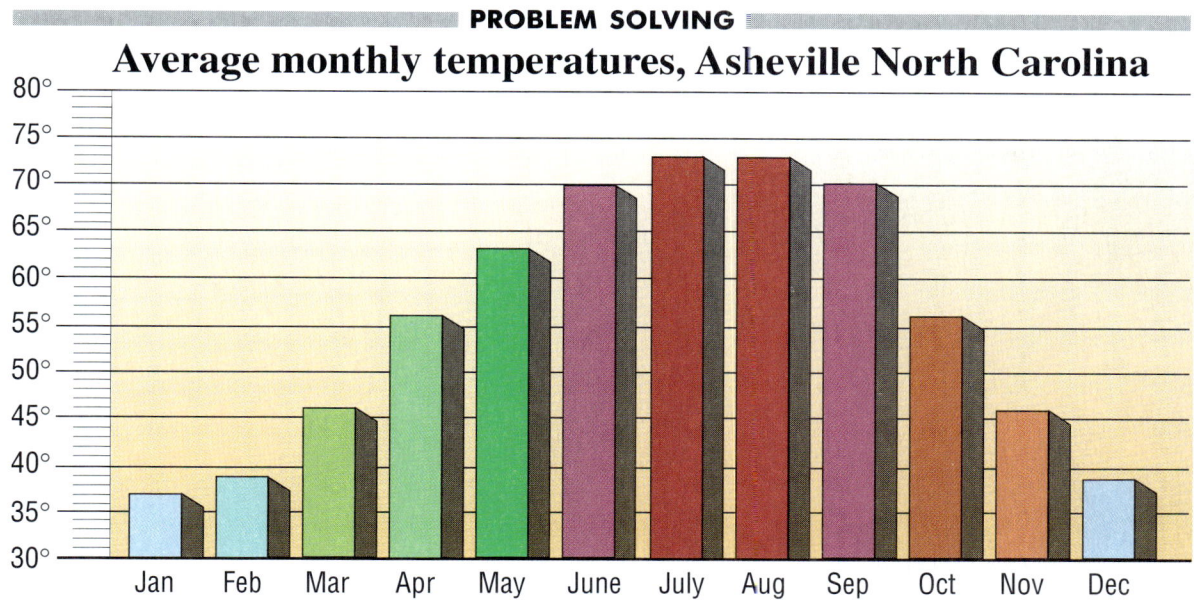

Average monthly temperatures, Asheville North Carolina

Use the graph to solve.

16. Which months had an average temperature below 60°F?

17. Describe the pattern of temperatures shown on the graph.

18. Between which two consecutive months is the greatest temperature increase? decrease?

19. **IN YOUR WORDS** Explain how it is possible to have days in July when the temperature is above 80°F, even though the average monthly temperature is 73°F.

20. **CREATE YOUR OWN** Write a word problem using information from the graph. Trade with a partner and solve.

What's the Problem?

Problem Solving: Writing Problems

READ ABOUT IT

Here are some interesting facts about the lives of Wolfgang Amadeus Mozart and Ludwig van Beethoven.

Wolfgang Amadeus Mozart
(1756 – 1791)

Mozart is recognized as one of the great composers and performers of all time. He started composing in 1761.

By 1768 Mozart had composed several sonatas and symphonies, as well as an opera.

Mozart played his first major concert in 1762.

Mozart composed 20 operas, 41 symphonies, 17 piano sonatas, and 512 other works.

Ludwig van Beethoven
(1770 – 1827)

Beethoven brought a new appreciation of and flexibility to music.

Beethoven conducted his Ninth Symphony for the first time in 1821. He could not hear it because he had lost his hearing two years earlier.

Beethoven wrote 9 symphonies, 9 concertos, 32 piano sonatas, 1 opera, and many other works.

🟦 TALK ABOUT IT

Beethoven lived from 1770 to 1827. In 1821 he conducted his Ninth Symphony for the first time. Unfortunately, he could not hear the audience's applause because he had lost his hearing 2 yr earlier. How old was Beethoven when he lost his hearing?

1. The problem above was written from the facts about Beethoven. Solve it. How many steps did you use?

2. Think about the problem. Turn it around to create a problem that asks a different question.

Now take the facts on page 148 and create some other problems.

3. Use the facts about Mozart to create a subtraction problem.

4. Use the facts about Mozart to create a problem involving dates.

5. Use the facts about Beethoven to create a subtraction problem.

6. Some problems include extra facts. Change one of the problems you have created into a problem that includes too much information.

🟦 WRITE ABOUT IT

Work with a partner. Use the facts about Mozart and Beethoven. Trade with a partner to solve.

7. Write a problem involving ages.

8. Write an estimation problem.

9. Write an addition problem that can be solved using mental math.

10. Write a problem that compares Beethoven and Mozart.

MIDCHAPTER CHECKUP

LANGUAGE & VOCABULARY

Describe ways in which these units of measure could be classified or sorted. Then use your methods to sort the units.

| mile | foot | ounce | pint | ton | pound |
| inch | gallon | yard | quart | cup | fluid ounce |

QUICK QUIZ

Choose the best unit (*in., ft, yd,* or *mi*). *(pages 142–143)*

1. A car is about 10 ▨ long.
2. Complete: 8,000 ft = 1 mi ▨ ft

Add or subtract. *(pages 142–143)*

3.   2 yd 1 ft 7 in.
   + 3 yd 2 ft 9 in.

4.   6 ft 3 in.
   − 4 ft 9 in.

Choose best estimate of capacity. *(pages 144–145)*

5. a bathtub: 40 fl oz or 40 gal

Complete. *(pages 144–145)*

6. $3\frac{3}{4}$ lb = ▨ oz
7. $4\frac{1}{2}$ qt = ▨ pt

Choose the most likely temperature. *(pages 146–147)*

8. for baking cookies   a. 120°F   b. 375°F   c. 500°F

9. in your classroom   a. 70°F   b. 50°F   c. 120°F

Solve. *(pages 148–149)*

10. During 30 summer days, the temperature in Sun City rose above 100°F on 10 days. During this same period there were 4 days of rain whose total rainfall was 1.8 in.

    a. For how many days are the weather facts given?
    b. Which facts would you use to write a problem about the number of days the temperature did not rise above 100°F?
    c. Based on the facts given, write a problem concerning rainfall in Sun City. Then solve it.

LEARNING LOG

Write the explanation in your learning log.

1. Your friend thinks the ceiling in your classroom is 20 ft high. Explain two ways in which you can prove whether your friend is right or wrong.

2. Explain why 3 ft 4 in. can be written as 40 in., and as 1 yd 4 in.

MATH AMERICA

In 1849, the sidewheel boat, *Francis Skiddy,* carried passengers on the Hudson River. It traveled from Albany to New York City. Find a river on a map of your state. Use the map's scale to estimate the actual length of the river.

BONUS

A tiny bug is at the bottom of a 30-ft hole. Each day the bug is able to crawl up 3 ft. But each night the bug slides down 1 ft. When will the bug reach the top of the hole?

Metric Units of Length

The metric system of measurement was created by French scientists in the 1790s. The **meter (m)** is the basic unit of length.

DID YOU KNOW...?
The meter was defined as 1 ten-millionth of the distance from the North Pole to the equator along Earth's surface

Smaller and larger units of length are based on the meter.

1 *deci*meter (dm) = 1 tenth of a meter
1 *centi*meter (cm) = 1 hundredth of a meter
1 *milli*meter (mm) = 1 thousandth of a meter

The picture shows the size of the units that are smaller than the meter.

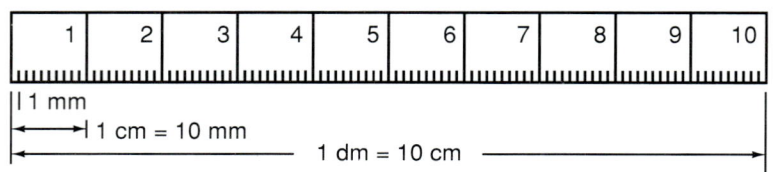

1 m = 10 dm. The distance from the floor to a doorknob is about 1 m.

Longer distances are usually measured in kilometers.

1 *kilo*meter (km) = 1,000 meters

You can walk 1 km in about 12 min.

If you know how to multiply and divide by powers of 10, you can make a MATH CONNECTION to change metric units.

Multiply to change from a larger unit to a smaller unit.

Divide to change from a smaller unit to a larger unit.

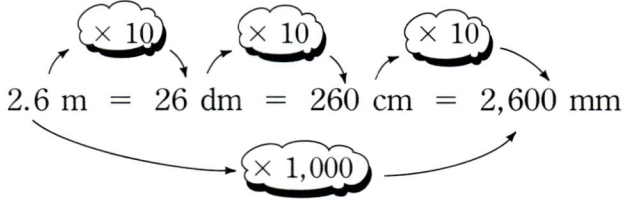

2.6 m = 26 dm = 260 cm = 2,600 mm

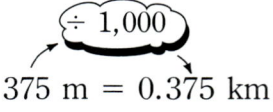

375 m = 0.375 km

GUIDED PRACTICE

Complete.

1. 2 m = ▨ cm
   2.5 m = ▨ cm

2. 5,000 mm = ▨ m
   5,250 mm = ▨ m

3. 70 dm = ▨ m
   71 dm = ▨ m

LESSON 5–6

Add or subtract.

4. 5 cm 3 mm
    + 3 cm 8 mm

5. 4 m 25 cm
    − 2 m 60 cm

6. 6 km
    − 3 km 750 m

PRACTICE

Complete.

7. 20 mm = ___ cm   **8.** 7 km = ___ m   **9.** 8 dm = ___ cm

10. 4,000 mm = ___ m   **11.** 30 cm = ___ mm   **12.** 45 dm = ___ m

13. 4.26 m = ___ cm   **14.** 8.5 m = ___ mm   **15.** 30 cm = ___ dm

16. 75 cm = ___ m   **17.** 25 m = ___ km   **18.** 1 km = ___ mm

Add or subtract.

19. 5 m 795 mm
     + 2 m 350 mm

20. 2 m 43 cm
     − 1 m 87 cm

21. 3 km
     − 1 km 4 m 6 cm 9 mm

Complete. Choose >, <, or =.

22. 140 mm ___ 1.4 m   **23.** 3,759 m ___ 4 km   **24.** 75 cm ___ 7.5 dm

ESTIMATE Which is the most likely measurement? Write *a* or *b*.

25. the length of a car   a. 4 m   b. 1 m

26. the height of a seat   a. 1.5 m   b. 0.5 m

27. the thickness of a dime a. 1 mm   b. 1 cm

PROBLEM SOLVING

28. The distance from the North Pole to the equator is 10,000 km. About how many kilometers is the distance around Earth?

29. The metric system was adopted in France in 1795. It was not legal to use the metric system in the United States until 1866. How many years later was that?

30. IN YOUR WORDS In what ways is the metric system easy to work with?

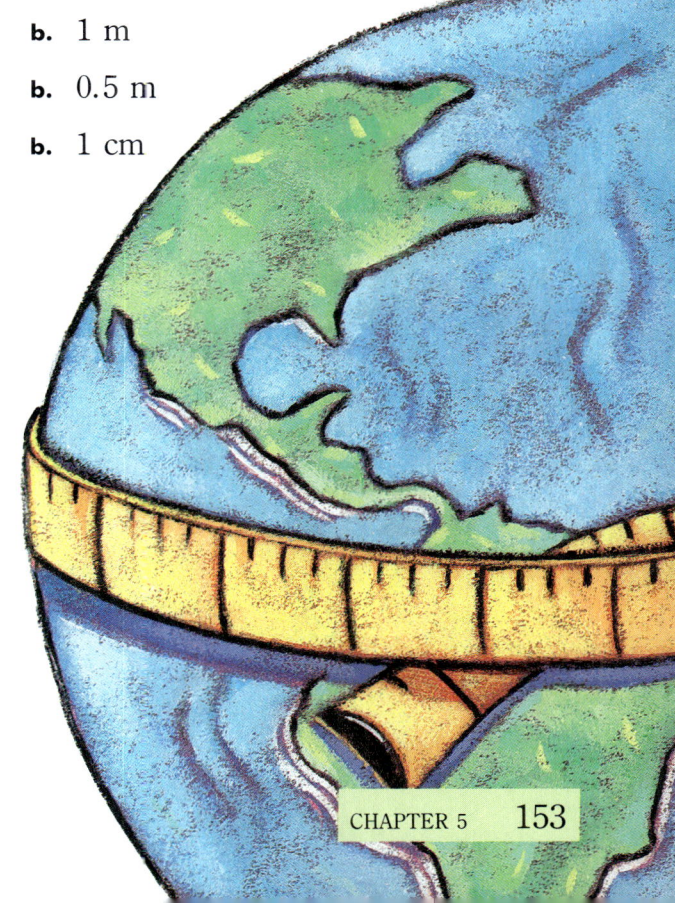

CHAPTER 5   153

Perimeter

Basketball courts are not all the same size. The largest basketball court can be 94 ft long by 50 ft wide. The smallest can be 74 ft long and 42 ft wide. What is the distance around the largest basketball court?

The distance around a figure is its **perimeter (P)**.

You can find the perimeter of a figure by adding the lengths of its sides.

$$
\begin{aligned}
P \text{ rectangle} &= 2 \text{ lengths} + 2 \text{ widths} \\
&= (2 \times l) + (2 \times w) \\
&= (2 \times 94) + (2 \times 50) \\
&= 188 + 100 \\
&= 288
\end{aligned}
$$

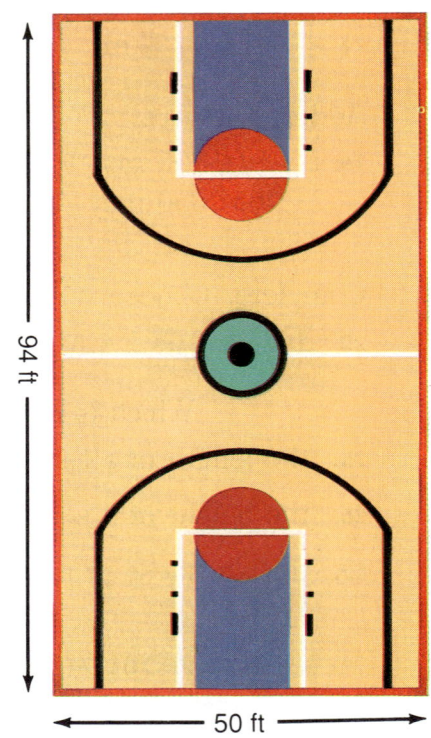

The perimeter of the largest basketball court is 288 ft.

CRITICAL THINKING Write a rule for the perimeter of a square.

GUIDED PRACTICE

Find the perimeter.

1. Square, 6 in. on each side.
2. Triangle, 9 ft, 9 ft, 6 ft.
3. Figure with sides 9 in., 7 in., 6 in., 15 in., 1 ft.
4. Rectangle, 2.3 mi by 1.4 mi.

5. **THINK ALOUD** Explain two different ways to find the perimeter of this star if each side is 2 in.

PRACTICE

Find the perimeter. Use a calculator when needed.

6. rectangle: 23 ft by 12 ft

7. 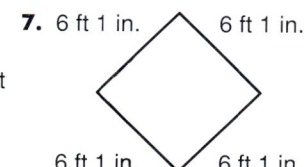 rhombus with all sides 6 ft 1 in.

8. trapezoid: top 10 in., sides 4 in., bottom 1 ft 4 in.

9. 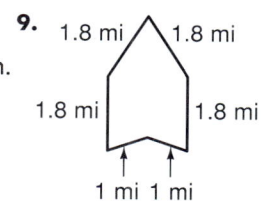 figure with sides 1.8 mi, 1.8 mi, 1.8 mi, 1.8 mi, 1 mi, 1 mi

Find the perimeter of the rectangle.

10. length, 26 yd; width, 13 yd

11. length, 38 ft; width, 10.5 ft

12. a square with a side of 20 in.

13. length, 14 yd; width, 14 ft

Find the value of n. Draw a rectangle to help you.

	Length	Width	Perimeter		Length	Width	Perimeter
14.	15 mi	5 mi	n	15.	22 yd	13 yd	n
16.	6 in.	n	22 in.	17.	n	7 ft	80 ft
18.	11 yd	n	46 yd	19.	n	1,000 ft	1 mi

CALCULATOR Find the rectangle with the greater perimeter. Write a or b.

20. a. length, 19.5 mi; width, 12.75 mi
b. length, 17.75 mi; width, 14.5 mi

21. a. length, 9 yd 2 ft; width, 1 ft 7 in.
b. length, 17 ft 4 in.; width, 3 yd

PROBLEM SOLVING

CHOOSE Choose pencil and paper, mental math, or calculator to solve.

22. A basketball's diameter is 9.5 in. About how many basketballs could be lined up along the perimeter of a 94-ft by 50-ft basketball court?

23. A field goal is worth 2 or 3 points. A free throw earns 1 point. Find ten combinations of 10 points that a player could earn.

Critical Thinking

Work with a partner. Draw a square whose sides are 4 in. long. Draw a circle that is 4 in. across its center.

1. Do you think the square and the circle have the same perimeter?

2. How could you use string to find the perimeter of each figure? Try it.

3. Try it again with other pairs of circles and squares. Compare your results with the results from another group.

Metric Units of Capacity

The basic metric units of capacity are the **milliliter (mL)**, **liter (L)**, and **kiloliter (kL)**.

    1 L = 1,000 mL
    1 kL = 1,000 L

To change between these units, multiply or divide by 1,000.

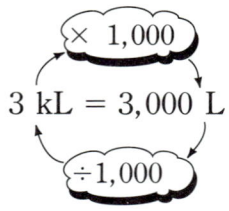
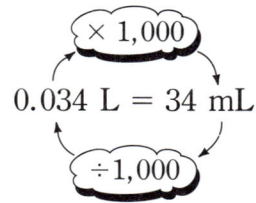

This holds about 1 kL of water.

The prefixes *deci* and *centi* are used to form other metric units of capacity.

    1 m = 10 decimeters,
so 1 L = ▨ **deciliters (dL)**.

    1 m = 100 centimeters,
so 1 L = ▨ **centiliters (cL)**.

CRITICAL THINKING How many milliliters are there in 1 deciliter?

GUIDED PRACTICE

Complete.

1. 2 L = ▨ mL
2. 4 kL = ▨ L
3. 5 mL = ▨ L
4. 1 dL = ▨ L
5. 5 L = ▨ cL
6. 6,500 L = ▨ kL

PRACTICE

ESTIMATE Choose the more likely measurement. Write *a* or *b*.

7. the capacity of a bathtub   a. 10 L   b. 400 L
8. the amount of soup in a bowl   a. 200 mL   b. 2,000 mL
9. the amount of gas in a car   a. 56 cL   b. 56 L
10. the amount of water in a sink   a. 35 L   b. 5 kL
11. the amount of medicine in a dropper   a. 5 mL   b. 100 mL

This holds about 1 L.

This glass holds about 250 mL.

Complete.

12. 4.2 kL = ▨ L
13. 3,000 mL = ▨ L
14. 9 L = ▨ mL
15. 4.5 L = ▨ dL
16. 0.1 L = ▨ mL
17. 8 L = ▨ cL
18. 7 mL = ▨ L
19. 15 L = ▨ mL
20. 922 L = ▨ kL

Complete. Choose >, <, or =.

21. 1 L ▨ 100 mL
22. 1 kL ▨ 1,000 L
23. 50 L ▨ 0.5 kL
24. 10 mL ▨ 1 cL
25. 3,500 mL ▨ 4 L
26. 0.5 dL ▨ 50 mL

NUMBER SENSE Use a different unit to name the same amount.

27. 100 cL
28. 1.5 L
29. 86 dL
30. 123 mL
31. 9.05 kL

MIXED REVIEW Solve.

32. 3 m = ▨ km
33. 17 cm = ▨ m
34. 8 mm = ▨ cm
35. 1.4 m = ▨ dm

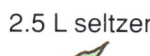

Fruit Punch Recipe

3.5 L unsweetened pineapple juice
500 mL fresh strawberries quartered
900 mL orange juice
300 mL mint leaves
300 mL lemon juice
2.5 L seltzer

PROBLEM SOLVING
Use the recipe to answer the questions.

36. If you double the recipe, will 2 L of orange juice be enough?

37. Will the punch fit in a bowl that holds 7 L?

38. How many 200-mL servings does this recipe make?

39. a. Estimate how many milliliters of punch a person might drink during a party.

b. If you want to be sure to have enough, how many liters of punch should you make for 20 people?

CHAPTER 5 157

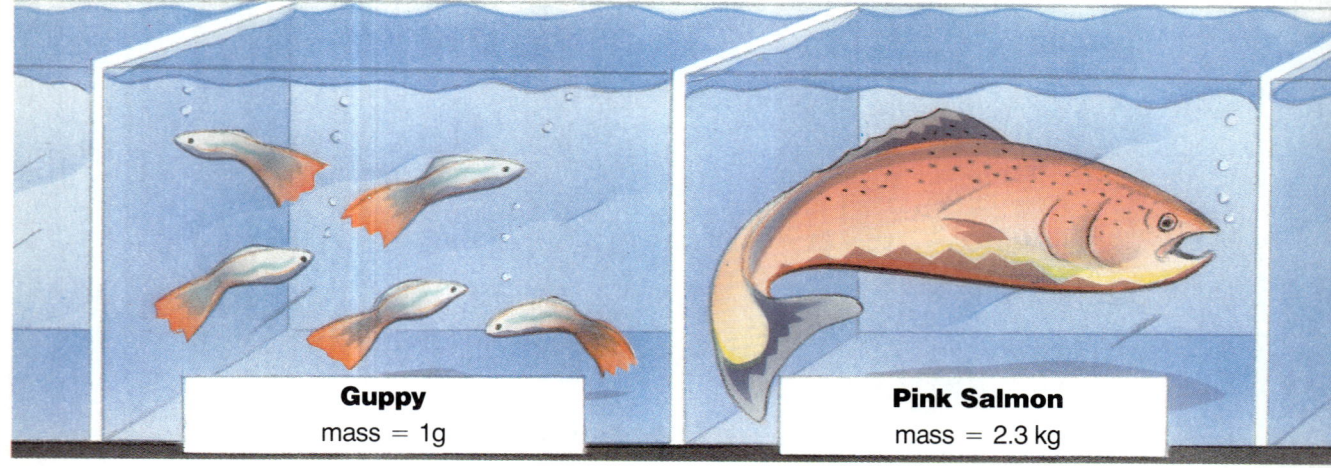

Metric Units of Mass

The **milligram (mg)**, **gram (g)**, **kilogram (kg)**, and **metric ton (t)** are standard units of mass.

1 g = 1,000 mg
1 kg = 1,000 g
1 t = 1,000 kg

You can change kilograms to grams by multiplying by 1,000.

2.3 kg × 1,000 = 2,300 g

The mass of the pink salmon is about 2,300 times that of the guppy.

Other units of mass are the **decigram (dg)** and the **centigram (cg)**.

1 g = 10 dg = 100 cg

- Many vitamin and medicine dosages are measured in milligrams.
- A thumbtack has a mass of about 1 g.
- Your math book has a mass of about 1 kg.
- A small car has a mass of about 1 metric ton.

The chart shows a special relationship among length, capacity, and mass in the metric system.

Length of a Side of a Cube	Capacity	Mass of Water
1 cm	1 mL	1 g
1 dm = 10 cm	1 L	1 kg

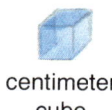
centimeter cube

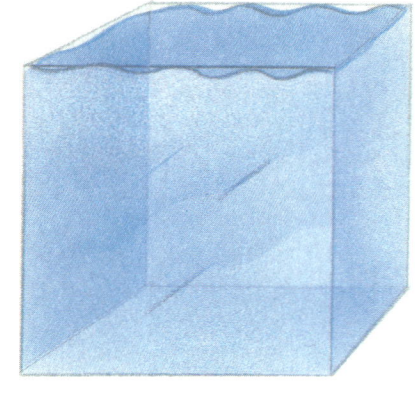
decimeter cube

GUIDED PRACTICE

Which unit of mass would you use to measure each?
Choose *mg*, *g*, *kg*, or *t*.

1. a stamp
2. a bagel
3. a whale
4. a bowling ball

Complete.

5. 4 g = ▨ mg
6. 1.3 kg = ▨ g
7. 852 kg = ▨ t

PRACTICE

Choose the best unit for measuring mass (*mg*, *g*, or *kg*).

8. A pencil is about 5 ▨.
9. A loaf of bread is about 0.75 ▨.
10. A toothbrush is about 8,000 ▨.
11. A woman is about 53 ▨.
12. A pair of running shoes is about 900 ▨.

ESTIMATE Choose the more likely measurement. Write *a* or *b*.

13. the mass of an apple   a. 2.5 kg   b. 250 g
14. the mass of an elephant   a. 6,000 kg   b. 6,000 mg
15. the mass of a pickup truck   a. 2 t   b. 2 kg
16. the mass of a quarter   a. 6 cg   b. 6 g

Complete.

17. 8 kg = ▨ g
18. 14 t = ▨ kg
19. 14 mg = ▨ g
20. 395 mg = 0.395 ▨
21. 50 cg = 5 ▨
22. 0.43 kg = 430 ▨

23. **MENTAL MATH** How many milligrams are there in 1 kg?

MIXED REVIEW Solve.

24. 4,700 mL = ▨ L
25. 2.56 m = ▨ cm
26. 3 kL = ▨ L

PROBLEM SOLVING

27. Would 40 newborn guppies, each about 3 mm long, measure more than or less than 1 dm if they were lined up end to end?

28. The Chinook is the largest species of salmon. How many more grams does a 10-kg Chinook salmon weigh than 4 pink salmon, each weighing 2.3 kg?

29. To lay their eggs, salmon swim upstream, sometimes as far as 3,200 km. About how far does a salmon have to swim per day to travel 3,200 km in 3 months?

Problem Solving: Elapsed Time

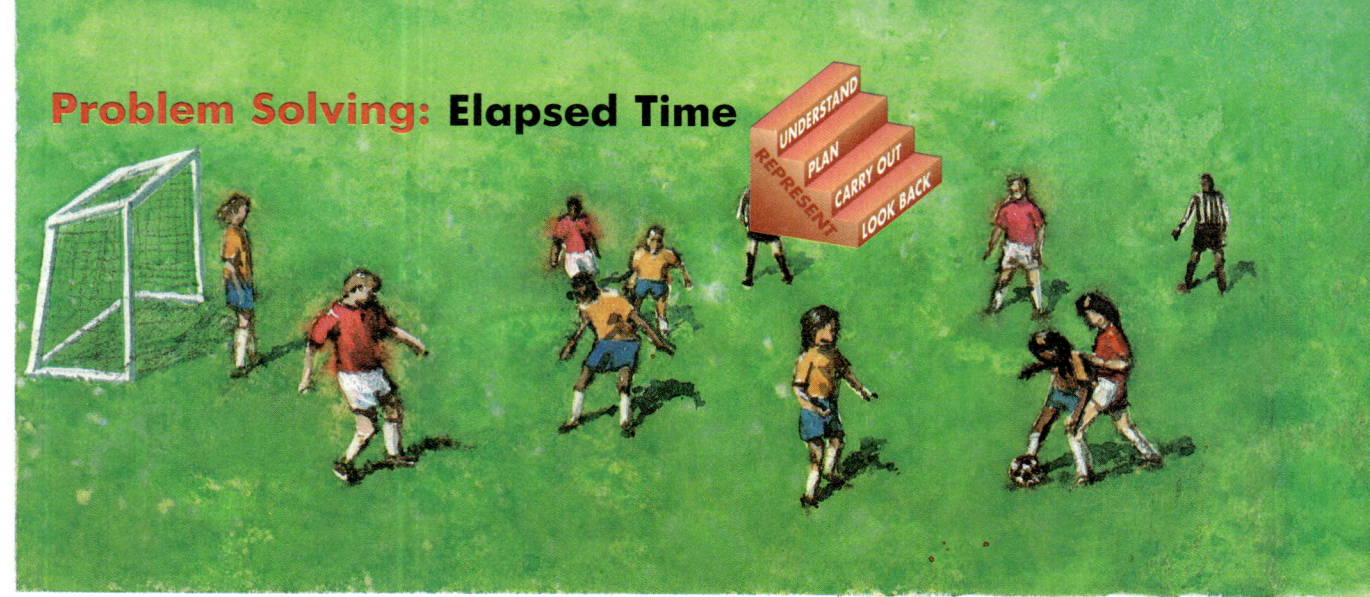

The Crocodiles arrived at the field at 7:15 A.M. for the soccer play-offs. Their game begins at 10:05 A.M. How long must they wait until their game starts?

You can subtract to find out.

$$\begin{array}{r} 10:05 \text{ A.M.} \rightarrow \overset{9}{\cancel{10}} \text{ h } \overset{65}{\cancel{05}} \text{ min} \\ 7:15 \text{ A.M.} \rightarrow -7 \text{ h } 15 \text{ min} \\ \hline 2 \text{ h } 50 \text{ min} \end{array}$$

60 min = 1 h, so 60 + 5 = 65 min.

They must wait 2 h 50 min until their game starts.

The Antelopes game started at 11:50 A.M. and ended at 1:25 P.M. How long did their game last?

$$\begin{array}{r} 1:25 \text{ P.M.} \rightarrow 13 \text{ h } 25 \text{ min} \\ 11:50 \text{ A.M.} \rightarrow -11 \text{ h } 50 \text{ min} \end{array}$$

Before subtracting to find the difference, you need to add 12 h to 1:25 to make it 13:25. Why?

Now you can solve the problem.

The Lions-Cougars game began at 4:15 P.M. and lasted 1 h 50 min. At what time did the game end?

- At what time did the game start?
- Do you need to add or subtract to solve?
- How would you write 65 min, using hours and minutes?

Now you are ready to solve.

160 LESSON 5-10

GUIDED PRACTICE

1. Read the passage. Then answer the questions.

   The Eagles won their play-off game against the Tigers at 3:35 P.M. Both teams will play again tomorrow morning at 9:45.

   a. At what time did the Eagles finish their game against the Tigers?

   b. Will they have to wait more than or less than 12 h for their next play-off game?

   c. How long must they wait from the end of their first play-off game against the Tigers until the start of their second play-off game tomorrow?

PRACTICE

2. Shina had soccer practice for 2 h 20 min on Saturday. Practice ended at 4:00 P.M. When did practice start?

3. Ricky was supposed to play in a game that began at 11:50 A.M. He arrived 1 h 10 min late. At what time did he arrive?

4. José was so tired after his game that he went to sleep at 7:30 P.M. How long can he sleep if his alarm goes off at 6:15 A.M.?

5. **CREATE YOUR OWN** Write a word problem that has the answer 2 h 45 min.

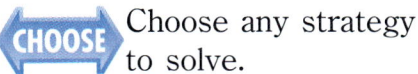 Choose any strategy to solve.

6. Sofia's game took $1\frac{3}{4}$ h. Then she took 50 min for lunch. At what time did her game start if she finished lunch at 1:25 P.M.?

7. The Hawks scored 1, 3, 5, 4, 4, 2, 3, 4, 3, 2, 1, 0, 6, and 4 points in their games this season. How many points did they average per game?

8. Linda's coach told her to practice this play: run up the field 5 yd, turn right and run 10 yd; run up the field 5 yd, turn left and run 5 yd. How far and in what direction would Linda's ending place be from where she started?

9. The winners of the play-off games were the Alligators, the Crocodiles, the Antelopes, and the Eagles. What combinations could play in the next round of games?

CHAPTER 5   **161**

Decision Making: Building a Fence

Suppose you are on a committee planning a split rail fence to go around a rectangular playground 32 ft long and 40 ft wide. The drawing below shows how a split-rail fence is set up.

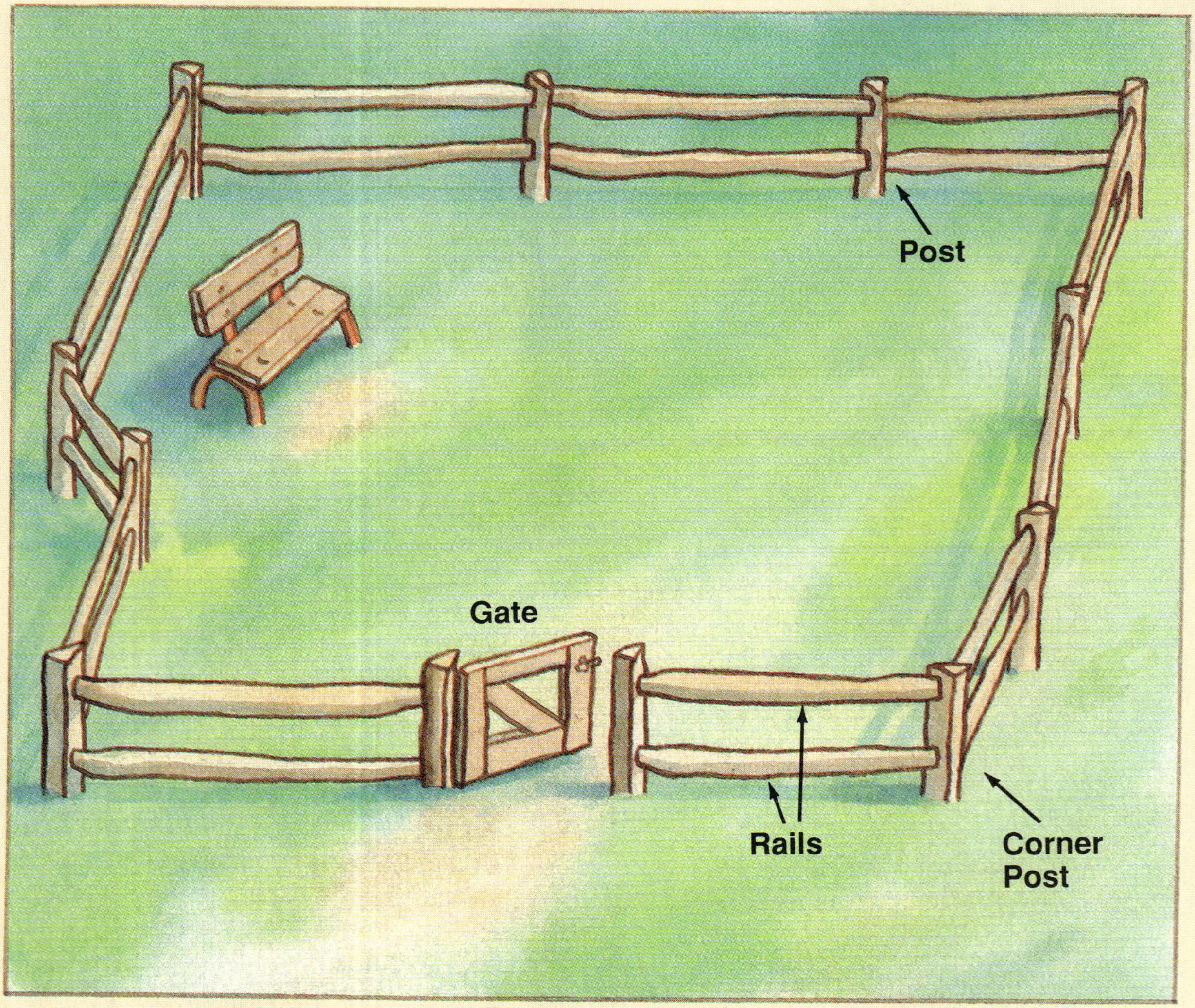

Two long pieces of wood, or rails, run between two wooden posts. The ends of the rails are slipped into holes, or notches, on either side of the posts. Corner posts are cut so that the rails form an angle.

162 LESSON 5-11

Work in a group. A diagram and a calculator may be helpful.

1. How many corner posts will you need? Using the price list shown, how much will the corner posts cost?

2. You will need a post every 8 ft. How many other posts will you need? How much will these cost?

3. The fence will need at least 1 gate. How many gates will you include in your plan? What will be your cost?

4. How many rails will you need to complete the fence? How much will the rails cost? (Remember: 2 rails go between each post.)

5. What is the total cost of the project as planned? Will your committee qualify for a free bench?

Do-It-Yourself Fencers

Split Rail Kit
(price per unit)

Split rails (8 ft)	$6.00
Posts	$8.00
Corner posts	$12.00
Gate set	$25.00

All pieces come ready to install. Free park bench included with orders over $400.

6. With your group, pick an area near your school or in your community that would look good with a split-rail fence.

Note: You don't have to build a rectangular fence, but it must have straight sides and square corners.

7. Measure the area or estimate its size. Draw a diagram of the area you plan to fence in.

8. Decide how many gates you will need and where they should be.

9. Make a list of all of the materials you will need and their cost.

Note: You can cut rails to fit between posts less than 8 ft apart, but you still have to buy 8-foot rails.

10. Find the total cost of the project.

11. Your budget is $450. Can you afford your fence as planned? If not, how will you adjust your plan to lower your costs?

Mixed Review

Write in standard form.
1. fifty-six thousand, thirty-eight
2. six and thirty thousandths
3. 7 + 0.5 + 0.003
4. 300,000 + 7,000 + 900 + 80

Write the value of the underlined digit.
5. 2<u>7</u>5,689
6. <u>3</u>,480,725
7. 2,74<u>1</u>,983
8. 24.63<u>5</u>
9. 65.<u>3</u>7
10. <u>6</u>98.45

Compare. Write >, <, or =.
11. 267,385 ■ 267,358
12. 680,375 ■ 68,375
13. 2,123 ■ 2,132
14. 0.5 ■ 0.499
15. 0.2 ■ 0.200
16. 0.07 ■ 0.7

Find the answer.
17. 84,732 + 6,457
18. 679 × 84
19. 241.08 ÷ 57.4
20. 0.45 × 27
21. 88,089 − 3,095
22. 972 ÷ 54
23. 1.632 ÷ 8
24. 19.777 + 0.958
25. 905 × 63
26. 114.144 − 97.79
27. 7.52 × 67
28. 7,814 ÷ 73
29. 19,925 + 6,048
30. 576,796 − 97,807
31. 1.589 ÷ 0.3
32. 50.005 − 4.6
33. 195 × 893
34. 5.9 − 4.023
35. 4.886 + 53.7 + 6.08 + 24.033
36. 2,783 + 486 + 21 + 36,459

PROBLEM SOLVING

CHOOSE Choose estimation, calculator, or paper and pencil to solve. Use the chart as needed.

37. What was the total mileage for the Texas trip?

38. Which day's expenses were the greatest? the least?

39. About how much was the total cost of the trip to Texas?

Texas Trip Expense

Day	Miles	Meals	Motels	Gas
1	385	$75.82	$58.00	$26.00
2	467	$56.75	$55.00	$29.50
3	553	$92.95	$70.00	$33.00

40. Mrs. Singh gave the cashier $40 for a bill of $32.65. How much change did she receive?

41. The odometer read 54,324 mi when the Singhs left home. If the odometer read 57,118 when they returned home, how many miles did they travel?

42. Each day on the trip, Anita drove 75 of the miles traveled. How far did she drive?

43. If the Singhs drove an average of 55 mi/h, how long did they drive the first day?

44. The Singhs bought 15 T-shirts. Each T-shirt costs $12.95. How much did they spend on T-shirts?

45. Jason had $39.50. He loaned his brother $10 and spent $8.95. How much did he have left?

46. The distance from Houston to El Paso, 756 mi, is 30 more than 3 times the distance from Houston to Dallas. How far is it from Houston to Dallas?

47. Jean counted 304 American cars, 176 foreign cars, and 37 motorcycles. How many vehicles did she count?

48. **CREATE YOUR OWN** Use the information in the chart to write a word problem. Give it to a partner to solve.

49. Use the chart to find the average cost of the Texas trip per day.

CHAPTER CHECKUP

LANGUAGE & VOCABULARY

Write these terms in order from least to greatest. Tell what each prefix means.

meter　　　decimeter　　　millimeter　　　kilometer　　　centimeter

TEST

CONCEPTS

Choose the more likely measurement.
(pages 142–147, 152–153, and 156–159)

1. the width of a field　　　　　　　　　　　　a. 30 ft　　　b. 30 yd
2. the weight of a cat　　　　　　　　　　　　a. 10 lb　　　b. 10 oz
3. the capacity of a soup bowl　　　　　　　　a. 12 fl oz　b. 3 qt
4. the temperature of a person with the flu　　a. 74°F　　　b. 101°F
5. the length of a pencil　　　　　　　　　　　a. 18 cm　　 b. 15 mm
6. the amount of milk in a glass　　　　　　　 a. 240 cL　　b. 240 mL
7. the mass of an orange　　　　　　　　　　　a. 2,100 mg　b. 210 g

SKILLS

Complete. *(pages 142–145, 152–153, and 156–159)*

8. 5,000 yd = 2 mi ▓ yd　　　　　　9. 4 t = ▓ lb
10. 6 gal = ▓ qt　　　　　　　　　　11. 6 dm = ▓ mm
12. 8,000 L = ▓ kL　　　　　　　　　13. 2,795 mg = ▓ g

Add or subtract. *(pages 142–143 and 152–153)*

14.　　4 ft 7 in.　　　　　　　　　　15.　　8 ft 1 in.
　　+ 2 ft 8 in.　　　　　　　　　　　　　− 5 ft 9 in.

16.　　4 m 286 mm　　　　　　　　　 17.　　5 m 65 cm
　　− 1 m 453 mm　　　　　　　　　　　　+ 4 m 55 cm

Find the perimeter. *(pages 154–155)*

18.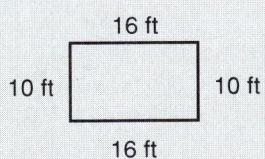

19.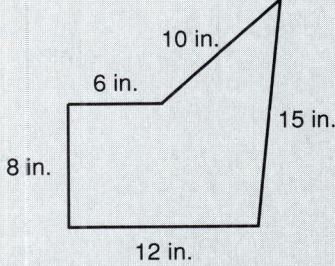

Find the perimeter of each rectangle. *(pages 154–155)*

20. length: 12 yd, width: 8 yd

21. a square with a side of 11 in.

Solve. *(pages 160–161)*

22. Tawanna attended a concert. The concert started at 8:30 P.M. and lasted until 10:10 P.M.
   a. At what time did the concert start?
   b. Which operation would you use to find how long the concert lasted?
   c. How long did the concert last?

Write a word problem. *(pages 148–149)*

23. Write a word problem that involves adding the lengths 4 ft 7 in. and 2 ft 8 in. Then solve the problem.

LEARNING LOG

Write the explanation in your learning log.

1. Explain how knowing the relationships between the metric units of length can help you remember the metric units of capacity and mass.

2. Do you think we need both customary and metric units? Defend your choice.

3. Explain why two rectangles that have different lengths and widths can have the same perimeter.

EXTRA PRACTICE

Choose the best unit (*in, ft, yd,* or *mi*) to measure.
(pages 142–143)

1. The distance from New York to Los Angeles is about 2,500 ▇.
2. The length of a pen is about $\frac{1}{2}$ ▇.

Choose the more likely measurement.
(pages 142–147, 152–153, and 156–159)

3. the capacity of a mug   a. $\frac{1}{2}$ pt   b. 25 fl oz
4. the weight of a student's desk   a. 20 lb   b. $\frac{1}{4}$ t
5. the temperature of ice cream   a. 36°F   b. 12°F
6. the height of your bedroom ceiling   a. 2.75 m   b. 27.5 cm
7. the amount of punch in a punchbowl   a. 2 kL   b. 15 L
8. the mass of a sixth grader   a. 140 kg   b. 35 kg

Complete. *(pages 142–145, 152–153, and 156–159)*

9. $2\frac{1}{2}$ yd = ▇ in.
10. $2\frac{1}{2}$ qt = ▇ c
11. $3\frac{1}{4}$ t = ▇ lb
12. 40 oz = ▇ lb ▇ oz
13. 10 pt = ▇ qt
14. 7,000 ft = 1 mi ▇
15. 50 dm = ▇ cm
16. 3.25 km = ▇ m
17. 0.4 L = ▇ mL
18. 246 L = ▇ kL
19. 12 t = ▇ kg
20. 327 mg = ▇ g

Add or subtract. *(pages 142–143 and 152–153)*

21.   5 yd 1 ft 8 in.
   + 4 yd 1 ft 8 in.

22.   6 yd 2 in.
   − 2 yd 20 in.

23.   2 m 476 mm
   + 3 m 964 mm

24.   8 m 46 cm
   − 3 m 54 cm

Find the perimeter. *(pages 154–155)*

25. 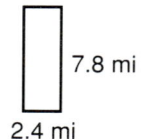 7.8 mi, 2.4 mi

26. 7 in., 9 in., 8 in., 4 in., 8 in.

27. a rectangle with length 8 ft, width 14 ft
28. a square with a side of 15 mi

Solve. *(pages 148–149 and 160–161)*

29. Write a word problem in which you use the times 8:45 A.M. and 3:20 P.M. Then solve it.

30. Dustin started mowing the lawn at 1:15 P.M. He finished 2 h 50 min later. When did he finish?

ENRICHMENT

MILITARY TIME

The United States military uses a 24-hour clock.

12-hour clock	24-hour clock
7:00 A.M.	0700
7:30 A.M.	0730
11:00 A.M.	1100
7:00 P.M.	1900

1200 h + 700 h

1. Make a schedule of the things you do during a normal day from the time you get up to the time you go to bed. Show all times using military time.

2. What advantages do you think there would be to everyone using 24-hour clocks? What disadvantages might there be?

Why is an inch as long as it is?

Use reference books in your classroom or library to find the answer to the question posed in the title.

Also read how other customary units of measure became standard units. Make a poster to show your findings.

METRIC vs. CUSTOMARY

- Use reference books to find out which countries have *not* adopted the metric system as their official system of measurement.

- Describe how you think the following people would be affected if the United States adopted the metric system as its official system of measurement.

  you
  your parents
  doctors
  builders
  food store managers
  farmers

ENRICHMENT 169

CUMULATIVE REVIEW

Solve each equation.

1. $n + 45 = 100$
   a. 45
   b. 55
   c. 65
   d. none of these

2. $8 \times n = 64$
   a. 7
   b. 8
   c. 9
   d. none of these

3. $n \div 9 = 63$
   a. 4
   b. 6
   c. 7
   d. none of these

Choose the best estimate.

4. 43×6
   a. 300
   b. 240
   c. 400
   d. none of these

5. 68×2
   a. 1,800
   b. 1,200
   c. 2,100
   d. none of these

6. 392×48
   a. 20,000
   b. 2,000
   c. 1,600
   d. none of these

7. $387 \div 8$
   a. 5
   b. 50
   c. 40
   d. none of these

8. $2.5 \div 0.4$
   a. 0.6
   b. 60
   c. 6
   d. none of these

9. $5,326 \div 88$
   a. 700
   b. 600
   c. 60
   d. none of these

Divide. Round to the nearest hundredth if necessary.

10. $12 \div 8$
    a. 0.67
    b. 1.33
    c. 1.5
    d. none of these

11. $43.2 \div 0.06$
    a. 720
    b. 7.2
    c. 72
    d. none of these

12. $9.756 \div 4.7$
    a. 2.07
    b. 2.08
    c. 2.76
    d. none of these

Find the average. Round to the nearest hundredth if necessary.

13. 3, 5, 7, 8, 9
    a. 7
    b. 6.2
    c. 6.4
    d. none of these

14. $77, $42, $58
    a. $59.00
    b. $42.00
    c. $88.50
    d. none of these

15. 4.28, 5.4, 2.3, 6.18
    a. 2.81
    b. 6.05
    c. 4.45
    d. none of these

CUMULATIVE REVIEW

PROBLEM SOLVING REVIEW

Remember the strategies and types of problems you have had so far. Solve.

Problem Solving Check List
- Choosing the operation
- Drawing a diagram
- Using estimation
- Using guess and check
- Too much information
- Too little information
- Multistep problems
- Looking back

1. The gas tank of Mr. Tallchief's car holds 16 gal. He uses 24 gal of gas each week driving back and forth to work. Gas costs $1.32 per gal.
   a. What is the cost of 1 gal of gas?
   b. What information is needed to find the cost of filling a gas tank?
   c. Mr. Tallchief's gas tank is empty. How much will it cost to fill it?

2. The Eagles play their last game 7 wk after their first game. The first game is on March 5. On what date do they play their last game?

3. Laura has 4 test scores that range from 80 to 92. The average of these scores is 88. Find one possible set of test scores for Laura.

4. The Bradys bought a refrigerator that cost $964.38 including tax. They plan to pay for this in 24 equal monthly payments. About how much will each payment be?

5. Attendance at opening day of the fair included 7,438 adults and 4,395 children. Admission is $3 per person. How much was paid for admission on opening day?

6. A train travels at an average speed of 40 mi/h. At this rate, how far can the train travel in 30 h?

7. How many cuts are needed to saw a 12-ft board into 2-ft pieces?

8. Mei has gymnastics every 4th day and a piano lesson every 7th day. She has gymnastics and a piano lesson today. In how many days will both activities be on the same day?

Tell whether each answer is *reasonable* or *unreasonable*. If *unreasonable*, explain why.

9. Each van can carry 8 people. How many vans are needed to carry the 20 members of the ski club?
   Answer: 2 R4 vans

10. Dan bought shoes for $27.95, jeans for $32.50, and a jacket for $74. About how much did he spend?
    Answer: about $130

TECHNOLOGY

PATHFINDERS

In the computer game "Metric Paths," players estimate length in centimeters. Try this paper-and-pencil version of the computer game to sharpen your estimation skills. Estimate the total length of each path. Then check your estimate by measuring.

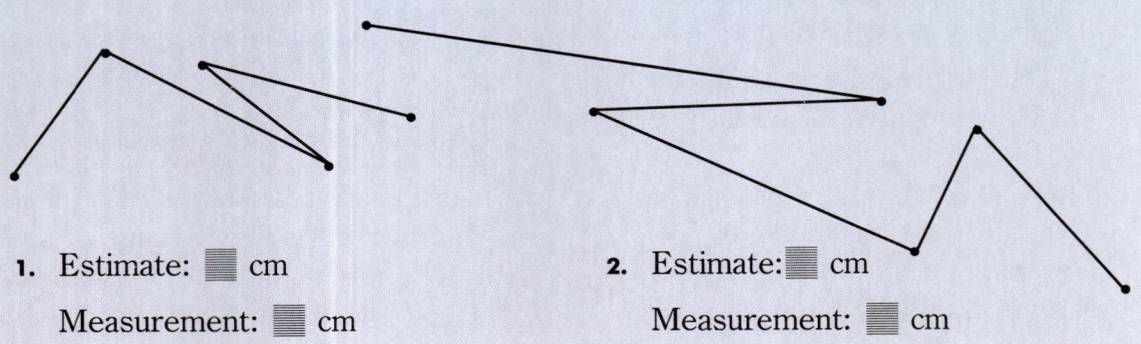

1. Estimate: ▓ cm

   Measurement: ▓ cm

2. Estimate: ▓ cm

   Measurement: ▓ cm

LIGHT WEIGHTS

Payne's Paper Clip Plant shipped 3 metric tons of tiny paper clips. If 3 paper clips weigh 1 gram, how many paper clips were in the shipment?

(*Hint:* 1 metric ton = 1,000 kg)

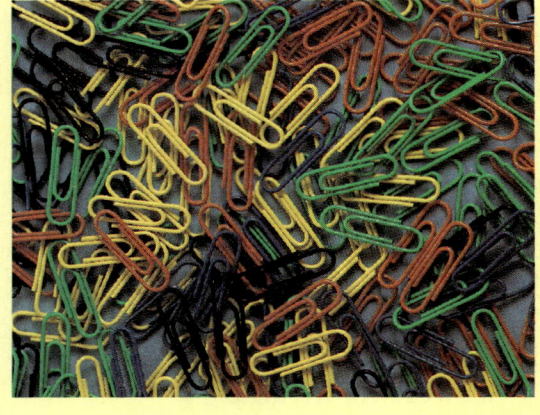

ROASTS

With an oven temperature of 175°C, it takes about 45 min per kilogram to cook a stuffed turkey. In hours and minutes, how long should it take to cook a stuffed turkey that weighs 7.6 kg?

Number Theory and Fraction Concepts 6

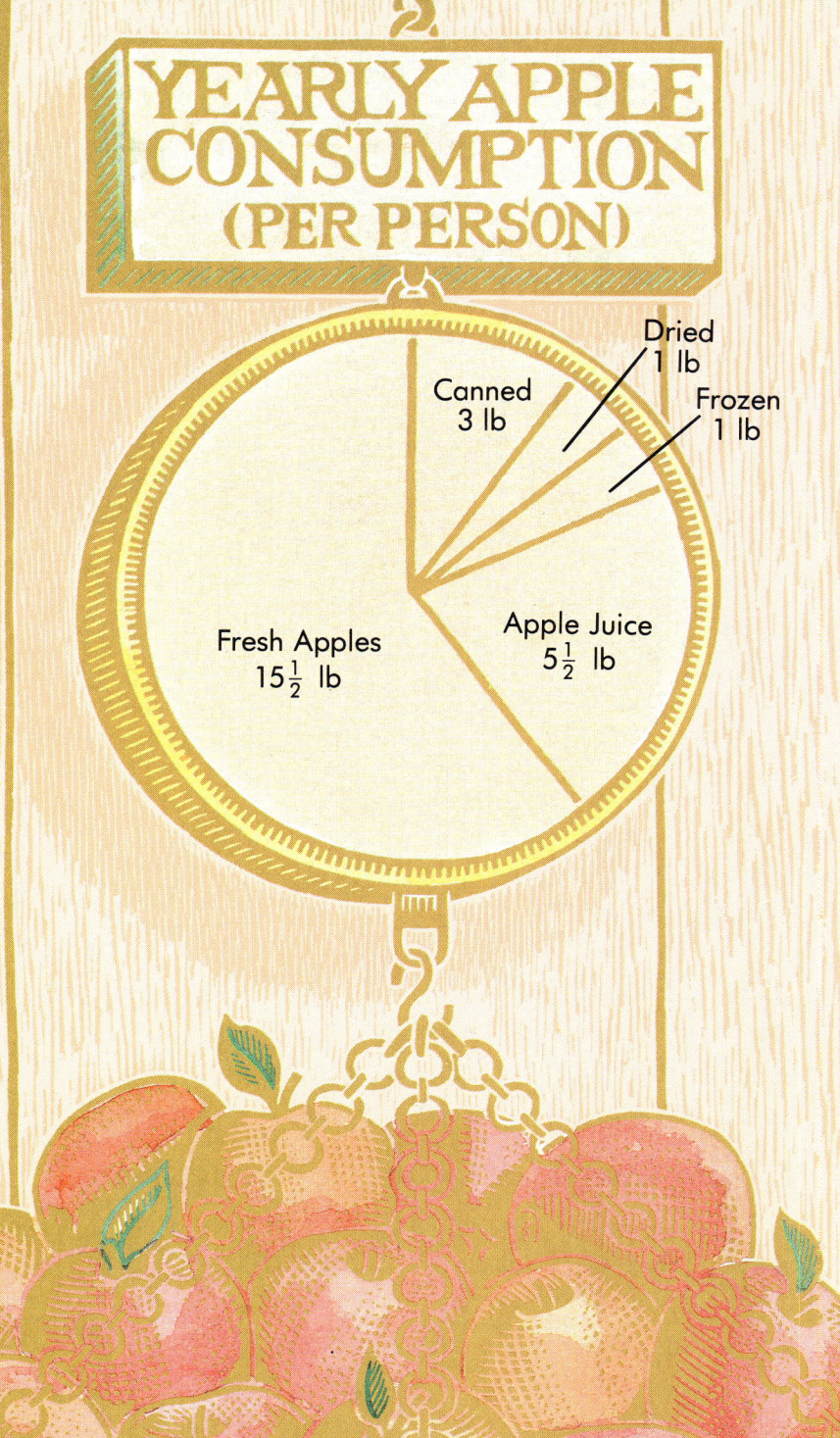

YEARLY APPLE CONSUMPTION (PER PERSON)

- Canned 3 lb
- Dried 1 lb
- Frozen 1 lb
- Fresh Apples 15 1/2 lb
- Apple Juice 5 1/2 lb

DID YOU KNOW…?
Over 7,000 varieties of apples have been grown in the United States! How many varieties can you name?

How many pounds of apples and apple products does the average person consume in a year? How much is that per week?

Take an apple survey! Ask 10 people to estimate how many pounds of apples and apple products they eat in a week. (Tell them that 3 medium apples weigh about 1 lb). Find the average of the estimates. How close is this average to the national average?

USING DATA
Collect
Organize
Describe
Predict

Least Common Multiples

Main Street is 100 blocks long, with a bus stop every 4 blocks and a subway station every 6 blocks. Bus stops are listed by street numbers in **multiples** of 4 and subway stations are listed by street numbers in multiples of 6.

Bus stops: 4 8 12 16 20 24 28 32 36 40 . . .
Subway stops: 6 12 18 24 30 36 42 . . .

What is the first street at which you can switch from the bus to the subway?

Find street numbers that are common to both lists.

> 4 8 **12** 16 20 **24** 28 32 **36** 40 . . .
> 6 **12** 18 **24** 30 **36** 42 . . .

The **common multiples** of 4 and 6 are 12, 24, 36, . . .

The **least common multiple (LCM)** of 4 and 6 is 12.

The first street at which you can switch from the bus to the subway is 12th Street.

Another example: Find the LCM of 5, 6, and 10. List multiples of each number.

> 0 5 10 15 20 25 **30** 35 40 45 50 55 **60** 65 . . .
> 0 6 12 18 24 **30** 36 42 48 54 **60** 66 . . .
> 0 10 20 **30** 40 50 **60** 70 . . .

We say that the first two common multiples are 30 and 60 and that the least common multiple is 30.

CRITICAL THINKING When you list the common multiples you do not include 0. Why not?

GUIDED PRACTICE

Complete. Do not include 0.

1. **a.** List the first ten consecutive multiples of 2.
   b. List the first five consecutive multiples of 5.
   c. List the first two common multiples of 2 and 5.
   d. What is the LCM of 2 and 5?

2. Find the LCM of 5 and 15.

174   LESSON 6–1

PRACTICE

Find the LCM. Use a calculator or pencil and paper.

3. 2, 6	**4.** 11, 3	**5.** 6, 9
6. 5, 6	**7.** 15, 25	**8.** 7, 35
9. 8, 6	**10.** 9, 27	**11.** 2, 9
12. 8, 32	**13.** 5, 12	**14.** 14, 6
15. 20, 12	**16.** 6, 24	**17.** 10, 18
18. 8, 15	**19.** 2, 3, 5	**20.** 3, 5, 8

List the first five even multiples. Do not include 0.

21. 3	**22.** 15	**23.** 7
24. 5	**25.** 9	**26.** 11

PROBLEM SOLVING

27. Local bus stops are listed on page 174. A new express bus will stop every 10 blocks, starting with 10th Street. Where can you transfer from the express bus to a local bus?

28. CREATE YOUR OWN Work with a partner to design a new transportation system for Main Street using the subway stations on page 174. Include many local bus stops, a limited number of express bus stops, and plenty of transfer points.

Mental Math

You can find the LCM of some numbers mentally.

Find the LCM of 8 and 10.

Think of multiples of the larger number, 10. For each one, ask yourself if it's also a multiple of 8. Stop if it is. That's the LCM.

Think: 10   20   30   40
         no   no   no   yes

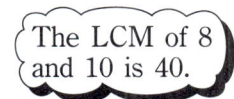

The LCM of 8 and 10 is 40.

Use mental math to find the LCM of each group of numbers.

1. 9, 15   **2.** 5, 7   **3.** 4, 5   **4.** 9, 12   **5.** 2, 5, 10

Exploring Divisibility

One whole number is **divisible** by another if it can be divided by that number without having a remainder. Multiples of a number are always divisible by that number.

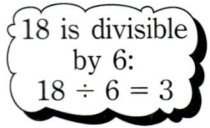

18 is divisible by 6: $18 \div 6 = 3$

Look at the first fifteen multiples of 2.

0	2	4	6	8
10	12	14	16	18
20	22	24	26	28

1. What pattern do you notice?

2. How can you tell if a number is divisible by 2 without actually dividing? We call a quick check like this a **divisibility rule**.

Look at the first twelve multiples of 5.

0	5	10	15	20	25
30	35	40	45	50	55

3. Make up a divisibility rule for 5.

Look at the first ten multiples of 10.

0	10	20	30	40
50	60	70	80	90

4. Make up a divisibility rule for 10.

- **A number is divisible by 2 if the last digit is even.**
- **A number is divisible by 5 if the last digit is 0 or 5.**
- **A number is divisible by 10 if the last digit is 0.**

Look at the first thirty-five multiples of 4.

0	4	8	12	16
20	24	28	32	36
40	44	48	52	56
60	64	68	72	76
80	84	88	92	96
100	104	108	112	116
120	124	128	132	136

5. What happens to your list after you reach 100?
6. 24 is a multiple of 4. Is 124 a multiple of 4?
7. 34 is not a multiple of 4. Is 134 a multiple of 4?
8. 16 is a multiple of 4. Are 216 and 516 multiples of 4? Explain how you can tell without dividing.

A number is divisible by 4 if the last two digits are divisible by 4.

9. Write *true* or *false*.
   a. Any number that is divisible by 2 is also divisible by 4.
   b. If a number is divisible by 10, it is always divisible by 5.

LESSON 6–2

Look at the first twenty multiples of 3.

10. For each multiple, add up the digits. What do you notice?

```
 0   3   6   9  12
15  18  21  24  27
30  33  36  39  42
45  48  51  54  57
```

Look at these numbers that are not divisible by 3.

11. For each one, add up the digits. What do you notice?

12. Make up a divisibility rule for 3.

```
17  53  13  59  88
41  29  32  25  73
93 127 136 214 301
```

> **A number is divisible by 3 if the sum of the digits is divisible by 3.**

Look at the first fifteen multiples of 6.

13. Is every multiple of 6 also a multiple of 3?

14. Is every multiple of 3 also a multiple of 6?

15. Start to list the multiples of 3 which are also multiples of 6. What kind of number are they?

16. To tell that a number is divisible by 6, you need to check that it is divisible by what two numbers?

```
 0   6  12  18  24
30  36  42  48  54
60  66  72  78  84
```

> **A number is divisible by 6 if it is divisible by 2 and by 3.**

Who am I? Use the numbers **2, 3, 4, 5, 6,** and **10**.

17. I go evenly into 10.

   To check whether a number is divisible by me, you need to look only at its last digit.

18. I go evenly into 100, but not into 10.

   To check whether a number is divisible by me, you need to look at its last two digits.

19. I do not go evenly into 10 or 100 or 1,000 or any power of 10.

   When checking if a number is divisible by me, you need to consider all its digits.

Primes and Composites

It took computer scientists $3\frac{1}{2}$ weeks and hundreds of computers to factor a number with 100 digits.

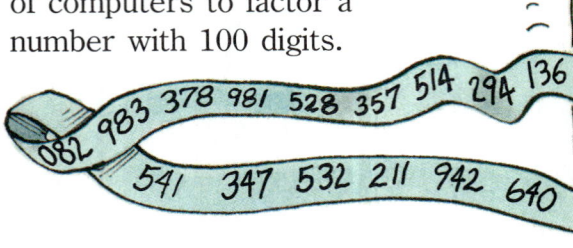

A **prime number** has exactly two factors, 1 and the number itself. The smallest five prime numbers are 2, 3, 5, 7, and 11.

A **composite number** has more than two factors. Since 4 has three factors, 1, 2, and 4, it is a composite.

Is the 100-digit number shown on the input tape prime or composite? Explain how you decided.

THINK ALOUD   The numbers 0 and 1 are neither prime nor composite. Why isn't 1 prime?

You can write every composite as the product of prime numbers. This is called **prime factorization.** You can use a **factor tree** to find the prime factors.

Here are factor trees for 48, 210, and 53.

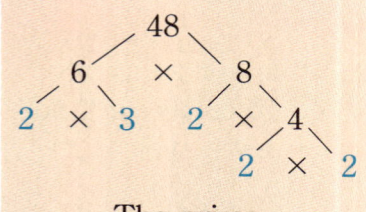

The prime factorization of 48 is $2 \times 2 \times 2 \times 2 \times 3$.

The prime factorization of 210 is $2 \times 3 \times 5 \times 7$.

53
53 cannot be split up any further.

53 is a prime number.

CRITICAL THINKING   Look at the factor tree for 48. Would you get the same prime factors for 48 if you started with 4×12 instead of 6×8? Explain.

GUIDED PRACTICE

Complete each factor tree.

1.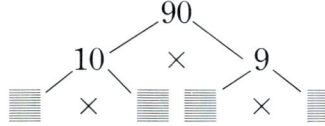
2.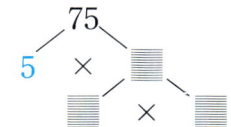
3.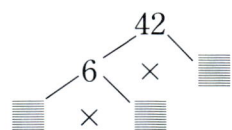

THINK ALOUD Is the number prime or composite? Explain.

4. 18   5. 29   6. 37   7. 28   8. 40   9. 41

PRACTICE

Is the number prime or composite?

10. 27   11. 17   12. 51   13. 47   14. 96   15. 254

Draw a factor tree for each number. Then write the prime factorization.

16. 36   17. 64   18. 81   19. 22   20. 144   21. 80
22. 84   23. 68   24. 1,000   25. 240   26. 675   27. 231

Use the numbers in the box to answer questions 28–30.

28. Which numbers have only 2's or 3's as prime factors?

29. Which numbers have only 2's or 5's as prime factors?

30. Which numbers have only 2's, 3's, or 5's as prime factors?

> 2, 5, 6,
> 9, 10, 15,
> 18, 20, 24,
> 27, 30, 42,
> 48, 50, 60

31. **MENTAL MATH** Suppose you know the prime factorization of 30. Explain how to find the prime factorization of 60, 150, and 270 mentally.

PROBLEM SOLVING

32. Think of the next consecutive number after the 100-digit number being inputted on page 178. What digit does it end in? Give one factor of that number other than 1 or the number itself.

33. Work with a partner to find the number between 1 and 50 that has the greatest number of prime factors. Count each time you repeat a prime. (5 × 5 counts as 2 factors.)

CHAPTER 6   179

Exponents

You can picture some numbers as squares or cubes.

You can show 25 as a 5 by 5 square.
25 is a **square number**.

You can show 125 as a 5 by 5 by 5 cube.
125 is a **cubic number**.

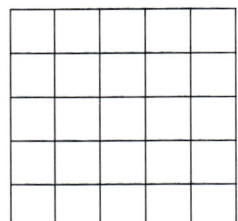 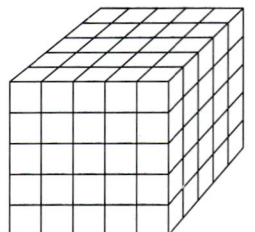

You can write 5×5 as 5^2.
Read 5^2 as *five squared* or
five to the second power.

You can write $5 \times 5 \times 5$ as 5^3.
Read 5^3 as *five cubed* or
five to the third power.

The **exponent** tells you how many times to use the **base** as a factor.

With Exponents	As a Product of Factors	In Standard Form
base→ 6^3 ←exponent	$6 \times 6 \times 6$	216

You can use exponents to write the prime factorization of a number.

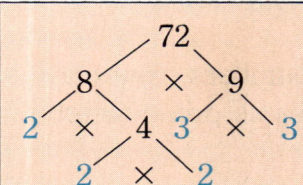

The prime factorization of 72 is
$2 \times 2 \times 2 \times 3 \times 3$ or $2^3 \times 3^2$

GUIDED PRACTICE

Read the number. Then write it as a product of factors.

1. 3^3 2. 1^2 3. 25^2 4. 12^3 5. 36^2

Complete. Write in standard form.

6. six squared = $6^\square$ = $\square \times \square$ = $\square$

7. $2 \times 2 \times 2 = 2^\square = \square$

PRACTICE

Write using exponents.

8. $7 \times 7 \times 7$ 9. 3×3 10. 17×17 11. 20×20 12. $15 \times 15 \times 15$

13. nine squared 14. nine cubed 15. eleven to the second power

Write in standard form. Use a calculator when needed.

16. 7^2 **17.** 12^2 **18.** 4^3 **19.** 10^3 **20.** 0^2 **21.** 1^3

Write the prime factorization, using exponents.

22. 18 **23.** 100 **24.** 40 **25.** 54 **26.** 24

CALCULATOR Use a calculator to write in standard form.

27. 8^3 **28.** 16^2 **29.** 13^3 **30.** 99^2 **31.** 21^3 **32.** 35^2

NUMBER SENSE Write *square number* or *cubic number*.

33. 4 **34.** 27 **35.** 8 **36.** 9 **37.** 16

PROBLEM SOLVING

38. Use exponents to write the number of playing spaces for each game.

39. Which two games have the same number of playing spaces?

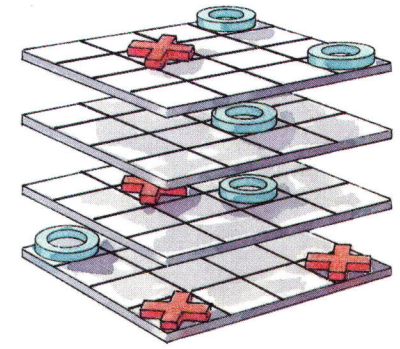

a form of
3-D tick-tack-toe

American checkers

Alquerque is an early form of checkers. A similar game is still played in Spain.

Critical Thinking

When you multiply a whole number by itself, the product is a square number. The number you started with is called the **square root**. Since $5^2 = 25$, the square root of 25 is 5. You can write $\sqrt{25} = 5$.

Write the square root for each. Discuss how you found it.

1. 4 **2.** 16 **3.** 9 **4.** 36 **5.** 49

6. Is the square root of 10 a whole number? Explain. If your calculator has a square root key, use it to find $\sqrt{10}$.

CHAPTER 6   181

Problem Solving Strategy: Making a Table

Ramona and Ian are studying cooking. They are testing various types of chiles.

> Chipotle hot ($2.50 for 2oz)
> Pasado cold (1oz for $1.75)
> Chiltecpin Wow!! ($2.25 for ½ oz)
> Ancho (68¢/oz) not so hot
> de Arbol pretty hot ($1.25 for 1oz bag)

Is it easy to tell from Ian's notes the differences between the kinds of chiles?

In what ways does organizing the data in a table help to compare the results of the testing?

In what other ways could Ian have organized the table?

THINK ALOUD What are some of the questions you need to ask yourself before making a table like this?

> Chiles — From Hot to Mild
>
> | Name | How Hot? | Price per Ounce |
> |---|---|---|
> | Chiltecpin | Very, very, hot | $4.50 |
> | de Arbol | Very hot | $1.25 |
> | Chipotle | Hot | $1.25 |
> | Ancho | Mild to Medium | $.68 |
> | Pasado | Mild to Medium | $1.75 |

GUIDED PRACTICE

Read the passage. Answer the questions.

One of the world's hottest chiles, the serrano chile, is about 2 in. long and $\frac{1}{2}$ in. wide. It is brick red in color. The pico de pajaro chile, deep red and very hot, grows to be about $\frac{1}{2}$ in. wide and $\frac{1}{4}$ to $\frac{3}{4}$ in. long. A much milder chile, the pasilla, is a blackish-brown chile about 7 in. long and 1 in. wide.

1. How can you organize the information in a table?

2. What headings should you use for the table?

3. Will it help to put the chiles in some kind of order?

4. Make the table. How does your table compare with those made by the other students?

LESSON 6-5

PRACTICE

Make a table. Answer the question.

5. A sauce recipe calls for 12 New Mexico chiles and 2 ancho chiles. The recipe makes 3 c of sauce. How many of each chile are needed to make 15 c of sauce?

6. Brian inspected chiles for a specialty food shop. He found 2 out of every 15 were of poor quality. How many poor quality chiles would he expect to find in a shipment of 120 chiles?

7. Bakersfield, California, averages 4.5 in. of rain per year. Fresno, California, where Fresno chiles are grown, averages about 10.5 in. How many years will it take for Bakersfield to get as much rain as Fresno gets in 3 years?

8. Mario, Ana, and Paul taste-tested three types of chiles. Mario liked the Bahamian. Ann liked the Tabasco. Paul liked both the Bahamian and the cayenne. Which chiles did each of them like and dislike?

CHOOSE Choose any strategy to solve.

9. Sue was testing seeds for a new type of chile plant. Four out of ten seeds did not sprout. At this rate, how many out of 100 seeds will not sprout?

10. It has been said that the habañero chile is 100 times hotter than the jalapeño and 50 times hotter than the serrano. Of the jalapeño and the serrano, which is the hotter chile? How could you describe its hotness?

11. Which would be the best buy? Three chiles for $1, five chiles for $1.59, or eight chiles for $1.99? Explain.

12. **CREATE YOUR OWN** Predict whether more people prefer spicy, hot foods or plain foods. Then survey 25 people and make a table to show your results.

CHAPTER 6   183

MIDCHAPTER CHECKUP

LANGUAGE & VOCABULARY

Choose the best word to complete the sentence.

least common multiple   divisible   prime
composite   prime factorization   exponent

1. A ? number has exactly two factors—itself and 1.

2. Composite numbers can be written as the product of prime numbers. It is called the ? of the number.

3. Even numbers are ? by 2.

4. The ? of 5 and 10 is 10.

5. Ten is a ? number because it has more than two factors.

QUICK QUIZ

1. List the first five multiples of 7. *(pages 174–175)*

2. What is the LCM of 9 and 12? *(pages 174–175)*

3. By which of these numbers—2, 3, 4, 5, 6, and 10—is 316 divisible? *(pages 176–177)*

Is the number prime or composite?
(pages 178–179)

Write using exponents.
(pages 180–181)

4. 39   5. 71   6. $4 \times 4 \times 4$ 7. seven squared

8. Write 8^2 in standard form. *(pages 180–181)*

9. Draw a factor tree for 120. Then write its prime factorization. *(pages 178–179)*

Solve. *(pages 182–183)*

10. After 3 weeks of saving, Jonah can afford to buy 7 pieces of track for his racing set. At that rate, how long does he need to save to buy 35 sections of track?

LEARNING LOG

Write the answers in your learning log.

1. Describe the least common multiple of two consecutive numbers.

2. Describe what the multiples of 2, 4, 6, and 8 all have in common.

MATH AMERICA

Among the tourist attractions in Utah are Bryce Canyon, the Rainbow Bridge, and Capitol Reef National Park. Plan a route for tourists to use when visiting your state. Choose at least three places of interest for them to visit.

BONUS

The number 5 can be written as the sum of two prime numbers.

$5 = 2 + 3$

How many numbers from 1 through 20 can be written as the sum of two prime numbers? Try it!

Picture This

Non-Numerical Graphing

READ ABOUT IT

Line graphs show changes that take place over time. This line graph shows how a swimmer's speed changes from the beginning to the end of a 200-meter race.

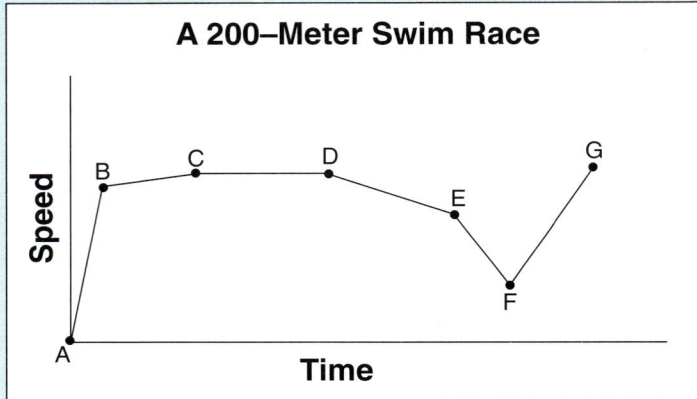

- Between points A and B, the swimmer was diving into the water.
- Between points B and C, the swimmer began swimming a little faster with each stroke.
- From point C to point D, the swimmer swam steadily at the same speed.

TALK ABOUT IT

Answer the questions.

1. What began to happen to the swimmer's speed between points D and E? Why?
2. What happened from E to F? Why?
3. Explain what was happening from point F to the end of the race.

186 LESSON 6–6

The students in Dr. Lin's math class are conducting an experiment popping popcorn. The graph shows the results of the experiment.

The students knew that the volume of the popcorn would increase as it popped.

Here are the results of the experiment.

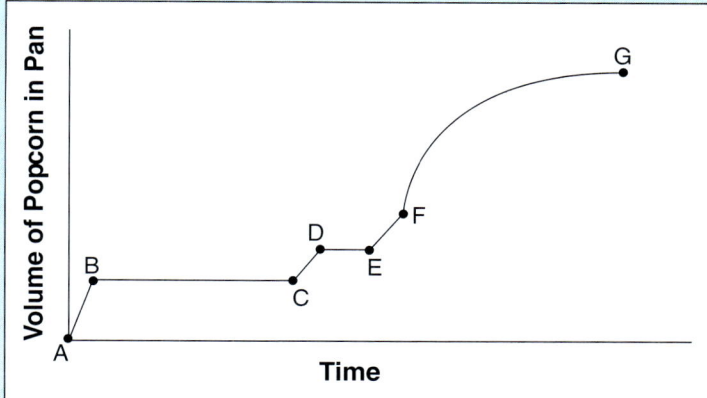

Use the graph to answer the questions.

4. What was done with the popcorn between points A and B?

5. What was happening between points B and C?

6. Why did the line start to go up at point C?

7. The line was horizontal again between points D and E. Why?

8. Describe what happened from point E to point G.

WRITE ABOUT IT

9. Choose one situation. Write a story and draw a graph.
   a. a hike
   b. driving in traffic
   c. driving in the city and then in the country
   d. a walk to school from your home
   e. the amount of talking you do in class all day

Meaning of Fractions

When Mount St. Helens erupted in 1980, about one eighth of the mountain's height was lost.

You can write a fraction to show a part of a region or measurement.

numerator: $\dfrac{1}{8}$
denominator:

If we think of the height of the mountain in 8 parts, 1 out of 8 was lost in the eruption.

You can write a fraction to represent a part of a set.

   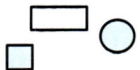

numerator: $\dfrac{4}{7}$ shaded items
denominator: items in set

The denominator tells how many equal-size parts are in the whole or how many items are in the set.

CRITICAL THINKING When writing a fraction for part of a set, the items can be different sizes. Why?

There are 850 active volcanos in the world. Three fourths of these are in the Ring of Fire.

GUIDED PRACTICE

Write the fraction for the part that is shaded.

1. 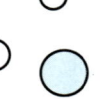   2. 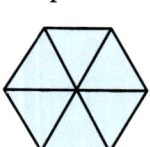   3. 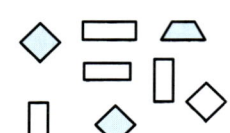   4.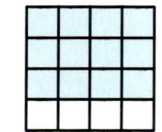

5. Write the fraction for *three out of five coins are pennies.* Then draw a picture to show it.

PRACTICE

Write the fraction for the part that is shaded.

6. 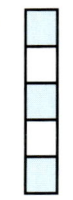   7. 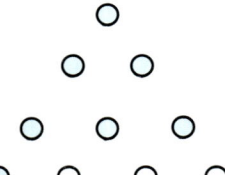   8. 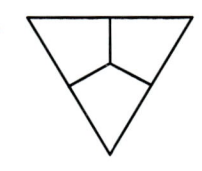   9.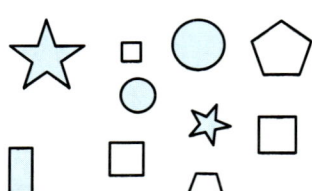

10. 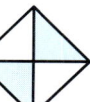 **11.**  **12.**  **13.**

Write the fraction. Then draw a picture to show it.

14. one eighth   **15.** three sevenths   **16.** four fifths

17. five sixths   **18.** zero halves   **19.** ten tenths

Write a fraction for the numbers in the statement.

20. 9 out of 10 students have pets.   **21.** 2 out of 5 classes are on a trip today.

22. 7 out of 7 students drink milk.   **23.** 3 out of 5 students ride bikes.

CRITICAL THINKING Part of the region or set is shown. Draw the whole set or region.

24. □ $\frac{1}{2}$ **25.** □□□□ □□□ $\frac{7}{15}$ **26.** ⌒ $\frac{1}{4}$ **27.** ◇ $\frac{2}{6}$

28. ○○○○ $\frac{1}{2}$ **29.** ◁▷ $\frac{1}{3}$ **30.** ▽ $\frac{2}{5}$ **31.** ▭ $\frac{3}{4}$

PROBLEM SOLVING

Use the information in the chart to solve.

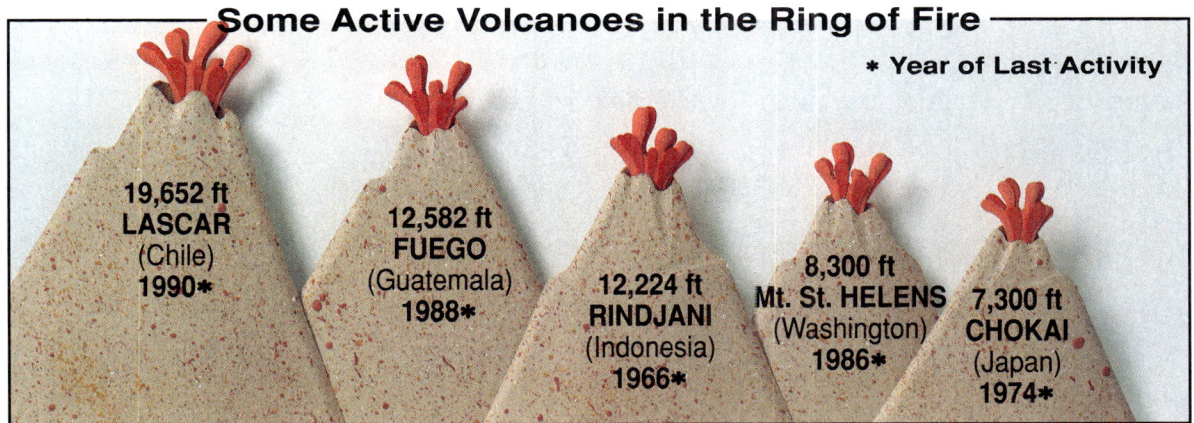

Some Active Volcanoes in the Ring of Fire
* Year of Last Activity

19,652 ft LASCAR (Chile) 1990*
12,582 ft FUEGO (Guatemala) 1988*
12,224 ft RINDJANI (Indonesia) 1966*
8,300 ft Mt. St. HELENS (Washington) 1986*
7,300 ft CHOKAI (Japan) 1974*

32. What fraction of the volcanoes listed in the chart are taller than 10,000 feet?

33. What fraction of the volcanoes listed in the chart have been inactive since 1980?

34. IN YOUR WORDS Which heights seem like rounded numbers? Explain. What place are they rounded to?

Equivalent Fractions

Two fractions that represent the same amount are **equivalent fractions.**

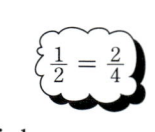

To find an equivalent fraction, you can multiply the numerator and denominator by the same number—as long as the number is not zero.

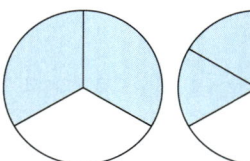 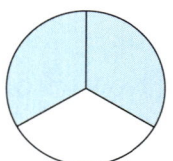 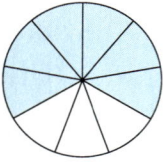 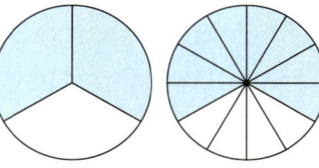

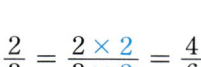

$$\frac{2}{3} = \frac{2 \times 2}{3 \times 2} = \frac{4}{6} \qquad \frac{2}{3} = \frac{2 \times 3}{3 \times 3} = \frac{6}{9} \qquad \frac{2}{3} = \frac{2 \times 4}{3 \times 4} = \frac{8}{12}$$

$\frac{2}{3} = \frac{4}{6} = \frac{6}{9} = \frac{8}{12} = \ldots$ is a **set of equivalent fractions**.

By multiplying in order by 2, by 3, and then by 4, you get the first three fractions that are equivalent to $\frac{2}{3}$. The dots show that the list could go on forever.

THINK ALOUD Suppose you multiplied the numerator and the denominator of $\frac{2}{3}$ by 7. How would you show that with a picture?

=== GUIDED PRACTICE ===

Name two equivalent fractions for the shaded part.

1. 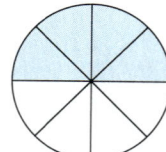 2. 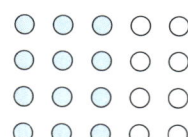 3. 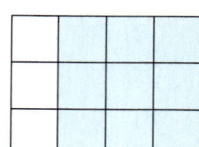 4.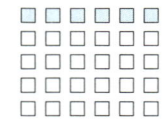

Complete.

5. $\frac{4}{7} = \frac{4 \times \blacksquare}{7 \times \blacksquare} = \frac{n}{35}$ 6. $\frac{3}{8} = \frac{9}{n}$ 7. $\frac{2}{5} = \frac{n}{10}$

=== PRACTICE ===

Write any three equivalent fractions for each.

8. $\frac{2}{5}$ 9. $\frac{2}{3}$ 10. $\frac{4}{9}$ 11. $\frac{3}{8}$ 12. $\frac{2}{7}$ 13. $\frac{5}{6}$

LESSON 6-8

Complete.

14. $\frac{1}{2} = \frac{n}{12}$ **15.** $\frac{1}{7} = \frac{n}{14}$ **16.** $\frac{2}{4} = \frac{n}{8}$ **17.** $\frac{2}{9} = \frac{n}{18}$ **18.** $\frac{3}{5} = \frac{n}{10}$

19. $\frac{6}{9} = \frac{n}{54}$ **20.** $\frac{3}{7} = \frac{21}{n}$ **21.** $\frac{2}{11} = \frac{n}{44}$ **22.** $\frac{2}{2} = \frac{n}{6}$ **23.** $\frac{4}{10} = \frac{n}{80}$

24. $\frac{6}{7} = \frac{42}{n}$ **25.** $\frac{5}{12} = \frac{n}{48}$ **26.** $\frac{8}{9} = \frac{24}{n}$ **27.** $\frac{10}{3} = \frac{n}{9}$ **28.** $\frac{4}{6} = \frac{n}{9}$

29. NUMBER SENSE Does adding the same amount to the numerator and denominator produce an equivalent fraction? Explain.

30. IN YOUR WORDS Look at $\frac{4}{14}, \frac{8}{28}, \frac{16}{56}$. Are these the first three fractions that are equivalent to $\frac{2}{7}$? Explain.

PROBLEM SOLVING
---

31. a. Fold a sheet of paper in half twice. Open the paper and shade one part. Write a fraction for the shaded part.

b. Refold the paper along the same lines. Fold it one more time. Write a fraction for the part you think is shaded. Unfold the paper to check.

32. Predict how many parts there will be if you fold a sheet of paper in half five times. Try it and check your answer.

33. This basic origami fold is used for many objects, including a star. Give two fractions for the part of the paper that is folded upward.

34. The diagram shows one stage in making origami flowers. The big triangle in front is what fraction of the sheet of paper? Give your answer in two different ways.

Origami is the Japanese art of paper folding. Different types of paper flowers can be made by cutting the petals various ways.

CHAPTER 6 191

Greatest Common Factors

Dairy Delight: 40 flavors
Frozen Fun: 24 flavors

The advertisers for Dairy Delight had to decide how many flavors of frozen yogurt each cone would represent. Since they wanted to use only whole cones, they needed to choose a number that goes into both 24 and 40 evenly.

Factors of 24: 1, 2, 3, 4, 6, 8, 12, 24
Factors of 40: 1, 2, 4, 5, 8, 10, 20, 40

The **common factors** of 24 and 40 are 1, 2, 4, and 8. One cone could represent 1, 2, 4, or 8 flavors.

The **greatest common factor (GCF)** of 24 and 40 is 8. If one cone represents 8 flavors, the pictograph uses the fewest possible cones.

THINK ALOUD When listing factors of a number, what system can you use to be sure you include every factor?

GUIDED PRACTICE

1. List the following to find the GCF of 21 and 35.
   a. factors of 21
   b. factors of 35
   c. common factors of 21 and 35
   d. GCF of 21 and 35

2. **THINK ALOUD** Explain how to find the GCF of 12, 20, and 36.

PRACTICE

List all the factors.

3. 15   4. 23   5. 18   6. 47   7. 2
8. 1   9. 21   10. 100   11. 38   12. 72

For each number, list all the factors. Then find the common factors and the GCF.

13. 9, 15   14. 16, 24   15. 27, 36   16. 15, 55   17. 12, 60

18. 22, 55   **19.** 16, 30   **20.** 30, 45   **21.** 21, 42, 49   **22.** 16, 32, 48

CRITICAL THINKING Give an example to prove that the statement is false.

23. All odd numbers have only two factors.

24. All even numbers have at least three factors.

25. The GCF of two numbers is always greater than 1.

PROBLEM SOLVING

26. Both stores will sell fruit ice this summer as well as the frozen yogurt flavors shown on page 192. Frozen Fun will add 12 flavors and Dairy Delight will add 14 flavors.

  a. How many flavors of frozen yogurt and fruit ice combined will each store sell then?

  b. If a new pictograph were made for the summer, what is the greatest number of flavors one cone could represent?

  c. How many cones would be shown for each store?

27. IN YOUR WORDS Use the results from Exercise 26. If Frozen Fun also adds 9 flavors of ice milk, would a pictograph comparing total flavors still be as effective an advertisement for Dairy Delight? Explain.

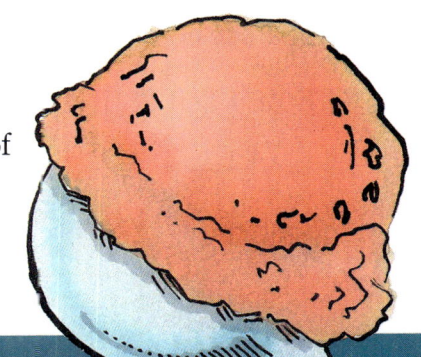

Mental Math

Sometimes you can find the GCF of two numbers mentally.

To find the GCF of 18 and 30:

List all the factors of 18.   1, 2, 3, 6, 9, 18

Starting with the largest one, mentally check each factor of 18 to see whether it is also a factor of 30. Stop if it is. That's the GCF.

18? no   9? no   6? yes   *The GCF of 18 and 30 is 6.*

Use mental math to find the GCF for each group of numbers.

1. 10, 20   **2.** 7, 28   **3.** 12, 16   **4.** 25, 40   **5.** 15, 25, 30

Lowest Terms

The ancient Egyptian measurement system was based on the human body. The basic unit of measure was the width of a finger. It was called a digit.

1 palm = 4 digits
1 hand = 5 digits
1 cubit (elbow to finger tip) = 28 digits

A palm was $\frac{4}{28}$ of a cubit.

You can reduce $\frac{4}{28}$ to **lowest terms.** Divide the numerator and the denominator by the greatest common factor (GCF).

The GCF of 4 and 28 is 4.

$$\frac{4}{28} = \frac{4 \div 4}{28 \div 4} = \frac{1}{7}$$

Divide the numerator and the denominator by 4.

A palm was $\frac{1}{7}$ of a cubit.

CRITICAL THINKING How could you draw a picture to show that $\frac{4}{28} = \frac{1}{7}$?

A fraction like $\frac{5}{28}$ is in lowest terms because the only common factor of 5 and 28 is 1.

THINK ALOUD Why do you need to use a common factor when reducing a fraction? Why would you use the GCF?

This ancient carving is from the Hathor Temple in Egypt. The picture symbols, or hieroglyphs, at left represent ideas and sounds. The Egyptians called their writing *mdw ntr* (phonetically, *medew netjer*) meaning the words of the god.

GUIDED PRACTICE

Complete.

1. $\frac{4}{10} = \frac{2}{n}$
2. $\frac{7}{14} = \frac{n}{2}$
3. $\frac{16}{24} = \frac{2}{n}$
4. $\frac{11}{99} = \frac{n}{9}$
5. $\frac{15}{35} = \frac{3}{n}$

6. **THINK ALOUD** Describe the steps you would follow to write any fraction in lowest terms.

PRACTICE

Write the missing number.

7. $\frac{6}{12} = \frac{n}{2}$
8. $\frac{8}{24} = \frac{n}{3}$
9. $\frac{18}{27} = \frac{n}{3}$
10. $\frac{6}{15} = \frac{n}{5}$
11. $\frac{12}{28} = \frac{n}{7}$

12. $\frac{6}{8} = \frac{n}{4}$
13. $\frac{21}{36} = \frac{n}{12}$
14. $\frac{8}{12} = \frac{2}{n}$
15. $\frac{9}{36} = \frac{1}{n}$
16. $\frac{15}{30} = \frac{1}{n}$

Write the fraction in lowest terms.

17. $\frac{14}{32}$
18. $\frac{30}{100}$
19. $\frac{24}{40}$
20. $\frac{21}{48}$
21. $\frac{12}{42}$
22. $\frac{8}{10}$

23. $\frac{21}{33}$
24. $\frac{18}{30}$
25. $\frac{9}{24}$
26. $\frac{13}{39}$
27. $\frac{51}{33}$
28. $\frac{81}{108}$

IN YOUR WORDS Is the fraction in lowest terms? Explain why or why not.

29. $\frac{12}{25}$
30. $\frac{3}{42}$
31. $\frac{7}{56}$
32. $\frac{6}{11}$
33. $\frac{12}{84}$
34. $\frac{8}{15}$

Choose whether to multiply or to divide. Write the missing number.

35. $\frac{4}{30} = \frac{n}{15}$
36. $\frac{1}{3} = \frac{11}{n}$
37. $\frac{4}{5} = \frac{n}{20}$
38. $\frac{17}{51} = \frac{n}{3}$
39. $\frac{2}{5} = \frac{16}{n}$

PROBLEM SOLVING

Solve. Write fraction answers in lowest terms.

40. The ancient Roman mile was equal to 1,000 paces. Suppose a Roman walked 600 paces. What part of a Roman mile would she have covered?

41. The barleycorn is another historical unit of measure. Fifteen barleycorns placed side by side measured 5 in. What fraction of an inch is 1 barleycorn?

42. **IN YOUR WORDS** People used their own hands and paces to measure. What problems do you think this caused?

43. Cloth was once measured in ells. The width of an ell was 54 in. in France and 45 in. in England. Name the shortest width that measured a whole number of ells in both countries.

CHAPTER 6 195

Mixed Numbers

Amazing as it seems, a centipede has been timed at $3\frac{1}{2}$ min in the 100-meter dash. Suppose the race were timed with a stopwatch that beeped every half-minute.

The stopwatches show $3\frac{1}{2}$ min in 7 half-minute segments.

$$3\frac{1}{2} \text{ min} = \frac{7}{2} \text{ min}$$

$3\frac{1}{2}$ is a **mixed number.** $\frac{7}{2}$ is a fraction greater than 1.

You can rename a fraction that is greater than 1, such as $\frac{7}{3}$, as a mixed number.

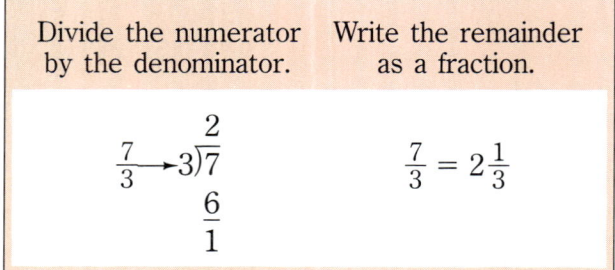

THINK ALOUD Why does dividing by 3 give you the number of wholes? How do you know that the 1 left over is $\frac{1}{3}$?

You can also rename a mixed number, such as $5\frac{3}{4}$, as a fraction.

Multiply the whole number by the denominator of the fraction.	Add the product and the numerator of the fraction.	Write the sum over the denominator.
$5\frac{3}{4} \rightarrow 5 \times 4 = 20$	$20 + 3 = 23$	$5\frac{3}{4} = \frac{23}{4}$

THINK ALOUD Why does multiplying the number of wholes by 4 give the number of fourths? Why must you add 3?

GUIDED PRACTICE

Write a mixed number and a fraction for the shaded amount.

1.   2.   3.

Complete.

4. $6\frac{1}{2} = \frac{}{2}$ 5. $\frac{15}{7} = \frac{1}{7}$ 6. $4\frac{3}{5} = \frac{}{5}$ 7. $\frac{22}{3} = 7\frac{1}{}$ 8. $\frac{26}{4} = \frac{}{4}$

PRACTICE

Rename as a mixed number or as a whole number.

9. $\frac{6}{3}$ 10. $\frac{8}{4}$ 11. $\frac{12}{5}$ 12. $\frac{16}{3}$ 13. $\frac{9}{4}$ 14. $\frac{10}{7}$

15. $\frac{24}{4}$ 16. $\frac{19}{8}$ 17. $\frac{15}{3}$ 18. $\frac{18}{6}$ 19. $\frac{27}{8}$ 20. $\frac{36}{5}$

21. $\frac{41}{8}$ 22. $\frac{29}{2}$ 23. $\frac{56}{9}$ 24. $\frac{64}{7}$ 25. $\frac{31}{17}$ 26. $\frac{53}{32}$

Rename as a fraction.

27. $2\frac{1}{3}$ 28. $5\frac{2}{3}$ 29. $1\frac{5}{9}$ 30. $8\frac{6}{7}$ 31. $3\frac{9}{10}$ 32. $9\frac{1}{6}$

33. $10\frac{5}{12}$ 34. $8\frac{7}{8}$ 35. $3\frac{1}{20}$ 36. $4\frac{3}{25}$ 37. $13\frac{2}{9}$ 38. $12\frac{1}{4}$

39. **IN YOUR WORDS** Sometimes a fraction greater than 1 can be renamed as a whole number rather than as a mixed number. Explain how this is possible.

PROBLEM SOLVING

40. During a race, a runner drank $\frac{5}{3}$ c of water. Explain whether this is more than or less than 1 pt.

41. The beginning and end markers for a race are $2\frac{1}{2}$ mi apart. How many other places must be marked to separate the course into $\frac{1}{2}$-mi sections? How many sections will the course have?

42. In 1984 Ernesto Canto, of Mexico, set a world record for the 20,000-m walk. His time was 1 hr 18 min 40.0 s. About how many minutes did he take per kilometer?

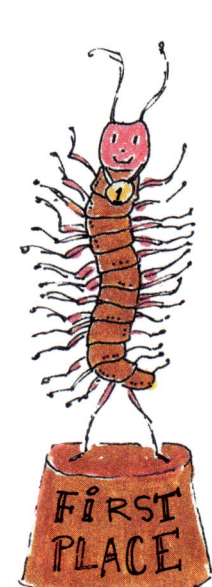

CHAPTER 6 197

Comparing and Ordering Fractions and Mixed Numbers

Asia covers about $\frac{3}{10}$ of Earth's land area. Africa covers about $\frac{1}{5}$. Which continent is larger?

You can compare fractions with different denominators by using equivalent fractions.

Compare $\frac{3}{10}$ and $\frac{1}{5}$.

Use the least common denominator (LCD) to write equivalent fractions.		Compare numerators to compare fractions.
$\frac{3}{10} = \frac{3}{10}$ $\frac{1}{5} = \frac{2}{10}$	The LCD is the LCM of the denominators.	$\frac{3}{10} > \frac{2}{10}$ so $\frac{3}{10} > \frac{1}{5}$ 3 > 2

Since $\frac{3}{10} > \frac{1}{5}$, Asia is larger than Africa.

You can use renaming to help you compare some fractions.

Compare $\frac{37}{5}$ and $7\frac{3}{8}$.

Write fractions as mixed numbers. Compare the whole numbers.	You need to compare the fractions.	Compare.
$\frac{37}{5} = 7\frac{2}{5}$ $7\frac{3}{8} = 7\frac{3}{8}$	$7\frac{2}{5} = 7\frac{16}{40}$ $7\frac{3}{8} = 7\frac{15}{40}$	$7\frac{16}{40} > 7\frac{15}{40}$ so $7\frac{2}{5} > 7\frac{3}{8}$

The fraction $\frac{37}{5}$ is greater than the mixed number $7\frac{3}{8}$.

THINK ALOUD Explain how to use equivalent fractions to put three or more fractions in order from least to greatest.

GUIDED PRACTICE

Compare. Write >, <, or =.

1. $\frac{3}{4}$ ▇ $\frac{7}{8}$
2. $\frac{5}{2}$ ▇ $\frac{17}{7}$

Write in order from least to greatest.

3. $\frac{7}{9}$, $\frac{5}{6}$, $\frac{2}{3}$
4. $\frac{31}{12}$, $\frac{19}{6}$, $2\frac{5}{9}$

LESSON 6–12

PRACTICE

Compare. Write >, <, or =.

5. $\frac{1}{3}$ ▨ $\frac{3}{8}$
6. $\frac{6}{18}$ ▨ $\frac{3}{9}$
7. $\frac{6}{12}$ ▨ $\frac{13}{24}$
8. $\frac{1}{6}$ ▨ $\frac{2}{9}$
9. $\frac{4}{10}$ ▨ $\frac{6}{15}$

10. $\frac{3}{2}$ ▨ $\frac{14}{9}$
11. $\frac{4}{1}$ ▨ $\frac{8}{4}$
12. $\frac{9}{5}$ ▨ $\frac{8}{10}$
13. $\frac{5}{2}$ ▨ $2\frac{1}{2}$
14. $\frac{11}{10}$ ▨ $1\frac{1}{5}$

15. $\frac{7}{9}$ ▨ $\frac{3}{4}$
16. $8\frac{2}{6}$ ▨ $8\frac{1}{3}$
17. $\frac{21}{7}$ ▨ $3\frac{1}{2}$
18. $3\frac{1}{3}$ ▨ $\frac{8}{3}$
19. $9\frac{3}{7}$ ▨ $9\frac{2}{5}$

Write the fractions in order from least to greatest.

20. $\frac{2}{3}, \frac{7}{12}, \frac{3}{4}$
21. $\frac{2}{5}, \frac{7}{10}, \frac{4}{15}$
22. $1\frac{5}{9}, 1\frac{2}{3}, \frac{25}{18}$
23. $\frac{11}{5}, 2\frac{1}{4}, \frac{19}{8}$

PROBLEM SOLVING

Use the fractions to order the continents from largest to smallest. Match each continent to a section of the circle graph.

Fraction of Earth's Land Area

24. North America $\frac{17}{100}$
25. Australia $\frac{1}{20}$
26. Europe $\frac{7}{100}$
27. Asia $\frac{3}{10}$
28. Antarctica $\frac{9}{100}$
29. Africa $\frac{1}{5}$
30. South America $\frac{3}{25}$

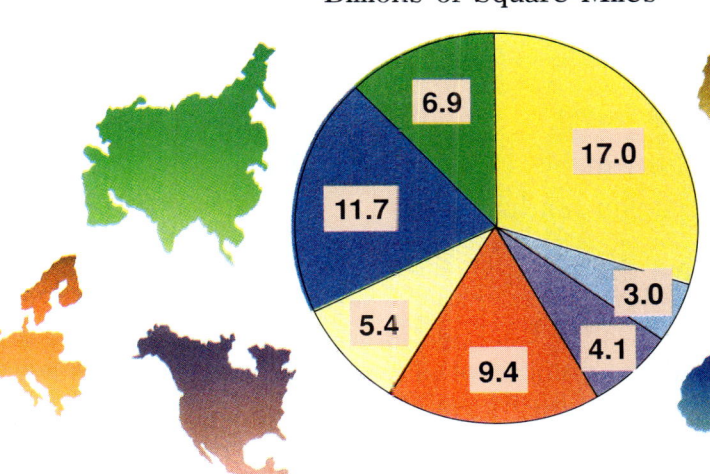

Area of Continents in Billions of Square Miles

Estimate

Use number sense to choose the letter which shows where $\frac{21}{40}$, $\frac{11}{24}$, $\frac{15}{14}$, $\frac{3}{47}$, and $\frac{19}{20}$ each belongs on the number line. Think about whether the fraction is close to 0, close to $\frac{1}{2}$, or close to 1.

```
       0           1/2        1
   ←──┼──────────┼──┼─────┼───┼──→
      (A)      (B)  (C)   (D)  (E)
      >0       <1/2  >1/2  <1   >1
```

CHAPTER 6   199

Fractions and Decimals

The world's smallest bird is the bee hummingbird of Cuba. When fully grown, it is about two inches long and weighs about seven hundredths of an ounce.

You can write this weight in two different ways:

fraction form $\frac{7}{100}$ oz

decimal form 0.07 oz

The largest hummingbird, the giant hummingbird of the Andes, weighs 0.75 oz.

If you can read and write fractions and decimals, you can use a **MATH CONNECTION** to change a fraction with a denominator of 10, 100, or 1,000 to a decimal, or vice versa.

2.03 → Read: *two and three hundredths*   Write: $2\frac{3}{100}$

$\frac{149}{1,000}$ → Read: *one hundred forty-nine thousandths*   Write: 0.149

Other examples:

Word Name	Fraction	Decimal
six and seven tenths	$6\frac{7}{10}$	6.7
thirty-five thousandths	$\frac{35}{1,000}$	0.035

THINK ALOUD Look at the number of zeros in the fraction form and the number of places to the right of the decimal point in the decimal form. What do you notice?

GUIDED PRACTICE

Complete.

1. $\frac{2}{10} = 0.\blacksquare$
2. $7.035 = 7\frac{\blacksquare}{\blacksquare}$
3. $5\frac{6}{100} = 5.\blacksquare$
4. $0.18 = \frac{\blacksquare}{\blacksquare}$

5. Write **three hundred and twenty-five thousandths**:
   a. as a mixed number
   b. as a decimal.

PRACTICE

Write as a decimal.

6. $\frac{7}{10}$
7. $\frac{215}{1,000}$
8. $5\frac{87}{100}$
9. $30\frac{9}{1,000}$
10. $\frac{10}{10}$
11. $\frac{255}{100}$

A hummingbird may beat its wings as many as 80 times per second. Hummingbirds must feed often—some captive hummingbirds eat every 15 or 20 min during the day.

Write as a fraction or mixed number.

12. 0.2 **13.** 7.684 **14.** 3.1 **15.** 71.04 **16.** 0.045 **17.** 8.88

Write as a decimal and as a fraction or mixed number.

18. one and sixty thousandths

19. twenty and fourteen hundredths

20. five and five tenths

21. forty and nine tenths

22. forty and nine hundredths

23. forty-nine hundredths

24. one hundred one thousandths

25. one hundred and one thousandth

PROBLEM SOLVING

CHOOSE Choose estimation, mental math, pencil and paper, or calculator to solve.

Write *true* or *false*. Use the information on pages 200 and 201.

26. The bee hummingbird's weight is approximately 0.1 oz.

27. The giant hummingbird weighs $\frac{3}{4}$ oz.

28. At its fastest, a small hummingbird beats its wings a little less than 500 times in 1 min.

29. Some hummingbirds eat about every 0.3 h.

30. It would take about 12,000 bee hummingbirds to equal the weight of an 85-lb child.

Exploring Writing Fractions as Decimals

This lesson explores how to write any fraction as a decimal.

Write each of the fractions in Set 1 as a decimal.
1. Why is it so easy to write these fractions as decimals?

2. Write another fraction that is easy to change to a decimal.

If you can find equivalent fractions, you can use a MATH CONNECTION to write some fractions as decimals.

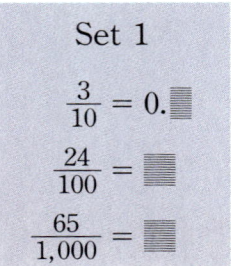

Set 1

$\frac{3}{10} = 0.\blacksquare$

$\frac{24}{100} = \blacksquare$

$\frac{65}{1,000} = \blacksquare$

Look at the fractions in Set 2.
3. Why are these fractions harder to write in decimal form than those in Set 1?

4. Write each fraction in Set 2 as a decimal by first rewriting it as an equivalent fraction with a denominator of 10, 100, or 1,000.

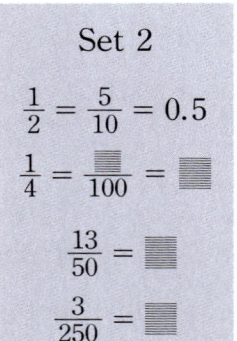

Set 2

$\frac{1}{2} = \frac{5}{10} = 0.5$

$\frac{1}{4} = \frac{\blacksquare}{100} = \blacksquare$

$\frac{13}{50} = \blacksquare$

$\frac{3}{250} = \blacksquare$

Now look at the fractions in Set 3.
5. Can you change these fractions to equivalent fractions with a denominator of 10, 100, or 1,000? Explain.

Set 3

$\frac{5}{9}, \frac{2}{3}, \frac{5}{12}, \frac{1}{7}$

For fractions like those in Set 3, a different approach is needed. You can use paper folding to discover it.
6. Fold 1 piece of paper into 8 parts. Shade 1 part.
   a. What fraction of the paper is shaded?
   b. You split one whole into $\blacksquare$ parts, or divided 1 by $\blacksquare$.
   c. You have just shown that $\frac{1}{8} = \blacksquare \div \blacksquare$.

202   LESSON 6-14

7. Take 3 pieces of paper that are the same size but different colors. Fold each one into 8 parts. Each paper represents one whole.

   a. Cut out one part of each of the 3 papers. If you put the parts together, what fraction of a whole do you get?

   b. You can think of the 3 cut parts as one person's share. Altogether, you are sharing the 3 papers among ▊ people, or dividing 3 by ▊.

   c. You have just shown that ▊ = ▊ ÷ ▊.

8. The fraction $\frac{5}{8}$ corresponds to what division example? Use paper folding to check yourself.

Write the division for each fraction.

9. $\frac{5}{9} = ▊ \div 9$

10. $\frac{3}{7} = ▊ \div ▊$

11. $\frac{12}{5} = ▊ \div ▊ = ▊\frac{▊}{5}$

12. Is there a division problem for any fraction? Explain.

> **You can think of every fraction as division:**
> **denominator)numerator.**

 Choose pencil and paper or calculator to solve.

13. Divide to find the decimal form of the fraction. Try these.

    $\frac{1}{2}, \frac{1}{4}, \frac{1}{8}, \frac{6}{125}, \frac{7}{16}$

14. None of those had a remainder. Give another fraction that will have no remainder.

15. Now divide these. Round decimals to the nearest hundredth.

    $\frac{1}{6}, \frac{2}{3}, \frac{4}{11}, \frac{5}{12}, \frac{1}{7}$

    Example:

    ```
        0.833    Rounds to 0.83
    6)5.000
      4 8
        20
        18
         20
         18
          2
    ```

Problem Solving Strategy: Drawing a Diagram

The local dog show has 38 dogs entered. As they wait to be shown, the dogs are kept in waiting areas. Each area holds 6 dogs. How many waiting areas are needed?

This problem can be solved by dividing. But the answer to a division problem depends on the question asked.

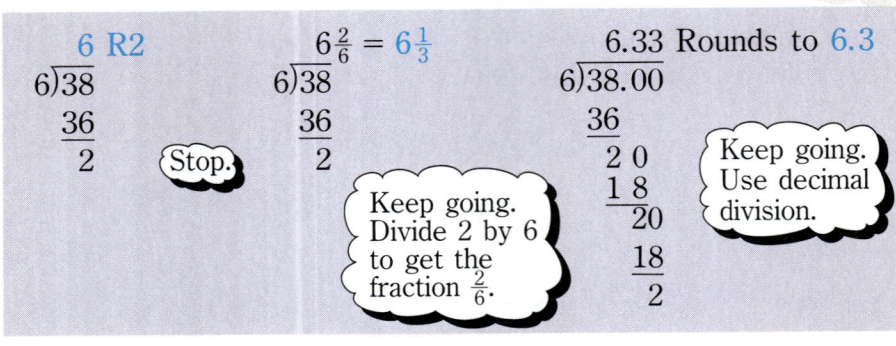

Drawing a diagram can help you decide which number answers the question.

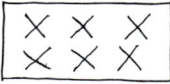 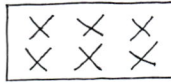 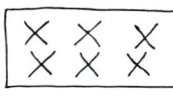 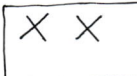
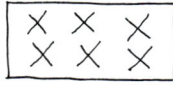 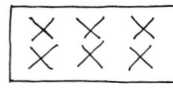 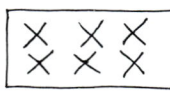

- Are 6 waiting areas enough to hold all the dogs?
- If not, how many more waiting areas are needed?
- Does a mixed number or decimal answer make sense?

Some other questions:

- Which number tells us how many waiting areas are totally full?
- Which number tells us the number of dogs in the last waiting area?
- Which number tells us exactly how much space is used?

GUIDED PRACTICE

1. There were 195 people at the show. Each section of seating holds 30 people. The sections were filled in order.

   a. How many seats are in each section?
   b. Draw a diagram that shows how the people were distributed among the sections.
   c. How many people sat in the last section?

PRACTICE

Solve. Draw a diagram.

2. There are 13 German shepherds in the show. How many vans were needed to transport them if 5 dogs could ride in 1 van?

3. The judging for the dog show takes place in 4-hour sessions. How many judging sessions can take place in 11 hours?

4. There are 450 dogs in the county dog show. If no more than 20 dogs can be shown in each judging arena, what's the smallest number of arenas that could be used?

5. A dog food company donated 1 bag of dog food for every 10 dogs entered in the show. How many bags did the company donate for the 38 dogs entered this year?

CHOOSE Choose any strategy to solve.

6. Of the 38 dogs in the dog show, there are 13 German shepherds, 8 golden retrievers, 7 poodles, 5 cocker spaniels, and 5 collies. Each breed has a separate arena for the judging. How many arenas are needed?

7. More Labrador retrievers than golden retrievers are registered with the American Kennel Club. There are fewer poodles than Labrador retrievers. There are more cocker spaniels than Labrador retrievers. Which breed has the greatest number registered?

8. Tickets to the show are $3 for each adult and $2 for each child. There are 110 adults and 85 children at the show. How much money was collected?

9. The 38 dogs in the show this year are what fraction of the 51 dogs entered last year? Choose the closest estimate.

   a. $\frac{3}{5}$ b. $\frac{3}{4}$ c. $\frac{9}{10}$

Mixed Review

Round to the place of the underlined digit.
1. 6̲7
2. 2̲48
3. 3̲,125
4. 1,9̲68
5. 16.59̲7
6. 138̲.63
7. 3.421̲3
8. 0.01̲74

Write in standard form.
9. 7 and 3 hundredths
10. 5 + 0.7 + 0.004
11. 8 thousandths
12. three million, two

Put the numbers in order from the least to the greatest.
13. 7,639; 76,390; 70,369; 76,039
14. 0.104; 0.014; 0.140; 1.40
15. 2,571; 2,715; 27,175; 7,215
16. 7.3; 7.31; 0.73; 73

Find the answer.
17. 142.1 × 805
18. 17,293 + 3,965
19. 239.802 − 43
20. 68,437 − 12,639
21. 819 ÷ 5
22. 409 × 78
23. $.98 + $13.23
24. 1.26 × 40
25. 7,063 ÷ 7
26. 51.84 ÷ 7.2
27. 509 × 40
28. 1,998.5 + 72.965
29. 0.45 × 0.9
30. 16 − 1.507
31. 11,455 ÷ 395
32. 56,875 − 8,794
33. 0.625 ÷ 2.5
34. 230,050 − 18,276
35. 32.96 + 7.72 + 5.9 + 74.08
36. 9,846 + 38 + 23,485 + 254

PROBLEM SOLVING

CHOOSE Choose estimation, calculator, mental math, or paper and pencil to solve. Use the chart.

37. Which trail is 2.2 mi longer than another trail?

38. Jack hiked to Mirror Lake and back. Annette hiked to Nevada Falls and back. How much farther did Annette hike than Jack?

39. Ann's family hiked to Half Dome and back. Did they walk at least 15 mi?

40. If your walking speed is 4 mi/h, about how long will it take you to hike to Happy Isles and back?

41. The bus tickets to Yosemite cost $8.75 each. What was the cost for a group of 138 students?

42. Five Boy Scouts' packs weighed a total of 96.5 lb. What was the average weight of a pack?

43. How much longer is the longest hiking trail in the chart than the shortest trail?

Yosemite National Park Hiking Trails

Location	Length
Vernal Falls	1.3 mi
Nevada Falls	3.5 mi
Little Yosemite Valley	4.3 mi
Half Dome	7.8 mi
Happy Isles	0.9 mi
Mirror Lake	2.0 mi

44. What two other ways can the information on the chart be organized?

45. What is the average length of a trail in the chart?

46. Sara took three different roundtrip hikes at Yosemite National Park. She walked a total of 8.4 mi. What trails did she walk?

47. The upper falls of Yosemite Falls measures 1,430 ft. The height of the lower falls is 320 ft. The cascade between the falls measures 675 ft. What is the falls' total height?

CHAPTER 6   207

CHAPTER CHECKUP

LANGUAGE & VOCABULARY

Explain the meanings in your own words. Use the fraction $\frac{8}{6}$.

1. numerator
2. denominator
3. equivalent fractions
4. lowest terms
5. mixed number

TEST

CONCEPTS

Is the number prime or composite? *(pages 178–179)*

1. 27
2. 49
3. 2
4. 53

Use 2, 3, 4, 5, 6, and 10. *(pages 176–177)*

5. By which numbers is 570 divisible?

Write a fraction for the shaded part. *(pages 188–189)*

6.

Write as a decimal. *(pages 200–201)*

7. $\frac{348}{1,000}$
8. $2\frac{6}{100}$

Write as a fraction or mixed number. *(pages 200–201)*

9. 0.7
10. 5.29

SKILLS

Find the LCM.
(pages 174–175)

11. 7, 8
12. 5, 15

Write the prime factorization.
(pages 178–179)

13. 250

Write using exponents. *(pages 180–181)*

14. 14 × 14
15. 2 cubed

Write in standard form. *(pages 180–181)*

16. 4^3
17. 15^2

List the factors and the GCF. *(pages 192–193)*

18. 16, 36
19. 7, 21

Write an equivalent fraction in lowest terms. *(pages 194–195)*

20. $\dfrac{5}{35}$ 21. $\dfrac{24}{46}$ 22. $\dfrac{18}{48}$

Write as a mixed or whole number. *(pages 196–197)*

23. $\dfrac{19}{4}$ 24. $\dfrac{27}{3}$ 25. $\dfrac{17}{5}$ 26. $\dfrac{34}{9}$

Compare. Write >, <, or =. *(pages 198–199)*

27. $\dfrac{2}{5}$ ▨ $\dfrac{3}{8}$ 28. $\dfrac{10}{3}$ ▨ $3\dfrac{1}{4}$ 29. $\dfrac{4}{9}$ ▨ $\dfrac{7}{12}$ 30. $2\dfrac{5}{6}$ ▨ $2\dfrac{3}{4}$

PROBLEM SOLVING

Solve. *(pages 182–183 and 204–205)*

31. A packing box holds 60 notebooks. There are 357 notebooks ready to be boxed.
   a. How many notebooks can fit in 1 box?
   b. Draw a diagram that would help you find the number of notebooks in the last box.
   c. How many notebooks are in the last box?

32. Mrs. Quan planted 36 marigolds in rows. She put an equal number of plants in each row. Name all the possible numbers of plants she could have in each row.

LEARNING LOG

Write the answers in your learning log.

1. One number is a factor of another number. Describe the greatest common factor of the two numbers.

2. Your friend said that all mixed numbers are greater than fractions. Is that thinking correct? Explain.

3. You wrote a fraction in lower terms by dividing. You did not use the GCF. What do you notice about your answer?

EXTRA PRACTICE

Find the LCM. *(pages 174–175)*

1. 8, 12   **2.** 5, 9   **3.** 6, 30

By which numbers—2, 3, 4, 5, 6, and 10—is each divisible? *(pages 176–177)*

4. 80   **5.** 441   **6.** 700

Write the prime factorization. *(pages 178–179)*

7. 33   **8.** 72   **9.** 360

Write in standard form. *(pages 180–181)*

10. 3^3 **11.** 1^3 **12.** 5^2 **13.** 7^3 **14.** 45^2 **15.** 19^3

Write a fraction for the shaded part. *(pages 188–189)*

16. 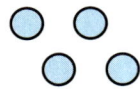   **17.**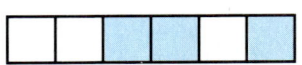

Find the GCF. *(pages 192–193)*

18. 11, 33   **19.** 8, 20   **20.** 7, 13   **21.** 1, 15   **22.** 2, 8   **23.** 9, 17

Complete. *(pages 190–191 and 194–195)*

24. $\frac{5}{8} = \frac{n}{16}$ **25.** $\frac{15}{4} = \frac{45}{n}$ **26.** $\frac{8}{24} = \frac{1}{n}$ **27.** $4\frac{3}{4} = \frac{n}{4}$ **28.** $5\frac{2}{3} = 3\frac{n}{3}$

Write the fraction or mixed number in lowest terms. *(pages 194–195 and 196–197)*

29. $\frac{6}{10}$ **30.** $\frac{5}{25}$ **31.** $\frac{9}{21}$ **32.** $\frac{70}{100}$ **33.** $\frac{10}{5}$ **34.** $\frac{16}{3}$

Compare. Write >, <, or =. *(pages 198–199)*

35. $\frac{7}{9}$ ▧ $\frac{4}{5}$ **36.** $2\frac{2}{3}$ ▧ $\frac{14}{6}$ **37.** $\frac{3}{4}$ ▧ $\frac{5}{8}$ **38.** $1\frac{5}{12}$ ▧ $\frac{11}{8}$

Write as a decimal. *(pages 200–201)*

39. $\frac{7}{10}$ **40.** $3\frac{85}{100}$ **41.** $9\frac{8}{1,000}$ **42.** $5\frac{5}{100}$ **43.** $4\frac{1}{1,000}$ **44.** $\frac{12}{10}$

Solve. *(pages 182–183 and 204–205)*

45. Martha has only dimes and quarters in her pocket. She has 9 coins that total $1.35. How many of each coin does she have?

46. A banquet will have 347 people attending. How many tables are needed if 8 are seated at a table?

ENRICHMENT

SIEVE OF ERATOSTHENES

The ancient scholar Eratosthenes was born in Cyrene (in modern-day Libya) and spent most of his working life in Alexandria, Egypt.

Find out what the Sieve of Eratosthenes is and construct one to identify all the prime numbers less than 100.

Be A Game Designer

Choose one of the games described below. Work with a partner to design and make the materials needed to play it. Check your work by playing the game.

Game 1: Roll-a-Factor

Materials: 2 number cubes, each face should have a different number.

Take turns rolling the 2 number cubes. Score a point each time one number rolled is a factor of the other number rolled.

Game 2: Fraction Challenge

Materials: 36 cards, each with a fraction or mixed number.

Deal 18 cards to each player. Each player places their cards in a stack face down.

Players then turn over their top cards, and the player with the greater fraction takes both cards. If the fractions are equivalent, players turn over their next cards, and the player with the greater fraction takes all 4 cards.

Continue play. The player with more cards at the end of the game is the winner.

Describe any changes you need to make to improve your game. Trade games with another group and then try their game.

CUMULATIVE REVIEW

Choose the most likely measurement.

1. the length of your classroom
   a. 25 ft
   b. 100 in.
   c. 25 yd
   d. none of these

2. the weight of a hamburger
   a. 2 lb
   b. 200 oz
   c. 5 oz
   d. none of these

3. the temperature of cold milk
   a. 0°F
   b. 40°F
   c. 80°F
   d. none of these

4. the height of a car
   a. 2.5 m
   b. 100 cm
   c. 1.5 m
   d. none of these

5. the capacity of a thimble
   a. 200 mL
   b. 2 L
   c. 2,000 mL
   d. none of these

6. the mass of a car
   a. 500 kg
   b. 1 t
   c. 2,000 g
   d. none of these

Find the perimeter of each.

7. (rectangle: 15 ft by 8 ft)
   a. 23 ft
   b. 120 ft
   c. 38 ft
   d. none of these

8. (triangle: 6 in., 10 in., 14 in.)
   a. 24 in.
   b. 42 in.
   c. 30 in.
   d. none of these

9. a square with a side of 7 mi
   a. 14 mi
   b. 28 mi
   c. 49 mi
   d. none of these

Compute.

10. $70 \overline{)45{,}395}$
    a. 647 R5
    b. 648 R35
    c. 6,485
    d. none of these

11. $37 \overline{)7{,}586}$
    a. 205 R1
    b. 206 R36
    c. 25 R1
    d. none of these

12. $52 \overline{)8{,}405}$
    a. 16 R13
    b. 178 R29
    c. 161 R33
    d. none of these

13. $\$3.27 \times 48$
    a. $156.94
    b. $39.24
    c. $156.96
    d. none of these

14. 2.68×3.4
    a. 91.12
    b. 1.876
    c. 6.08
    d. none of these

15. 0.36×0.19
    a. 0.0684
    b. 0.684
    c. 6.84
    d. none of these

PROBLEM SOLVING REVIEW

Remember the strategies and types of problems you have had so far. Solve.

Problem Solving Check List
- Choosing the operation
- Drawing a diagram
- Making a table
- Using estimation
- Too much information
- Too little information
- Multistep problems
- Looking backward

1. Jose had softball practice for 2 h 45 min on Saturday. Practice ended at 11:30 A.M.
   a. How long did softball practice last?
   b. What operation would you use to find when practice started?
   c. When did practice start?

2. During a craft fair Mrs. Tao sold 46 handpainted sweatshirts for $34 each. About how much did Mrs. Tao receive for the sweatshirts?

3. Video games are usually priced at $37.75 each. How much would you save by buying two at the sale price of 2 for $59?

4. A recipe calls for $\frac{3}{4}$ c of raisins. Michael plans to double the recipe. What other information does Michael need to determine whether he has enough raisins?

5. The low temperature Monday was −5°F and the high temperature was 12°F. Chris said that this was a difference of 7°F. Is Chris' statement reasonable? Explain.

6. Fruit drinks are sold in 4 flavors: strawberry, banana, tropical fruit, and orange. They are sold in 3 sizes: small, medium, and large. How many different combinations of flavor and size are available?

7. Deanna paid $8.57 for a pizza. She paid with a $10-bill. The cashier gave her change in the fewest bills and coins possible. Name the bills and coins Deanna received.

8. The perimeter of Toya's bedroom is 44 ft. It is 12 ft wide. Draw a diagram to show the dimensions of Toya's room.

Read the passage and then write the word problems.

The Science Club meets every Wednesday from 3:20 P.M. to 4:45 P.M. Its members include 24 sixth graders, 33 seventh graders, and 31 eighth graders. Each member must pay $3 for supplies and $4.50 to cover the cost of a field trip.

9. Write an addition problem using the information about the Science Club. Then solve it.

10. Write a multiplication problem using the information about the Science Club. Then solve it.

TECHNOLOGY

SUSPECTS

In the computer game "Number Stumper," players unravel clues to determine composite numbers. Try this paper-and-pencil puzzle to sharpen your logic skills.

List three possible "suspects" for each set of clues.

1. Factor: 3
   Multiple: 240
   Not a factor: 16
   Not a multiple: 24

2. Factor: 4
   Multiple: 240
   Not a factor: 32
   Not a multiple: 32

3. Factor: 3
   Multiple: 420
   Not a factor: 8
   Not a multiple: 30

FIRST TIMERS

The play, *The Terrifics,* is now at the local theater. Of the 336 audience members Friday, 294 were seeing the play for the first time. In Sunday's audience of 324 people, 285 were first-time viewers. On Monday, the fraction was 288 of 360. On which night did the greatest fraction of the audience see the play for the first time?

NOT EQUAL

Find the fraction in each group that does not belong.

1. $\frac{150}{200}$ $\frac{165}{220}$

   $\frac{180}{280}$ $\frac{111}{148}$

2. $\frac{180}{480}$ $\frac{135}{370}$

   $\frac{216}{576}$ $\frac{126}{336}$

214   TECHNOLOGY

Adding and Subtracting Fractions 7

DID YOU KNOW...?
The fastest train in the world is the Japanese maglev, or magnetically levitating, train. It doesn't move along tracks — it speeds along 4 inches above the ground!

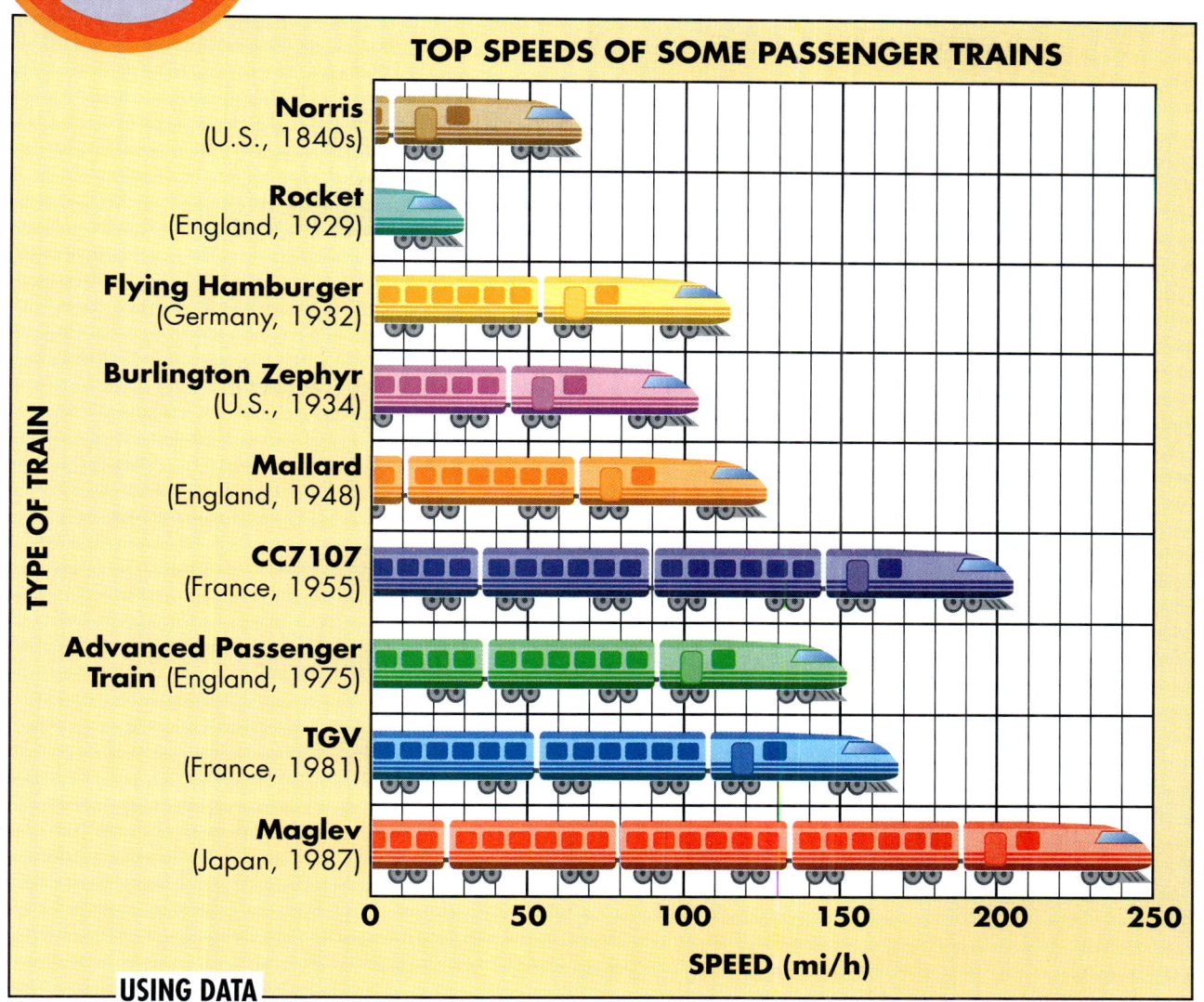

USING DATA
- Collect
- Organize
- Describe
- Predict

Work with a partner. Take turns making up and answering three questions about the data in the graph.

Find out more about the Japanese maglev train. Share your group's information with the class.

CHAPTER 7 215

Adding and Subtracting Fractions: Same Denominators

Through the Jobs for Youth program, George got a job painting the community center in his neighborhood.

To make orange paint, George mixed $\frac{5}{8}$ gal of red paint and $\frac{5}{8}$ gal of yellow paint. How much orange paint did he make?

To find out, add $\frac{5}{8} + \frac{5}{8}$.

   5 eighths + 5 eighths = 10 eighths

$\frac{5}{8} + \frac{5}{8} = \frac{10}{8} = 1\frac{2}{8} = 1\frac{1}{4}$ *Write the answer in lowest terms.*

George mixed $1\frac{1}{4}$ gal of orange paint.

Another example:

   3 fourths − 1 fourth = 2 fourths

$\frac{3}{4} - \frac{1}{4} = \frac{2}{4} = \frac{1}{2}$

THINK ALOUD Explain how and why each answer was reduced to lowest terms. Why was $1\frac{1}{4}$ given as a mixed number?

====== GUIDED PRACTICE ======

Add or subtract. Answer in lowest terms.

1. 3 eighths
    + 2 eighths

2. 8 ninths
    − 4 ninths

3. $\frac{5}{6}$
    $- \frac{1}{6}$

4. $\frac{5}{8}$
    $+ \frac{3}{8}$

5. $\frac{7}{12} + \frac{11}{12}$

====== PRACTICE ======

Add or subtract.

6. $\frac{1}{8}$
    $+ \frac{4}{8}$

7. $\frac{7}{9}$
    $- \frac{3}{9}$

8. $\frac{5}{16}$
    $+ \frac{6}{16}$

9. $\frac{3}{7}$
    $+ \frac{2}{7}$

10. $\frac{7}{8}$
    $- \frac{2}{8}$

11. $\frac{5}{10}$
    $- \frac{2}{10}$

Add or subtract. Write the answer in lowest terms.

12. $\frac{5}{6}$
    $+ \frac{5}{6}$

13. $\frac{8}{9}$
    $- \frac{2}{9}$

14. $\frac{7}{10}$
    $- \frac{2}{10}$

15. $\frac{3}{4}$
    $- \frac{1}{4}$

16. $\frac{7}{12}$
    $- \frac{1}{12}$

17. $\frac{7}{10}$
    $+ \frac{9}{10}$

18. $\frac{7}{8} + \frac{5}{8}$ **19.** $\frac{5}{8} - \frac{3}{8}$ **20.** $\frac{9}{16} + \frac{7}{16}$ **21.** $\frac{13}{16} - \frac{7}{16}$ **22.** $\frac{11}{12} - \frac{1}{12}$

23. $\frac{5}{10} - \frac{3}{10}$ **24.** $\frac{9}{12} + \frac{5}{12}$ **25.** $\frac{7}{8} + \frac{7}{8}$ **26.** $\frac{3}{4} + \frac{3}{4}$ **27.** $\frac{11}{16} - \frac{5}{16}$

28. $\frac{5}{10} + \frac{2}{10} + \frac{1}{10}$ **29.** $\frac{8}{12} + \frac{4}{12} + \frac{5}{12}$ **30.** $\frac{5}{8} + \frac{2}{8} - \frac{1}{8}$ **31.** $\frac{5}{8} - \frac{2}{8} + \frac{1}{8}$

MIXED REVIEW Find the answer.

32. $0.7 + 0.5$ **33.** $40 - 17.87$ **34.** $2{,}572 \div 37$ **35.** 92.4×0.07

MENTAL MATH Look for fractions with a sum of 1. Add.

36. $\frac{7}{8} + \frac{1}{8} + \frac{5}{8}$ **37.** $\frac{6}{11} + \frac{8}{11} + \frac{5}{11}$ **38.** $\frac{1}{6} + \frac{4}{6} + \frac{5}{6}$ **39.** $\frac{98}{100} + \frac{75}{100} + \frac{2}{100} + \frac{26}{100}$

40. IN YOUR WORDS Reta added $\frac{2}{8} + \frac{5}{8}$ and got the answer $\frac{7}{16}$. How would you explain to Reta the mistake he made?

PROBLEM SOLVING

41. Bonnie bought twice as many gallons of blue paint as red paint. She bought equal amounts of green paint and yellow paint. How many whole gallons of each color paint did she buy if she bought 12 gal in all?

42. If you mixed $\frac{3}{8}$ qt of red paint and $\frac{3}{8}$ qt of blue paint to make purple paint, how much paint would you have in all?

43. CREATE YOUR OWN Write a word problem for the equation $\frac{7}{8} - \frac{3}{8} = n$. Trade with a partner and solve.

Adding and Subtracting Mixed Numbers: Same Denominators

When *Homesick, My Own Story* begins, Jean is 9 yr 11 months, or $9\frac{11}{12}$ yr, old. She sets sail for America almost 1 yr 7 months, or $1\frac{7}{12}$ yr, later. About how old is she then?

To find out, add $9\frac{11}{12} + 1\frac{7}{12}$.

Add the fractions.	Add the whole numbers.	Rename.
$9\frac{11}{12}$	$9\frac{11}{12}$	$10\frac{18}{12} = 10 + 1\frac{6}{12}$
$+ 1\frac{7}{12}$	$1\frac{7}{12}$	$= 11\frac{6}{12} = 11\frac{1}{2}$
$\frac{18}{12}$	$10\frac{18}{12}$	

Jean is about $11\frac{1}{2}$ yr old.

Estimate to check that your answer is reasonable. $9\frac{11}{12} + 1\frac{7}{12} \Rightarrow 10 + 2 = 12$ $11\frac{1}{2}$ is close to 12.

GUIDED PRACTICE

Explain the steps you follow to add or subtract.

1. $6\frac{1}{5} + 3\frac{2}{5}$
2. $4\frac{7}{9} - 2\frac{4}{9}$
3. $15\frac{3}{10} - 8\frac{3}{10}$
4. $8\frac{2}{3} + 7\frac{1}{3} + 3$

PRACTICE

Add or subtract. Write the answer in lowest terms.

5. $7\frac{3}{16} + 4\frac{9}{16}$
6. $5\frac{3}{10} + \frac{7}{10}$
7. $6\frac{9}{15} - 4\frac{4}{15}$
8. $16\frac{5}{6} - 8\frac{1}{6}$
9. $2\frac{5}{12} + 13\frac{3}{12}$
10. $15\frac{7}{9} - 8$

Add or subtract. Write the answer in lowest terms.
Estimate to be sure your answer is reasonable.

11. $3\frac{9}{10} - 1\frac{5}{10}$

12. $15\frac{2}{5} - 7\frac{2}{5}$

13. $8\frac{15}{16} + 3\frac{5}{16}$

14. $9\frac{11}{15} + 7\frac{10}{15}$

15. $6\frac{1}{8} + 4\frac{3}{8}$

16. $\frac{3}{4} + 1\frac{2}{4}$

17. $5\frac{7}{16} + 9\frac{5}{16}$

18. $10\frac{2}{3} - 7\frac{2}{3}$

19. $2\frac{3}{7} + 11\frac{2}{7}$

20. $19\frac{7}{12} - 7\frac{5}{12}$

21. $9\frac{7}{8} - 6\frac{4}{8}$

22. $13\frac{5}{6} - 2\frac{1}{6}$

23. $7\frac{3}{12} + 2\frac{5}{12} - 3\frac{2}{12}$

24. $4\frac{4}{10} + 5\frac{3}{10} - 2\frac{5}{10}$

25. $8\frac{5}{6} - \frac{3}{6} + 1\frac{4}{6}$

Find the value of *n*.

26. $8\frac{5}{6} - n = 5\frac{4}{6}$

27. $n + 6\frac{1}{4} = 11\frac{3}{4}$

28. $13\frac{2}{5} - n + 2\frac{3}{5} = 5\frac{1}{5}$

MIXED REVIEW Find the answer.

29. $387 + 2{,}848$

30. $\frac{7}{8} - \frac{1}{8}$

31. 167×48

32. $720 \div 8$

PROBLEM SOLVING

33. Jean's family drove across the U.S. to reach their home. After driving for $21\frac{2}{4}$ days, they called to say that they would arrive in another $1\frac{3}{4}$ days. How long was the trip?

34. Jean solved problems about trains. If a train traveled 55 mi/h and made five 15-min stops, about how long would it have taken to go the 800 mi from Hankow to Peking?

Mental Math

You can add or subtract mixed numbers and fractions mentally by counting on or back to make the next whole.

Add $5\frac{3}{4} + \frac{3}{4}$.   Think: $5\frac{3}{4} + \frac{1}{4} = 6$; $6 + \frac{1}{2} = 6\frac{1}{2}$.

Subtract $5\frac{2}{5} - \frac{3}{5}$.   Think: $5\frac{2}{5} - \frac{2}{5} = 5$; $5 - \frac{1}{5} = 4\frac{4}{5}$.

Count on or back to find the answer.

1. $2\frac{5}{9} + \frac{5}{9}$

2. $4\frac{9}{16} + \frac{9}{16}$

3. $3\frac{2}{5} - \frac{4}{5}$

4. $5\frac{1}{8} - \frac{5}{8}$

CHAPTER 7

Renaming Before Subtracting: Same Denominators

The *Solar Challenger*, a solar-powered airplane, has a wingspan of 47 ft. The largest wandering albatross ever measured had a wingspan of $11\frac{5}{6}$ ft. How much longer is the *Solar Challenger*'s wingspan than the albatross's?

Subtract to find out.

To subtract, you must rename to get sixths.		Subtract.
$47 = 46\frac{6}{6}$ $-11\frac{5}{6} = -11\frac{5}{6}$	Rename 47 as $46\frac{6}{6}$	$46\frac{6}{6}$ $-11\frac{5}{6}$ $\overline{35\frac{1}{6}}$

The *Solar Challenger*'s wingspan is $35\frac{1}{6}$ feet longer.

THINK ALOUD Why was one whole traded for $\frac{6}{6}$ instead of $\frac{4}{4}$ or $\frac{5}{5}$?

Another example:

$$9\frac{5}{8} = 8\frac{13}{8}$$
$$-5\frac{6}{8} = -5\frac{6}{8}$$
$$\overline{3\frac{7}{8}}$$

=== GUIDED PRACTICE ===

Subtract. Answer in lowest terms.

1. $3 = 2\frac{}{5}$
   $-1\frac{3}{5} = -1\frac{3}{5}$

2. $9\frac{1}{8} = 8\frac{}{8}$
   $-2\frac{3}{8} = -2\frac{3}{8}$

3. 6
   $-5\frac{2}{3}$

4. $8\frac{3}{16} - 4\frac{9}{16}$

LESSON 7-3

PRACTICE

Subtract. Write the answer in lowest terms.

5. $9 - 6\frac{1}{4}$
6. $4 - 2\frac{5}{10}$
7. $8 - \frac{5}{8}$
8. $7 - 1\frac{1}{3}$
9. $6 - 3\frac{1}{2}$
10. $10 - 7\frac{3}{8}$

11. $7\frac{2}{4} - 4\frac{3}{4}$
12. $18\frac{5}{12} - 15\frac{7}{12}$
13. $14 - 9\frac{7}{8}$
14. $2\frac{3}{6} - \frac{5}{6}$
15. $16 - 7\frac{1}{4}$
16. $15\frac{3}{8} - 9\frac{7}{8}$

17. $8\frac{1}{5} - \frac{4}{5}$
18. $9 - 3\frac{2}{3}$
19. $4\frac{5}{7} - 3\frac{6}{7}$
20. $13\frac{7}{9} - 3\frac{8}{9}$

21. $13\frac{3}{10} - \frac{6}{10}$
22. $21\frac{5}{16} - 8\frac{11}{16}$
23. $15\frac{5}{9} - 12\frac{7}{9}$
24. $11\frac{1}{10} - 7\frac{3}{10}$

MIXED REVIEW Find the answer.

25. $\frac{4}{6} + \frac{5}{6}$
26. $80.1 - 67$
27. 0.074×0.2
28. $32.4 \div 8$

MENTAL MATH Use counting on to find the value of n.

29. $8\frac{2}{5} - n = 7\frac{2}{5}$
30. $9\frac{2}{5} - n = 8\frac{1}{5}$
31. $8\frac{1}{4} - n = 7\frac{3}{4}$
32. $7\frac{5}{8} - n = 6\frac{3}{8}$

PROBLEM SOLVING

33. The condor has an average wingspan of 9 ft. The Kalong fruit bat has an average wingspan of $5\frac{7}{12}$ ft. What is the difference in wingspans?

34. The marabou stork has an average wingspan of $11\frac{1}{2}$ ft. Is this more than the total length of 3 yardsticks? Explain.

35. **CREATE YOUR OWN** Write a subtraction word problem. The answer should be $3\frac{1}{2}$.

36. It took the *Solar Challenger* 5 h 30 min to fly from France to England at 30 mi/h. How far did the *Solar Challenger* travel?

CHAPTER 7

Exploring Fractions and Measurement

Make a number line on an 8-in. by 1-in. strip of paper. Use a ruler to mark inches.

Use a ruler to divide each inch into halves, then into fourths, and then into eighths.

1. How can you make the marks so that you can tell quickly whether they stand for half inches, quarter inches, or eighths of an inch?

2. If you were to divide each eighth of an inch in half, what part of an inch would each unit be?

Work in a group of four. Make a tape measure by taping your number lines together and numbering the inches.

3. Find $6\frac{1}{2}$ in. and $6\frac{4}{8}$ in. on your tape measure. What do you notice?

4. Take turns giving a mixed number and using the tape measure to name the mixed number in other ways. Keep a list. Compare your list with lists from other groups.

Measure to the nearest half inch.
Then tell how you measured.

5. the length of your shoe
6. your height
7. the width of your finger
8. the distance across your hand
9. the distance around your wrist
10. your armspan

The line above measures 2 in. to the nearest inch.
$2\frac{1}{2}$ in. to the nearest $\frac{1}{2}$ in.
$2\frac{1}{4}$ in. to the nearest $\frac{1}{4}$ in.
$2\frac{3}{8}$ in. to the nearest $\frac{1}{8}$ in.

11. Which measurement is closest to the actual length?

PRACTICE

Use a ruler to measure to the nearest $\frac{1}{2}$ in., $\frac{1}{4}$ in., and $\frac{1}{8}$ in.

12.
13.
14.
15.
16.
17.

ESTIMATE First estimate the length of each line. Then measure to the nearest $\frac{1}{8}$ inch. How close were you?

18.
19.
20.

Use a ruler to draw lines of the following lengths.

21. $\frac{1}{2}$ in.
22. $2\frac{3}{4}$ in.
23. $6\frac{5}{8}$ in.
24. between $3\frac{7}{8}$ in. and $4\frac{1}{8}$ in.

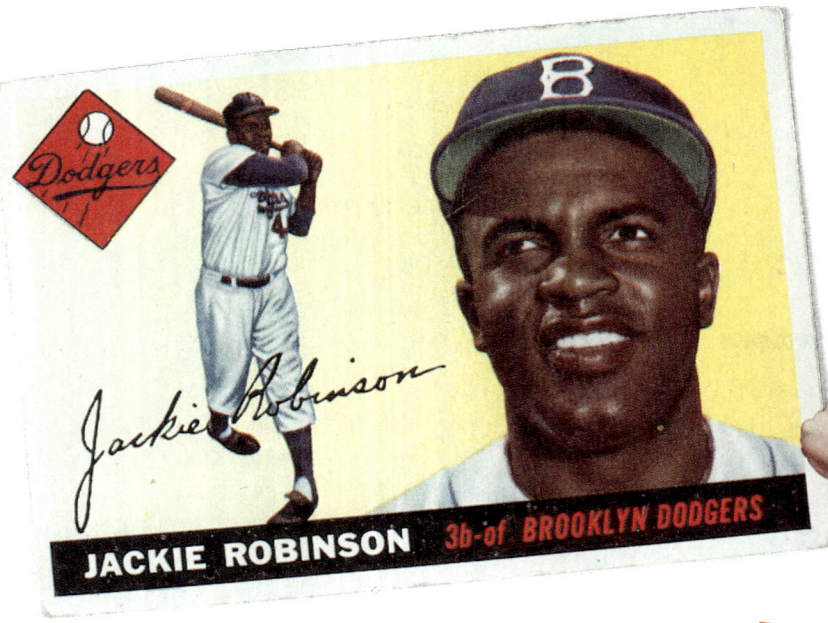

Problem Solving Strategy: Guess and Check

Carlos read 3 consecutive pages of *Thank You, Jackie Robinson*. The 3 page numbers had a sum of 90. Which 3 pages did Carlos read?

To answer the question, you need to find 3 consecutive numbers (numbers in order, such as 1, 2, 3) whose sum is 90.

You can solve this problem by using guess and check. Since $90 \div 3 = 30$, why does it make sense to begin with numbers close to 30?

Jackie Robinson was the first African American to play modern major league baseball.

1st guess

Try 30, 31, and 32.
$30 + 31 + 32 = 93$

Should the next guess use higher or lower numbers?

Use what you learned from your first guess to make a new guess.

2nd guess

Try 29, 30, and 31.
$29 + 30 + 31 = n$

Is the second guess correct?

GUIDED PRACTICE

Use guess and check to solve.

1. Find 2 consecutive even numbers (even numbers in order, such as 2 and 4) whose sum is 70.
   a. How many numbers are you looking for?
   b. How would you use the facts in the problem to make a first guess?
   c. What are the two numbers?

PRACTICE

Use guess and check to solve these problems.

2. Find 3 consecutive even numbers whose sum is 102.

3. Find 4 consecutive odd numbers whose sum is 264.

4. Winona read 5 consecutive pages in *A Wrinkle in Time*. The sum of the page numbers was 500. What pages did Winona read?

5. At a book fair, paperback books sold for $.99 or $1.99. Ken spent a total of $5.96 on paperback books. How many books did he buy?

6. Is the sum of 5 consecutive numbers an even or odd number? Explain. Give examples.

7. Can the sum of 3 consecutive numbers be 100? Explain. Give examples.

8. Lian gets either 3 or 5 points for reading a book. She has 59 points. What are two possibilities for the number of books she has read?

9. Todd read 9 pages of *A Cricket in Times Square*. The product of the page numbers he started and stopped on is 48. On what pages did he start and stop?

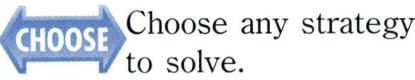 Choose any strategy to solve.

10. Akins has 35 books and Barry has 17. How many books must Akins give to Barry for the two boys to have the same amount?

11. Altogether, Kiah and Elizabeth have 153 books. Elizabeth has 9 more books than Kiah. How many books does Kiah have?

12. Nan has $1.15 in coins in her pocket but she doesn't have the exact change to buy a book for $1.00. What coins can Nan have?

MIDCHAPTER CHECKUP

LANGUAGE & VOCABULARY

Tell whether each sentence is true or false. If false, make the sentence true by changing the underlined word(s).

1. To add or subtract fractions with the same denominators, you add or subtract the <u>denominators</u>.

2. To subtract a mixed number from a whole number, it is necessary to rename the <u>mixed number</u>.

3. To write a sum that is a fraction in lowest terms, divide the numerator and denominator of the fraction by their <u>greatest common factor</u>.

QUICK QUIZ ✓

Add or subtract. Write the answer in lowest terms. *(pages 216–221)*

1. $\frac{3}{7} + \frac{1}{7}$
2. $\frac{11}{12} - \frac{7}{12}$
3. $\frac{3}{8} + \frac{7}{8}$

4. $3\frac{7}{9}$
   $-1\frac{5}{9}$

5. $4\frac{1}{6}$
   $+3\frac{5}{6}$

6. $2\frac{7}{16}$
   $+5\frac{13}{16}$

7. 8
   $-5\frac{2}{5}$

8. $7\frac{2}{9}$
   $-4\frac{4}{9}$

Use your ruler to measure to the nearest sixteenth inch. *(pages 222–223)*

9.

Use guess and check to solve. *(pages 224–225)*

10. Tawanna is thinking of 3 consecutive odd numbers whose sum is 75.
    a. What are consecutive odd numbers?
    b. How would you use the facts in the problem to make a first guess?
    c. What numbers is Tawanna thinking of?

Write the answers in your learning log.

1. You usually write fractions in lowest terms. When might you need an equivalent fraction that is not in lowest terms?

2. Compare trading in subtraction with fractions to trading in subtraction with whole numbers.

In 1869, the Cincinnati Red Stockings became the first professional baseball team in the United States. What are some local teams that you follow? Write their wins out of total games, and losses out of total games as fractions.

You have a jar that holds $\frac{3}{8}$ qt, a jar that holds $\frac{5}{8}$ qt, and a 5-gal bucket. How can you use these items to measure exactly $1\frac{3}{4}$ quarts of water?

Adding and Subtracting Fractions: Different Denominators

The first United States one-cent coin was made in Philadelphia in 1793. This coin was $\frac{15}{16}$ in. wide. Today's one-cent coin is $\frac{3}{4}$ in. wide. How much wider was the first cent than today's?

Since you know how to subtract fractions with the same denominator and how to write equivalent fractions, you can make a **MATH CONNECTION** to solve $\frac{15}{16} - \frac{3}{4}$.

Find the LCD (least common denominator).	Use the LCD to write equivalent fractions.	Subtract.
$\begin{array}{r} \frac{15}{16} \\ -\frac{3}{4} \\ \hline \end{array}$ The LCD of 4 and 16 is 16.	$\frac{15}{16} = \frac{15}{16}$ $-\frac{3}{4} = -\frac{12}{16}$	$\begin{array}{r} \frac{15}{16} \\ -\frac{12}{16} \\ \hline \frac{3}{16} \end{array}$

CRITICAL THINKING Will the LCD of two fractions always be the denominator of one of the fractions? Explain.

GUIDED PRACTICE

Add or subtract. Answer in lowest terms.

1. $\frac{1}{12} = \frac{}{12}$
 $+\frac{3}{4} = +\frac{}{12}$

2. $\frac{1}{2} = \frac{}{10}$
 $-\frac{4}{10} = -\frac{}{10}$

3. $\frac{9}{10}$
 $+\frac{3}{5}$

4. $\frac{6}{9} - \frac{1}{3}$

PRACTICE

Add or subtract. Write the answer in lowest terms.

5. $\frac{4}{5} - \frac{1}{10}$

6. $\frac{8}{9} - \frac{2}{3}$

7. $\frac{3}{4} + \frac{3}{8}$

8. $\frac{2}{7} - \frac{5}{21}$

9. $\frac{5}{12} + \frac{1}{3}$

10. $\frac{1}{4} + \frac{1}{12}$

11. $\frac{1}{8} + \frac{1}{2}$

12. $\frac{3}{5} - \frac{7}{20}$

13. $\frac{7}{12} + \frac{2}{6}$

14. $\frac{5}{6} - \frac{5}{18}$

15. $\frac{3}{4} - \frac{5}{12}$

16. $\frac{7}{8} - \frac{2}{16}$

17. $\frac{4}{5} + \frac{6}{25}$

18. $\frac{9}{16} - \frac{1}{4}$

19. $\frac{11}{24} + \frac{1}{6}$ **20.** $\frac{3}{7} - \frac{2}{14}$ **21.** $\frac{7}{20} + \frac{3}{5}$ **22.** $\frac{2}{5} - \frac{1}{15}$

23. $\frac{5}{6} + \frac{5}{12}$ **24.** $\frac{12}{25} - \frac{1}{5}$ **25.** $\frac{5}{6} - \frac{3}{6} + \frac{1}{3}$ **26.** $\frac{5}{6} + \frac{1}{6} - \frac{1}{3}$

MIXED REVIEW   Find the answer.

27. $8.7 - 6.89$ **28.** $2\frac{5}{9} + \frac{1}{9}$ **29.** $3.68 \div 100$ **30.** 0.7×0.8

NUMBER SENSE   Decide which is greater. Do not compute.

31. $\frac{4}{5} + \frac{1}{5}$ or $\frac{4}{5} + \frac{1}{6}$ **32.** $\frac{5}{6} + \frac{5}{8}$ or $\frac{5}{6} + \frac{3}{4}$ **33.** $\frac{3}{4} + \frac{9}{10}$ or $\frac{3}{4} + \frac{11}{12}$

MENTAL MATH   Complete the table.

34.

In	$\frac{1}{8}$	$\frac{3}{8}$	$\frac{5}{8}$	$\frac{7}{8}$	
Out	$\frac{1}{2}$	$\frac{3}{4}$	1		$1\frac{1}{2}$

35.

In	$\frac{1}{2}$	1	$1\frac{1}{2}$		$2\frac{1}{2}$
Out	$\frac{5}{6}$	$1\frac{1}{3}$	$1\frac{5}{6}$	$2\frac{1}{3}$	

PROBLEM SOLVING

CHOOSE Choose mental math or pencil and paper to solve.

HISTORY OF U.S. COINS—YEAR OF FIRST MINTS

1793 Cent, Half Cent
1794 Dollar, Half Dollar, Half Dime
1795 Eagle $10, Half Eagle $5
1796 Quarter, Dime

36. The half-cent coin was the size of today's nickel, or $\frac{13}{16}$ in. wide. How much wider was it than today's penny, which is $\frac{3}{4}$ in. wide?

37. In 1794, a half dozen apples cost $5\frac{1}{2}$¢. Suppose you paid with a half dollar. What is your change using the fewest coins possible?

38. Today's penny was first minted in 1909. How many years earlier was the first United States cent minted?

39. In 1794, what possible combinations of 11 coins were there that would total $1.28?

CHAPTER 7   229

Adding and Subtracting Fractions: Any Denominator

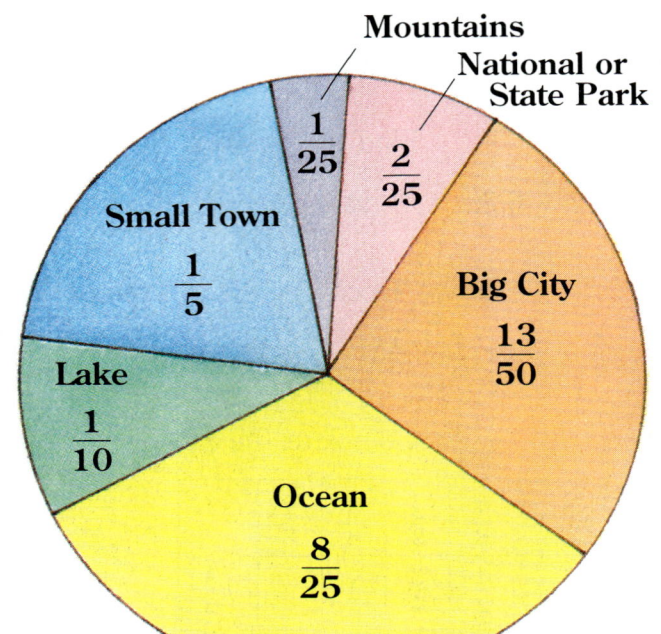

Where is your favorite place to vacation? In a survey, 8 out of every 25 people polled said they prefer to vacation near the ocean. One out of every 10 said they prefer to vacation near a lake. What fraction of the people polled prefer to vacation near the water?

To find out, add $\frac{8}{25} + \frac{1}{10}$.

Find the LCD (least common denominator).	Use the LCD to write equivalent fractions.	Add.
$\begin{array}{r}\frac{8}{25}\\+\frac{1}{10}\\\hline\end{array}$ The LCD of 25 and 10 is 50.	$\begin{array}{r}\frac{8}{25}=\frac{16}{50}\\+\frac{1}{10}=+\frac{5}{50}\\\hline\end{array}$	$\begin{array}{r}\frac{16}{50}\\+\frac{5}{50}\\\hline\frac{21}{50}\end{array}$

Of the people polled, $\frac{21}{50}$ prefer to vacation near the water.

THINK ALOUD Does $\frac{21}{50}$ represent more than half the people polled? Explain how you know.

Other examples:

$$\begin{array}{r}\frac{3}{4}=\frac{27}{36}\\-\frac{6}{9}=-\frac{24}{36}\\\hline\frac{3}{36}=\frac{1}{12}\end{array}$$

$$\begin{array}{r}\frac{5}{9}=\frac{10}{18}\\\frac{2}{6}=\frac{6}{18}\\+\frac{1}{2}=+\frac{9}{18}\\\hline\frac{25}{18}=1\frac{7}{18}\end{array}$$

LESSON 7-7

GUIDED PRACTICE

Explain how to find the LCD. Add or subtract.

1. $\dfrac{4}{5} - \dfrac{2}{3}$
2. $\dfrac{2}{7} + \dfrac{3}{4}$
3. $\dfrac{7}{10} - \dfrac{1}{4}$
4. $\dfrac{2}{3} - \dfrac{1}{2}$
5. $\dfrac{5}{9} + \dfrac{1}{9} + \dfrac{2}{3}$

PRACTICE

Add or subtract. Write the answer in lowest terms.

6. $\dfrac{3}{5} + \dfrac{1}{4}$
7. $\dfrac{4}{7} + \dfrac{2}{3}$
8. $\dfrac{2}{3} - \dfrac{1}{3}$
9. $\dfrac{5}{8} - \dfrac{1}{3}$
10. $\dfrac{7}{9} + \dfrac{5}{6}$
11. $\dfrac{5}{6} - \dfrac{3}{8}$

12. $\dfrac{5}{14} - \dfrac{4}{21}$
13. $\dfrac{4}{5} + \dfrac{1}{8}$
14. $\dfrac{11}{18} - \dfrac{3}{10}$
15. $\dfrac{1}{6} + \dfrac{5}{8}$
16. $\dfrac{9}{12} - \dfrac{3}{8}$
17. $\dfrac{5}{6} - \dfrac{2}{15}$

18. $\dfrac{2}{3} - \dfrac{5}{8}$
19. $\dfrac{8}{12} - \dfrac{1}{5}$
20. $\dfrac{1}{4} + \dfrac{5}{6}$
21. $\dfrac{6}{7} - \dfrac{1}{2}$

22. $\dfrac{5}{10} + \dfrac{2}{3}$
23. $\dfrac{1}{3} + \dfrac{1}{2} + \dfrac{1}{8}$
24. $\dfrac{2}{5} + \dfrac{1}{6} + \dfrac{7}{10}$
25. $\dfrac{3}{4} + \dfrac{2}{3} + \dfrac{4}{5}$

26. $\dfrac{2}{3} + \dfrac{3}{5} + \dfrac{1}{10}$
27. $\dfrac{8}{9} + \dfrac{1}{4} + \dfrac{1}{3}$
28. $\dfrac{3}{4} + \dfrac{2}{6} + \dfrac{1}{8}$
29. $\dfrac{5}{8} + \dfrac{1}{4} + \dfrac{3}{10}$

MIXED REVIEW Find the answer.

30. $1.32 \div 0.55$
31. $5{,}000 - 2{,}689$
32. $\$.89 \times 18$
33. $16{,}849 + 28{,}457$

ESTIMATE Is the answer closer to $\dfrac{1}{2}$, 1, or 2?

34. $\dfrac{9}{10} + \dfrac{4}{5}$
35. $\dfrac{9}{12} - \dfrac{3}{10}$
36. $\dfrac{1}{5} + \dfrac{1}{4}$
37. $\dfrac{1}{2} + \dfrac{5}{8}$
38. $\dfrac{7}{8} - \dfrac{1}{5}$

PROBLEM SOLVING

Use the circle graph on page 230. Write *true* or *false*.

39. More people surveyed liked to vacation in big cities than in all the other places put together.

40. The number of people who picked national or state parks was double the number who chose the mountains.

41. The sum of all the parts of the circle graph is greater than 1.

42. If 100 people were polled, 4 of them would probably choose the mountains as their favorite vacation spot.

CHAPTER 7   231

Adding and Subtracting Mixed Numbers: Any Denominator

It took Captain James Gallagher $3\frac{11}{12}$ days to fly the first nonstop flight around the world. Captain Walter Mullikin set a speed record around both poles with a $2\frac{1}{4}$-day flight.

How much longer did the first nonstop flight take?

You know how to subtract mixed numbers with the same denominators and how to write equivalent fractions. You can make a **MATH CONNECTION** to solve $3\frac{11}{12} - 2\frac{1}{4}$.

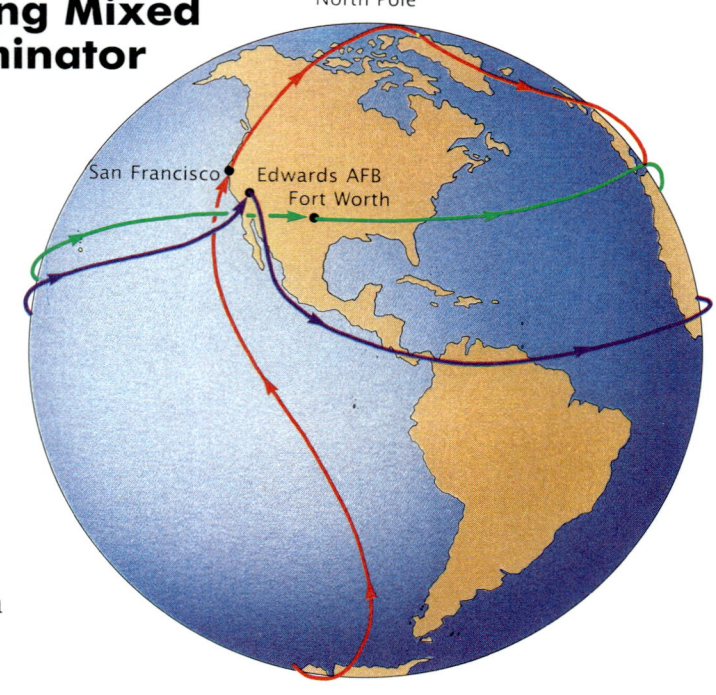

Use the LCD to write equivalent fractions.	Subtract. Write the answer in lowest terms.
$3\frac{11}{12} = 3\frac{11}{12}$ $-\ 2\frac{1}{4} = -\ 2\frac{3}{12}$	$3\frac{11}{12}$ $-\ 2\frac{3}{12}$ $1\frac{8}{12} = 1\frac{2}{3}$

■ 1st nonstop: 23,452 mi, $3\frac{11}{12}$ days, Capt. James Gallagher

■ Speed record around both poles: 26,382 mi, $2\frac{1}{4}$ days, Capt. Walter Mullikin

■ 1st non-stop, non-refueled flight 24,987 mi, 9 days, Dick Rutan and Jeana Yeager

The first nonstop flight took $1\frac{2}{3}$ days longer.

Estimate to see if the answer is reasonable.

$3\frac{11}{12} - 2\frac{1}{4} \Rightarrow 4 - 2 = 2 \quad 1\frac{2}{3}$ is close to 2.

CRITICAL THINKING What is $1\frac{2}{3}$ days in days and hours?

GUIDED PRACTICE

Add or subtract. Answer in lowest terms.
Check that your answer is reasonable.

1. $2\frac{2}{5}$
 $-\ 1\frac{1}{4}$

2. $15\frac{3}{4}$
 $+\ 2\frac{1}{6}$

3. $4\frac{7}{8}$
 $-\ 4\frac{4}{5}$

4. $6\frac{2}{3}$
 $+\ \frac{1}{2}$

5. $12\frac{1}{2} + 5\frac{3}{4}$

LESSON 7–8

PRACTICE

Add or subtract. Write the answer in lowest terms.

6. $7\frac{1}{3} + 5\frac{1}{4}$ **7.** $9\frac{7}{8} - 1\frac{3}{4}$ **8.** $21\frac{8}{9} - \frac{2}{5}$ **9.** $1\frac{1}{4} + 4\frac{11}{12}$ **10.** $5\frac{4}{5} + \frac{7}{8}$

11. $18\frac{5}{6} - 4\frac{3}{4}$ **12.** $23\frac{1}{6} + 25\frac{3}{8}$ **13.** $18\frac{7}{12} - 9\frac{5}{9}$ **14.** $12\frac{2}{3} - 9\frac{2}{5}$ **15.** $6\frac{7}{8} + \frac{2}{3}$

16. $9\frac{3}{4} + 3\frac{5}{10}$ **17.** $4\frac{6}{9} - 2\frac{1}{6}$ **18.** $5\frac{3}{10} + 2\frac{4}{5}$ **19.** $6\frac{7}{8} - 3\frac{3}{10}$ **20.** $5\frac{5}{6} - \frac{2}{3}$

Compute. Explain how to check that the answer is reasonable.

21. $14\frac{3}{4} - 8\frac{3}{5}$ **22.** $7\frac{5}{6} + 8\frac{1}{7}$ **23.** $8\frac{8}{11} - 6\frac{3}{5}$ **24.** $9\frac{3}{7} + 5\frac{4}{9}$

MIXED REVIEW   Find the answer.

25. 7.8×0.015 **26.** $\$2.49 \div 16$ **27.** $8.5 + 7.75$ **28.** $\$34.98 - \$17.89$

NUMBER SENSE   Explain how you know that the given answer is not correct. Do not compute.

29. $3\frac{3}{4} + 2\frac{1}{5} = 5\frac{4}{9}$ **30.** $9 - 7\frac{1}{4} = 2\frac{1}{4}$ **31.** $8\frac{1}{8} - 3\frac{3}{4} = 5\frac{4}{12} = 5\frac{1}{3}$

PROBLEM SOLVING

Use the graph on page 232.

32. How much longer did the nonstop, nonrefueled flight take than the first nonstop flight around the world?

33. **IN YOUR WORDS** Why do you think different numbers of miles are given for each of the flights? What could account for this?

Estimate

You can use rounding to find a range of sums.

	Round both addends down.	Round both addends up.
$3\frac{1}{2} + 2\frac{1}{4}$	$3 + 2 = 5$	$4 + 3 = 7$

The sum is between 5 and 7.

Find the range of the sum.

1. $5\frac{3}{5} + 7\frac{5}{6}$ **2.** $9\frac{7}{9} + 3\frac{1}{2}$ **3.** $17\frac{2}{7} + \frac{9}{10}$ **4.** $8\frac{1}{8} + 3\frac{3}{4}$

CHAPTER 7

Renaming Before Subtracting: Any Denominator

In 1985, Steve Cram of Great Britain ran a mile in about $3\frac{4}{5}$ min. Donald Davis of the United States ran a mile backward in about $6\frac{1}{10}$ min. How much longer did Davis take to run backward than Cram took to run forward?

You know how to rename to subtract mixed numbers with the same denominators and how to write equivalent fractions. You can make a **MATH CONNECTION** to subtract $3\frac{4}{5}$ from $6\frac{1}{10}$.

Use the LCD to write equivalent fractions.	You cannot subtract the fractions. Rename to get more tenths.	Subtract. Write the answer in lowest terms.
$6\frac{1}{10} = 6\frac{1}{10}$ $-3\frac{4}{5} = 3\frac{8}{10}$	$6\frac{1}{10} = 5\frac{11}{10}$ $3\frac{8}{10} = 3\frac{8}{10}$	$5\frac{11}{10}$ $-3\frac{8}{10}$ $\overline{2\frac{3}{10}}$

Davis took $2\frac{3}{10}$ min longer than Cram took.

Estimate to check that your answer is reasonable.

$6\frac{1}{10} - 3\frac{4}{5} \Rightarrow 6 - 4 = 2$ $2\frac{3}{10}$ is close to 2.

CRITICAL THINKING What is $2\frac{3}{10}$ min in minutes and seconds?

GUIDED PRACTICE

Subtract. Answer in lowest terms.

1. $4\frac{1}{6} = \quad 4\frac{1}{6} = \quad 3\frac{\square}{6}$
   $2\frac{1}{2} = -2\frac{3}{6} = -2\frac{\square}{6}$

2. $5\frac{1}{4} = \quad 5\frac{\square}{12} = \quad 4\frac{\square}{12}$
   $-1\frac{2}{3} = -1\frac{\square}{12} = -1\frac{\square}{12}$

3. $3\frac{1}{6}$
   $-1\frac{3}{4}$

4. 4
   $-3\frac{3}{8}$

234   LESSON 7-9

PRACTICE

Subtract. Write the answer in lowest terms.

5. $5\frac{1}{2} - 1\frac{2}{3}$
6. $7\frac{5}{12} - 3\frac{7}{9}$
7. $8\frac{1}{3} - 2\frac{5}{9}$
8. $10 - 1\frac{1}{8}$
9. $6\frac{2}{3} - 3\frac{3}{4}$
10. $5\frac{3}{5} - 2\frac{5}{6}$

11. $9\frac{1}{6} - 2\frac{4}{9}$
12. $15\frac{5}{8} - 1\frac{3}{4}$
13. $10 - 7\frac{9}{10}$
14. $25\frac{5}{12} - 3\frac{2}{3}$
15. $2\frac{1}{4} - \frac{5}{6}$
16. $8\frac{1}{2} - 6\frac{3}{4}$

17. $7\frac{2}{5} - 2\frac{1}{2}$
18. $10\frac{7}{12} - 7\frac{4}{5}$
19. $9\frac{2}{7} - 4\frac{2}{3}$
20. $8\frac{3}{5} - \frac{2}{3}$

21. $11\frac{2}{5} - 4\frac{7}{10}$
22. $16\frac{5}{12} - 7\frac{7}{8}$
23. $8 - 5\frac{3}{4}$
24. $4\frac{3}{8} - 2\frac{5}{6}$

25. $5\frac{1}{10} - 2\frac{1}{8}$
26. $18\frac{2}{5} - 14\frac{11}{20}$
27. $10 - \frac{5}{9}$
28. $8\frac{7}{12} - \frac{11}{15}$

MIXED REVIEW Find the answer.

29. 7.2×3.4
30. $5.67 \div 7$
31. $3 + 4.7$
32. $16\frac{1}{5} - 4\frac{4}{5}$

MENTAL MATH Use mental math to find the value of *n*. Explain your method.

33. $19 - n = 7\frac{1}{3}$
34. $n - 5\frac{7}{8} = 2\frac{6}{8}$
35. $n + 2\frac{2}{3} = 5\frac{1}{3}$
36. $2\frac{5}{8} + n = 7$

PROBLEM SOLVING

37. **CREATE YOUR OWN** Use this data about the Olympic javelin throw to write a word problem.

38. In 1948, Hungary won the Olympic gold medal in the women's long jump with a jump of 18 ft $8\frac{1}{4}$ in. In 1988, the United States won with a jump of 24 ft $3\frac{1}{2}$ in. How much longer was the 1988 jump?

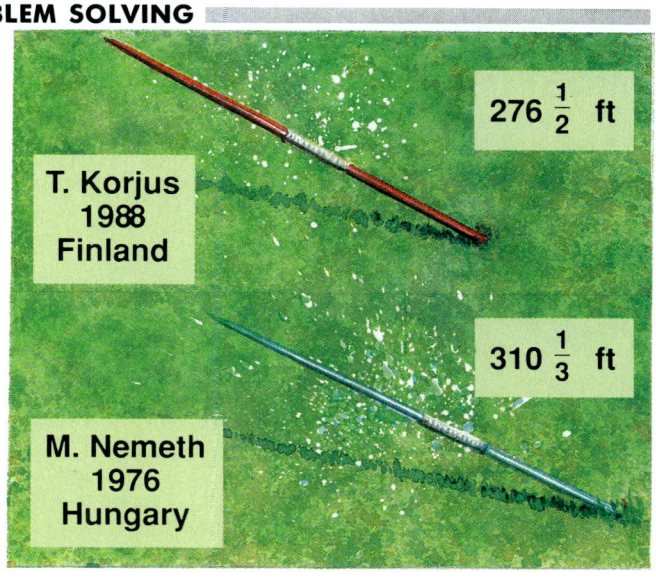

T. Korjus
1988
Finland

$276\frac{1}{2}$ ft

$310\frac{1}{3}$ ft

M. Nemeth
1976
Hungary

Problem Solving Strategy:
Using a Simpler Problem

Sometimes you can make a problem easier to solve by rewriting it. You can use simpler words and numbers and leave out unneeded information.

Original problem:

    A recipe from southeastern Asia for curried beef calls for $2\frac{1}{2}$ lb of beef. Dara wants to make the dish but only has $1\frac{1}{3}$ lb of beef. How much more beef does he need?

Simpler problem:

    A recipe calls for 2 lb of beef. Dara only has 1 lb of beef. How much more beef does he need?

To solve the simpler problem, you would subtract the whole numbers.

$2 \text{ lb} - 1 \text{ lb} = 1 \text{ lb}$

Now use the same method with the original problem.

$2\frac{1}{2} - 1\frac{1}{3}$
$\downarrow \quad \downarrow$
$2\frac{3}{6} - 1\frac{2}{6} = 1\frac{1}{6}$

Dara needs $1\frac{1}{6}$ lb more beef.

A spice seller in India.

GUIDED PRACTICE

Solve the simpler problem first. Then solve the original problem.

Original problem:

Simpler problem:

1. Amma has a recipe from Sri Lanka that makes 26 c of curry powder. She wants to give the curry powder as gifts, putting it into jars that hold $\frac{1}{2}$ c. How many jars will she need?

   Amma makes 2 c of curry powder. If she puts it into $\frac{1}{2}$-c jars, how many jars will she need?

2. Chef Lao's large oven is $57\frac{1}{4}$ in. wide. The width is $8\frac{1}{5}$ in. greater than the depth. How deep is the oven?

   Chef Lao's oven is 50 in. wide. The width is 10 in. greater than the depth. How deep is the oven?

PRACTICE

Write a simpler problem. Then use it to solve the original problem.

3. One Indian curry recipe uses $1\frac{3}{8}$ tsp cummin. If you wanted to double this recipe, how much cummin would you need?

4. Jamal has $7\frac{1}{2}$ c of rice. He used $\frac{3}{4}$ c in each of 2 curry dishes he made for a family gathering. How much rice does he have left?

5. Sara's recipe for shrimp curry uses $1\frac{1}{4}$ lb of shrimp. Tran's recipe uses $1\frac{3}{4}$ lb of shrimp. How much less shrimp is used in Sara's recipe than in Tran's?

6. Joshua found a curry lamb stew recipe that uses $1\frac{3}{4}$ lb green beans, $\frac{3}{4}$ lb of eggplant, $\frac{1}{4}$ lb green peppers, and $1\frac{1}{4}$ lb lamb. How much more vegetables than meat does the recipe use?

CHOOSE Choose any strategy to solve.

7. If curry powder costs $1.29 per can and curry paste costs $4.79 per jar, how much would 3 cans of powder and 2 jars of paste cost?

8. Rudyard served $\frac{1}{2}$ of his fish curry at lunch. He then served $\frac{1}{2}$ of what was left for dinner. What fraction of the fish curry was not served?

9. Victor and Neary worked together to collect 40 different curry recipes. Neary found 8 more than Victor. How many recipes each did they find?

10. Aun, Siko, Arlo, and Pete made curry sauces. Pete's was spicier than Siko's, but not as spicy as Aun's. Arlo's was spicier than Aun's. Whose curry was the spiciest? Whose was the mildest?

Investigating Stock Market Listings

Selling stocks is one method that companies use to raise money. If you buy stock, you buy a piece, or share, of the company.

Work with a partner.

- Find the daily stock market report in the financial section of a daily newspaper. It will look something like this.

STOCK MARKET REPORT

Name	Div	PE	Sales	High	Low	Last	Chg
Comcorp	.50	22	18	$33\frac{3}{4}$	$33\frac{1}{4}$	$33\frac{1}{4}$	$-\frac{1}{4}$
Leehar Inc.	.98	21	2,151	$60\frac{1}{2}$	$59\frac{3}{8}$	$60\frac{1}{4}$	$+\frac{1}{2}$

Look at the listing for Comcorp stock.

Dividend
portion of the company's profits paid to stockholders per year ($.50 per share of stock)

Sales
number of shares sold that day Shares are usually sold in lots of 100, so sales of 18 means 1,800 shares sold that day ($18 \times 100 = 1{,}800$).

Low
the lowest price in dollars, $33.25, for which one share of stock sold that day ($33\frac{1}{4} = \$33.25$).

PE
price/earnings ratio—stock price divided by the last 12 months' earnings

High
the highest price in dollars, $33.75, for which one share of stock sold that day ($33\frac{3}{4} = \$33.75$)

Last
the price the last batch of shares sold for that day ($33.25)

Change
the increase or decrease from the price at the end of the previous day. $-\frac{1}{4}$ means *down $.25*, so this stock lost $.25.

- Discuss with your partner what the listing for stock in Leehar Inc. means.

Choose a stock from the listings in your newspaper.

1. Make a chart to record the name of the company and the daily information given about it over a five-day period.

2. Keep track of your stock for five days. Record its daily listings.

3. How many shares of your stock were sold over the five days? (*Hint*: Remember to multiply by 100.)

4. What was the highest price your stock sold for during the five days? What was the lowest price?

5. Did the price of your stock increase or decrease over the five days? How much did it change?

6. Compare your findings with those of your classmates. Whose stock had the greatest increase?

CHAPTER 7 239

CHAPTER CHECKUP

LANGUAGE & VOCABULARY

Put these steps for subtracting $5\frac{1}{8} - 3\frac{3}{16}$ in order.

a. Subtract the whole numbers.
b. Rename the mixed number you are subtracting from.
c. Write the mixed number in lowest terms.
d. Subtract the fractions.
e. Find the least common denominator and write equivalent fractions.

TEST

CONCEPTS

Estimate each sum or difference. *(pages 218–219, 232–233)*

1. $3\frac{11}{16} + 5\frac{5}{16}$
2. $9\frac{2}{9} - 5\frac{7}{8}$

Find the range of the sum. *(pages 232–233)*

3. $8\frac{5}{8} + 4\frac{4}{7}$
4. $10\frac{1}{12} + 7\frac{15}{16}$

SKILLS

Add or subtract. Write the answer in lowest terms. *(pages 216–221, 228–234)*

5. $7\frac{5}{6} - 1\frac{1}{6}$
6. $2\frac{7}{10} + 5\frac{3}{10}$
7. $8\frac{2}{5} - 4\frac{4}{5}$
8. $9 - 3\frac{3}{4}$
9. $12\frac{7}{8} - 7\frac{1}{6}$
10. $7\frac{3}{10} + 3\frac{2}{5}$
11. $9\frac{2}{3} + 5\frac{3}{4}$
12. $8\frac{1}{4} - 2\frac{4}{5}$

13. $\frac{7}{9} - \frac{4}{9}$ 14. $\frac{5}{12} + \frac{11}{12}$ 15. $3\frac{7}{8} - 1\frac{3}{8}$ 16. $\frac{2}{5} + \frac{7}{20}$

17. $\frac{11}{12} - \frac{1}{3}$ 18. $\frac{3}{4} + \frac{9}{16}$ 19. $\frac{5}{8} - \frac{2}{9}$ 20. $\frac{3}{5} + \frac{1}{3}$ 21. $\frac{5}{6} + \frac{3}{4}$

Use your ruler to measure to the nearest sixteenth inch. *(pages 222–223)*

22.

PROBLEM SOLVING

23. Solve. *(pages 224–225 and 236–237)*

    On both Day 1 and Day 3 of her trip, Dawn rode her bicycle $14\frac{3}{4}$ mi. On Day 2 she rode $8\frac{1}{2}$ mi.

    a. For how many days did Dawn ride her bicycle?
    b. What simpler problem could you write to help you decide how to find the total distance Dawn rode?
    c. What is the total distance Dawn rode on the trip?

24. Admission to the museum is $2 for children under 12 and $3 for adults. The Redcloud family paid $13 for museum admission. What are the two possibilities for the number of admissions the Redclouds paid for children under 12?

25. Eugene lives $8\frac{1}{4}$ mi from school. He passes the tennis courts on his way to school. They are $2\frac{4}{5}$ mi from his house. How far are the tennis courts from the school?

LEARNING LOG

Write the answers in your learning log.

1. When would you multiply the denominators of two fractions to find the least common denominator?

2. When adding and subtracting fractions, when will the LCD be one of the denominators in the original example?

EXTRA PRACTICE

Add or subtract. Write the answer in lowest terms.
(pages 216–221, 228–235)

1. $4\frac{13}{15} - 1\frac{11}{15}$
2. $9\frac{9}{10} - 5\frac{7}{10}$
3. $3\frac{1}{12} + 2\frac{7}{12}$
4. $4\frac{3}{4} + 3\frac{1}{4}$
5. $6 - 2\frac{4}{5}$
6. $8\frac{1}{9} - 1\frac{5}{9}$
7. $5 - 4\frac{3}{8}$
8. $14\frac{5}{12} - 6\frac{7}{12}$
9. $\frac{3}{8} + \frac{1}{4}$
10. $\frac{11}{12} - \frac{2}{3}$
11. $\frac{2}{5} + \frac{9}{20}$
12. $\frac{3}{4} + \frac{7}{2}$
13. $\frac{2}{5} + \frac{3}{4}$
14. $\frac{1}{3} + \frac{3}{8}$
15. $\frac{7}{10} - \frac{1}{4}$
16. $\frac{5}{8} - \frac{3}{5}$
17. $5\frac{11}{12} - 2\frac{3}{4}$
18. $4\frac{5}{8} + 3\frac{5}{6}$
19. $8\frac{9}{10} - 4\frac{1}{3}$
20. $2\frac{1}{9} + 6\frac{4}{5}$
21. $5\frac{1}{4} - 3\frac{3}{8}$
22. $6\frac{2}{5} - 1\frac{5}{6}$
23. $7\frac{1}{3} - \frac{5}{9}$
24. $3\frac{3}{10} - 1\frac{3}{4}$
25. $9\frac{2}{3} - 8\frac{3}{4}$
26. $\frac{7}{10} + \frac{1}{10}$
27. $\frac{11}{16} - \frac{7}{16}$
28. $\frac{7}{8} - \frac{1}{8}$
29. $\frac{7}{12} + \frac{7}{12}$
30. $7\frac{5}{8} - 3\frac{5}{8}$
31. $8 - \frac{7}{10}$
32. $\frac{3}{4} - \frac{5}{16}$
33. $\frac{5}{6} - \frac{5}{24}$

Use a ruler to measure to the nearest sixteenth inch. *(pages 222–223)*

34. ├──────────────────────────────┤

Solve. *(pages 224–225 and 236–237)*

35. In one basketball game the Jets and the Eagles scored a total of 142 points. The Jets scored 12 points more than the Eagles. How many points did the Jets score?

36. Mr. Morris bought 2 packages of chicken for a barbeque. Each weighed $2\frac{3}{4}$ lb. He also bought $1\frac{5}{8}$ lb ground beef. How much meat did Mr. Morris buy?

242   EXTRA PRACTICE

ENRICHMENT

SHOPPING FOR FRACTIONS

Visit one of these types of stores:

    fabric   nursery
    produce   automotive
    hardware   sporting goods

Answer the questions.

- How are fractions used to describe items or amounts of items sold in that type of store?
- Do the fractions that were used have only certain denominators?

Write a problem whose solution uses addition or subtraction of fractions or mixed numbers. The problem should be one that a person at the store might have to solve. Trade with a partner and solve.

Magic Squares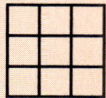

Work with a partner. Write each of these fractions on a small piece of paper or card:

$$\frac{1}{2}, \frac{1}{3}, \frac{1}{4}, \frac{1}{6}, \frac{1}{12}, \frac{2}{3}, \frac{3}{4}, \frac{5}{12}, \frac{7}{12}$$

Arrange the fractions to form a 3-by-3 magic square so that the sum of each row, column, and diagonal is the same.

Design another set of 9 fractions that can be arranged to form a 3-by-3 magic square. (*Hint:* Rewrite the fractions above with a common denominator. Look for a pattern.)

Pick's Formula

You can find the area, A, of this, or other complex shapes, drawn on dot paper by using a formula discovered by a man named Pick:

$A = \frac{b}{2} + i - 1$ b = number of points on polygon
                            i = number of points inside

$A = \frac{9}{2} + 8 - 1 = 11\frac{1}{2}$ square units

Work with a partner. Draw a shape on dot paper. Use Pick's formula to find the area of the shape.

CUMULATIVE REVIEW

Use compatible numbers to choose the best estimate.

1. $47\overline{)3{,}688}$
   a. 90
   b. 900
   c. 70
   d. none of these

2. $26.8 \div 3$
   a. 9
   b. 8
   c. 10
   d. none of these

3. $421.3 \div 23.6$
   a. 2
   b. 20
   c. 30
   d. none of these

Choose the appropriate customary unit.

4. A bedroom is about 12 ▬ long.
   a. yd
   b. in.
   c. mi
   d. none of these

5. An apple weighs about 4 ▬.
   a. oz
   b. T
   c. lb
   d. none of these

6. A soup bowl holds about 10 ▬.
   a. c
   b. fl oz
   c. pt
   d. none of these

Choose the appropriate metric unit.

7. A sixth grader is about 150 ▬ tall.
   a. m
   b. mm
   c. cm
   d. none of these

8. The mass of a dog is about 15 ▬.
   a. T
   b. mg
   c. g
   d. none of these

9. A car's gas tank holds about 60 ▬.
   a. kL
   b. L
   c. mL
   d. none of these

Find the LCM of the numbers.

10. 3 and 12
    a. 1
    b. 3
    c. 24
    d. none of these

11. 5 and 7
    a. 35
    b. 1
    c. 5
    d. none of these

12. 6 and 9
    a. 54
    b. 3
    c. 18
    d. none of these

PROBLEM SOLVING REVIEW

Remember the strategies and types of problems you have had so far. Solve.

Problem Solving Check List
- Choosing the operation
- Using an equation
- Using estimation
- Too much information
- Too little information
- Multistep problems
- Writing problems

1. During the last 4 yr, the total snowfall at High Peak was 72 in., 60 in., 80 in., and 78 in.
   a. For how many years is the total snowfall given?
   b. How would you find the average snowfall at High Peak for the last 4 yr?
   c. What was the average snowfall at High Peak during the last 4 yr?

2. In 1992, Mr. Meredith celebrated his 75th birthday. In what year was Mr. Meredith born?

3. In one month, Al's Discount Shop sold 72 stereos each priced at $495. About how much did this shop receive for all of these stereos?

4. The Ngus need 4 h 45 min to travel to their grandparents' house. They left at 1:30 P.M. What time did they arrive at their grandparents'?

5. A dolphin exhibit was toured by 810 people. They were admitted in groups of 24. How many people were in the last group?

6. Calendars at the bookstore regularly cost $7.95. During the first week of the new year, they are on sale for $5.50. How much did Mrs. Pei save by buying 3 calendars during the sale?

7. Leah decided to double her recipe for fruit punch since she invited 30 people to the party. The recipe calls for 750 mL of orange juice. Will she need more or less than 1 L of orange juice?

8. Dustin used part of his baby-sitting money to buy gifts for his sister. The gifts cost $12.83 and $9.50. He had $5.67 left. How much did Dustin earn baby-sitting?

Write a problem that could be solved using the equation. Then solve.

9. $n + 300 = 500$

10. $n - 75 = 25$

TECHNOLOGY

FRACTION PUZZLE

In the computer game "Fraction Hunt," players estimate, add, and subtract fractions. The fraction cross-number puzzle will help prepare you for the computer game.

Copy the cross-number puzzle. Use *only* the fractions in the cloud to fill the empty puzzle boxes.

$1\frac{2}{3}$	+		+		=	$4\frac{11}{12}$
+		−		+		
	−		−		=	$\frac{1}{12}$
=		=		=		
$3\frac{11}{12}$		$1\frac{11}{12}$		$1\frac{5}{6}$		

PRIME FIND

The greatest common factor of every number in the list below is a prime number. Use your calculator to find it.

391	527	697
323	119	901

MYSTERY FRACTION

The sum of the numerator and the denominator of a fraction is 520. The difference between them is 130. Written in lowest terms, the fraction is $\frac{3}{5}$.

Use your calculator and the guess and check strategy to help you find the mystery fraction.

Fractions: Multiplication and Division 8

DID YOU KNOW...?
A zoo elephant eats about 200 lb of food a day! An elephant in the wild eats more than 3 times as much food.

Recipes from the Zoo Kitchen

GORILLA ZOO STEW
16 carrots
16 oranges
12 apples
2 lb Monkey Chow
$\frac{1}{2}$ lb meat
2 lb primate-diet food
2 heads lettuce, any variety
4 eggs
4 yams
8 bananas
24–40 grapes
2 bunches celery
bales of hydroponic grass to taste
Toss all ingredients lightly. Divide among 4 trays. Serves 4 gorillas

BEAR FARE
$1\frac{1}{4}$ lb apples
$1\frac{1}{4}$ lb chicken
3 lb mackerel
15 lb Bear Chow
$1\frac{1}{2}$ lb hydroponic grass
Mix well. Serves 1 hungry polar bear

How would you adjust the Bear Fare recipe to serve 2 polar bears?

Predict the number of gorillas that could be fed with a recipe for Gorilla Zoo Stew that calls for 16 lb Monkey Chow.

Find out what a zoo elephant eats. Create a recipe for a meal that serves 2 zoo elephants.

USING DATA
Collect
Organize
Describe
Predict

CHAPTER 8

Multiplying Fractions: Using Arrays

In the design $\frac{3}{4}$ of the horizontal stripes are blue, and $\frac{1}{3}$ of the vertical stripes are red. Purple is created where the blue and the red stripes cross. What part of the design is purple?

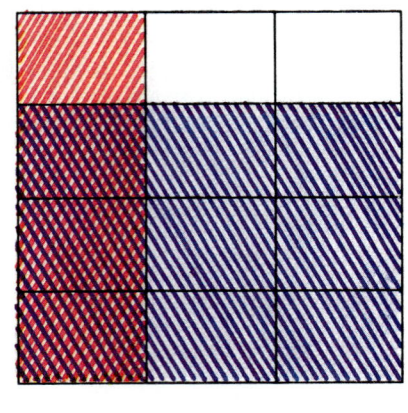

You can draw an array to show

$\frac{1}{3}$ of $\frac{3}{4}$, or $\frac{1}{3} \times \frac{3}{4}$.

First shade $\frac{3}{4}$ of the horizontal stripes.	Then shade $\frac{1}{3}$ of the vertical stripes.	Write a fraction to represent the part of the array that is double-shaded.
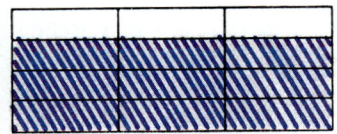		The stripes cross in 3 of the 12 parts. So, $\frac{1}{3} \times \frac{3}{4} = \frac{3}{12}$, or $\frac{1}{4}$.

CRITICAL THINKING How can you use the denominators to decide the number of parts in an array? How can you use the numerators to decide the number of double-shaded parts?

$$\frac{1}{3} \times \frac{3}{4} = \frac{1 \times 3}{3 \times 4} = \frac{3}{12}$$

1 column by 3 rows, or 1 × 3 parts, double-shaded

3 columns by 4 rows, or 3 × 4 parts, altogether

To multiply fractions, first multiply the numerators. Then multiply the denominators.

――――― GUIDED PRACTICE ―――――

Write the multiplication example shown by the array.

1.    2.

248 LESSON 8–1

Draw an array to find the product.

3. $\frac{1}{5} \times \frac{3}{4}$
4. $\frac{1}{3}$ of $\frac{2}{3}$
5. $\frac{3}{5} \times \frac{2}{3}$

PRACTICE

Write the multiplication shown by the array.

6.
7.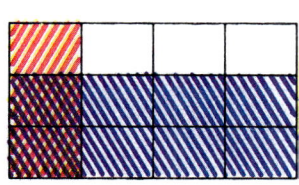
8.

Draw an array to find the product.

9. $\frac{1}{3} \times \frac{1}{2}$
10. $\frac{3}{8} \times \frac{1}{2}$
11. $\frac{1}{4} \times \frac{5}{8}$
12. $\frac{2}{5} \times \frac{2}{3}$
13. $\frac{3}{4} \times \frac{5}{6}$

Find the product.

14. $\frac{3}{7} \times \frac{2}{5}$
15. $\frac{5}{6} \times \frac{1}{9}$
16. $\frac{2}{11} \times \frac{3}{5}$
17. $\frac{5}{8} \times \frac{2}{3}$
18. $\frac{7}{12} \times \frac{1}{2}$

Write *true* or *false*.

19. $\frac{3}{4} \times \frac{2}{5} = \frac{5}{20}$
20. $\frac{1}{3}$ of $\frac{1}{4}$ is $\frac{1}{12}$.
21. $\frac{2}{3} \times \frac{1}{4} = \frac{2}{12}$
22. $\frac{1}{4}$ of $\frac{1}{3}$ is $\frac{1}{7}$.
23. $\frac{3}{4}$ of $\frac{3}{4}$ is $\frac{6}{8}$.
24. $\frac{1}{6}$ is $\frac{3}{6}$ of $\frac{1}{3}$.

25. **CRITICAL THINKING** When you multiply two fractions less than 1, is the product larger than both fractions, smaller than both fractions, or between the two fractions? Use examples to explain your answer.

PROBLEM SOLVING

26. Draw an array to show what part of a design is green if 4 of the 5 vertical stripes are blue, and 2 of the 3 horizontal stripes are yellow.

27. Enid worked $3\frac{1}{2}$ h drawing and cutting silkscreen designs and $2\frac{3}{4}$ h printing T-shirts. How long did she work in all?

28. If $\frac{2}{3}$ of Asabi's paint tubes are green and $\frac{3}{5}$ of the green tubes are blue-green, then what fraction of Asabi's paint tubes are blue-green?

29. What fraction multiplied by $\frac{1}{2}$ is $\frac{2}{8}$? Make an array to prove that your answer is correct.

CHAPTER 8 249

Multiplying Fractions

The string section makes up about $\frac{2}{3}$ of a symphony orchestra. Cellos make up about $\frac{1}{6}$ of the string section. About what part of a symphony orchestra are cellos?

Multiply to find $\frac{1}{6}$ of $\frac{2}{3}$.

Multiply the numerators. Multiply the denominators.	Reduce to lowest terms.
$\frac{1}{6} \times \frac{2}{3} = \frac{2}{18}$	$\frac{2}{18} = \frac{1}{9}$

Cellos are about $\frac{1}{9}$ of a symphony orchestra.

Since you know how to write a whole number as a fraction and you know how to multiply fractions, you can make a MATH CONNECTION to multiply a fraction by a whole number.

Write the whole number as a fraction.	Multiply the fractions.	Reduce to lowest terms.
$\frac{1}{10} \times 25 = \frac{1}{10} \times \frac{25}{1}$	$\frac{1}{10} \times \frac{25}{1} = \frac{25}{10}$	$\frac{25}{10} = \frac{5}{2} = 2\frac{1}{2}$

Other examples:

$\frac{3}{5} \times \frac{5}{8} = \frac{15}{40} = \frac{3}{8}$ $\frac{2}{5} \times \frac{2}{3} \times \frac{3}{4} = \frac{12}{60} = \frac{1}{5}$

GUIDED PRACTICE

1. **THINK ALOUD** Explain the steps needed to multiply $5 \times \frac{3}{7}$.

Multiply. Write the product in lowest terms.

2. $\frac{2}{3} \times \frac{3}{7}$ 3. $\frac{4}{5} \times \frac{3}{4}$ 4. $3 \times \frac{4}{9}$ 5. $\frac{2}{5} \times \frac{5}{6} \times \frac{1}{3}$

PRACTICE

Multiply. Write the product in lowest terms.

6. $\frac{1}{3} \times \frac{1}{3}$ 7. $8 \times \frac{1}{2}$ 8. $\frac{1}{2} \times \frac{4}{7}$ 9. $\frac{1}{3} \times \frac{1}{4}$ 10. $\frac{5}{6} \times 3$

11. $\frac{3}{4} \times \frac{1}{5}$ 　　　 12. $\frac{1}{3} \times \frac{2}{6}$ 　　　 13. $\frac{3}{8} \times 6$

14. $\frac{4}{5} \times \frac{5}{6}$ 　　　 15. $\frac{2}{3} \times \frac{9}{10}$ 　　　 16. $\frac{3}{5} \times \frac{1}{8}$

17. $\frac{2}{5} \times \frac{3}{7}$ 　　 18. $\frac{7}{8} \times \frac{1}{3}$ 　　 19. $\frac{9}{10} \times 5$ 　　 20. $\frac{9}{6} \times \frac{2}{3}$

21. $\frac{2}{3} \times \frac{1}{2} \times \frac{3}{10}$ 　　 22. $2 \times \frac{3}{4} \times \frac{2}{3}$ 　　 23. $\frac{5}{6} \times \frac{1}{5} \times n = \frac{1}{12}$

MIXED REVIEW Find the answer.

24. $\frac{3}{4} + \frac{2}{5}$ 　　 25. $\frac{7}{8} - \frac{5}{6}$ 　　 26. $3 + \frac{7}{8} + \frac{3}{4}$ 　　 27. $6 - \frac{3}{5}$

NUMBER SENSE Choose a value for n that makes the statement true. Use $\frac{5}{8}$, **1**, $\frac{3}{7}$, $\frac{8}{5}$, or **0**.

28. $\frac{5}{8} \times n = \frac{5}{8}$ 　　 29. $\frac{5}{8} \times n < \frac{5}{8}$ 　　 30. $\frac{5}{8} \times n > \frac{5}{8}$ 　　 31. $\frac{5}{8} \times n = 0$

PROBLEM SOLVING

Use the chart to answer the question.

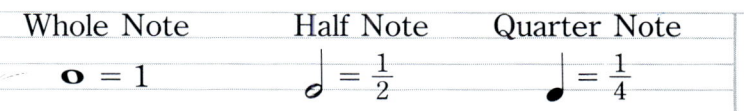

32. Putting a flag on the stem of a note halves the value of the note. What is the value of each note?

   a. ♪　　　b. 𝅘𝅥𝅯

33. Putting a dot at the right of a note adds on half the value of the note. What is the value of each note?

   a. 𝅗𝅥.　　　b. ♩.

Mental Math

To find $\frac{3}{4} \times 16$ mentally, picture splitting 16 into fourths.

Step 1: $\frac{1}{4}$ of $16 = 16 \div 4 = 4$　　　**Step 2**: $\frac{3}{4}$ of 16 is $3 \times 4 = 12$.

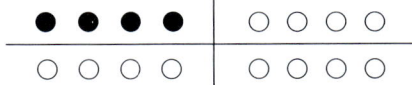　　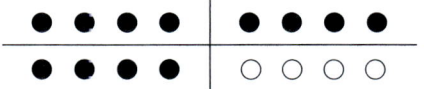

Use mental math to find the answer.

1. $\frac{1}{7}$ of 35 　 2. $\frac{5}{8} \times 32$ 　 3. $24 \times \frac{2}{3}$ 　 4. $\frac{7}{11}$ of 55 　 5. $\frac{4}{9} \times 81$

Multiplying Mixed Numbers

Claudia needs to make orange shakes for 14 people.

She figures out how to increase the recipe this way:
- 3 times the recipe serves 12.
- $\frac{1}{2}$ times the recipe serves 2.

So, $3\frac{1}{2}$ times the recipe serves 14.

Claudia's Orange Shake
$1\frac{1}{3}$ c orange juice
$\frac{1}{2}$ c powdered milk
$\frac{2}{3}$ tsp vanilla
Shake well. Serves 4

How much orange juice does Claudia need?
Multiply $3\frac{1}{2}$ times $1\frac{1}{3}$ c.

If you know how to write a mixed number as a fraction and you know how to multiply fractions, you can make a **MATH CONNECTION** to multiply mixed numbers.

Rename the mixed numbers as fractions.	Multiply. Write the product in lowest terms.
$3\frac{1}{2} \times 1\frac{1}{3} = \frac{7}{2} \times \frac{4}{3}$	$\frac{7}{2} \times \frac{4}{3} = \frac{28}{6} = \frac{14}{3} = 4\frac{2}{3}$

Claudia needs $4\frac{2}{3}$ c of orange juice.

GUIDED PRACTICE

Complete. Write the product in lowest terms.

1. $\frac{3}{4} \times 5\frac{1}{2} = \frac{3}{4} \times \frac{\rule{1em}{0.4pt}}{2} = \rule{1em}{0.4pt}$

2. $2\frac{1}{4} \times \frac{4}{5} = \frac{\rule{1em}{0.4pt}}{\rule{1em}{0.4pt}} \times \frac{4}{5} = \rule{1em}{0.4pt}$

Multiply. Write the product in lowest terms.

3. $12 \times \frac{5}{6}$ **4.** $2\frac{1}{3} \times 1\frac{1}{5}$ **5.** $\frac{2}{5} \times \frac{1}{2}$ **6.** $1\frac{3}{4} \times 3\frac{1}{3}$

PRACTICE

Multiply. Write the product in lowest terms.

7. $\frac{3}{4} \times 1\frac{1}{6}$ **8.** $\frac{2}{5} \times 8\frac{1}{2}$ **9.** $3\frac{1}{3} \times \frac{1}{5}$ **10.** $1\frac{1}{8} \times \frac{5}{6}$ **11.** $1\frac{3}{7} \times 1\frac{1}{5}$

12. $2\frac{2}{3} \times 2\frac{1}{2}$ **13.** $\frac{5}{9} \times \frac{2}{7}$ **14.** $1\frac{7}{8} \times \frac{2}{3}$ **15.** $6\frac{2}{3} \times 3$ **16.** $2\frac{1}{4} \times 3\frac{1}{3}$

17. $3\frac{3}{4} \times \frac{4}{5}$ **18.** $3 \times 2\frac{2}{9}$ **19.** $\frac{3}{7} \times 5\frac{1}{4}$ **20.** $2\frac{1}{5} \times 2\frac{1}{4}$ **21.** $12\frac{1}{2} \times \frac{1}{2}$

22. $2\frac{1}{2} \times 1\frac{2}{3}$ **23.** $\frac{4}{7} \times 1\frac{3}{5}$ **24.** $3\frac{1}{8} \times 2$ **25.** $2\frac{3}{4} \times \frac{2}{5}$ **26.** $3\frac{2}{3} \times 1\frac{1}{4}$

MIXED REVIEW Find the answer.

27. $2\frac{1}{8} + 3\frac{7}{8}$ **28.** $5\frac{5}{12} - 2\frac{11}{12}$ **29.** $3\frac{3}{5} + 2\frac{2}{3}$ **30.** $4\frac{5}{8} - 2\frac{3}{10}$ **31.** $6\frac{1}{6} - 2\frac{1}{2}$

32. CRITICAL THINKING
   a. Mike thought that $2\frac{1}{2} \times 3\frac{1}{3}$ was $6\frac{1}{6}$. How did he get that incorrect answer?
   b. Use the diagram to explain why $2\frac{1}{2} \times 3\frac{1}{3}$ must be more than $6\frac{1}{6}$.

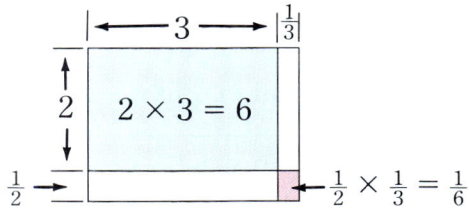

ESTIMATE Use rounding to the nearest whole number.

33. $4\frac{3}{4} \times 2\frac{1}{3}$ **34.** $5\frac{7}{8} \times 2\frac{9}{10}$ **35.** $\frac{5}{6} \times 3\frac{1}{4}$ **36.** $1\frac{3}{5} \times 4\frac{2}{7}$ **37.** $5\frac{2}{9} \times 4\frac{1}{8}$

PROBLEM SOLVING

38. How much lemon juice does Alvin need to make $2\frac{1}{2}$ times the recipe?

39. CREATE YOUR OWN Write a word problem that can be solved using the equation $5 \times \frac{1}{4} = 1\frac{1}{4}$.

40. Rewrite the recipe so that it makes enough dip to serve 9 people.

Mental Math

You can use the Distributive Property to multiply a mixed number by a whole number or by a fraction mentally.

$2\frac{2}{3} \times 9 = (2 \times 9) + (\frac{2}{3} \times 9) = 18 + 6 = 24$

Multiply each part of the mixed number by 9. Then add.

Use mental math to multiply.

1. $2\frac{1}{3} \times 6$ **2.** $1\frac{1}{5} \times 5$ **3.** $3 \times 5\frac{2}{3}$ **4.** $\frac{1}{4} \times 8\frac{1}{2}$ **5.** $\frac{1}{2} \times 2\frac{1}{2}$

Multiplying Fractions: Using the Short Cut

The protozoan *Monas stigmatica* can move 40 times its length in a second! If you were $4\frac{3}{4}$ ft tall, what would 40 times your height be?

$$40 \times 4\frac{3}{4} = \frac{40}{1} \times \frac{19}{4} = \frac{760}{4} = 190$$

Sometimes you can use a short cut to multiply fractions.

Divide both 40 and 4 by 4, the common factor.	Multiply the new numerators. Multiply the new denominators.
$\frac{\overset{10}{\cancel{40}}}{1} \times \frac{19}{\underset{1}{\cancel{4}}} = \blacksquare$	$\frac{\overset{10}{\cancel{40}}}{1} \times \frac{19}{\underset{1}{\cancel{4}}} = \frac{190}{1} = 190$

A protozoan is a one-celled organism.

Forty times your height would be 190 ft.

CRITICAL THINKING Explain why dividing both 40 and 4 by 4 doesn't change the size of the product.

Other examples using the short cut:

$\underset{1}{\cancel{\frac{2}{5}}} \times \frac{2}{\cancel{3}} \times \frac{\overset{3}{\cancel{9}}}{\underset{5}{\cancel{10}}} = \frac{6}{25}$ $\frac{\overset{4}{\cancel{12}}}{\underset{5}{\cancel{15}}} \times \frac{4}{5} = \frac{16}{25}$ $\frac{8}{15} \times 2\frac{1}{12} = \frac{8}{\underset{3}{\cancel{15}}} \times \frac{\overset{5}{\cancel{25}}}{\underset{3}{\cancel{12}}} = \frac{10}{9} = 1\frac{1}{9}$

═══════════════════ **GUIDED PRACTICE** ═══════════════════

Multiply. Use the short cut.

1. $1\frac{1}{2} \times 2\frac{1}{3}$ 2. $\frac{4}{5} \times \frac{10}{12}$ 3. $2\frac{1}{4} \times 10$ 4. $1\frac{7}{15} \times \frac{5}{11}$ 5. $\frac{2}{5} \times \frac{5}{9} \times \frac{1}{6}$

═══════════════════ **PRACTICE** ═══════════════════

Multiply. Use the short cut.

6. $35 \times \frac{5}{14}$ 7. $4\frac{1}{6} \times 2\frac{2}{5}$ 8. $\frac{8}{9} \times \frac{1}{4} \times \frac{6}{5}$ 9. $\frac{28}{15} \times 1\frac{9}{16} \times 2\frac{6}{7}$

IN YOUR WORDS Explain the mistake.

10. $\frac{\overset{5}{\cancel{10}}}{\underset{3}{\cancel{21}}} \times \frac{\overset{2}{\cancel{14}}}{2}$ 11. $\frac{7}{\underset{5}{\cancel{15}}} \times \frac{\overset{3}{\cancel{25}}}{8}$ 12. $\frac{\overset{2}{\cancel{10}}}{11} \times \frac{\overset{1}{\cancel{5}}}{3}$ 13. $\frac{\overset{3}{\cancel{3}}}{4} \times \frac{5}{\underset{3}{\cancel{18}}}$ 14. $3\frac{3}{4} \times 6\frac{\overset{1}{\cancel{1}}}{\underset{2}{\cancel{6}}}$

LESSON 8–4

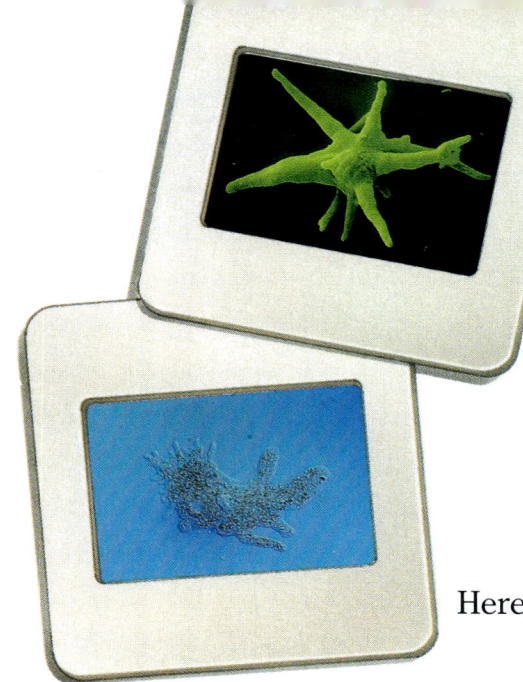

An amoeba moves by extending fingerlike pseudopods (false feet).

Reciprocals

The length of the amoeba, another type of protozoan, ranges from $\frac{1}{100}$ in. to $\frac{1}{10}$ in. How many times do you need to magnify the size of a $\frac{1}{10}$-in. amoeba for it to appear as 1 in.?

To find out, you can write $\frac{1}{10} \times \blacksquare = 1$.

$$\frac{1}{10} \times 10 = \frac{1}{10} \times \frac{10}{1} = \frac{10}{10} = 1$$

You need to magnify its size 10 times.

Here are some other pairs of numbers with a product of 1.

$$\overset{1}{\underset{1}{\frac{3}{2}}} \times \overset{1}{\underset{1}{\frac{2}{3}}} = 1 \qquad \overset{1}{\underset{1}{\frac{4}{7}}} \times 1\frac{3}{4} = \overset{1}{\underset{1}{\frac{4}{7}}} \times \overset{1}{\underset{1}{\frac{7}{4}}} = 1$$

THINK ALOUD What relationship do you see between the two fractions in each pair?

> **Any two numbers whose product is 1 are reciprocals. To find the reciprocal of a fraction, invert the numerator and the denominator.**

THINK ALOUD Describe how you can find the reciprocal of a whole number and of a mixed number.

=== GUIDED PRACTICE ===

Write the reciprocal.

1. 2
2. $\frac{3}{8}$
3. $\frac{1}{10}$
4. $2\frac{1}{2}$
5. $\frac{5}{6}$
6. 9

=== PRACTICE ===

Find the value of n.

7. $7 \times \frac{1}{n} = 1$
8. $n \times \frac{1}{100} = 1$
9. $\frac{5}{8} \times n = 1$
10. $n \times \frac{5}{17} = 1$
11. $4\frac{1}{2} \times \frac{2}{n} = 1$
12. $(n + \frac{1}{4}) \times \frac{4}{33} = 1$
13. $2\frac{1}{4} \times 2\frac{1}{2} \times n = 1$

Write the reciprocal.

14. 11
15. $\frac{5}{12}$
16. $1\frac{4}{5}$
17. $3\frac{1}{3}$
18. 20
19. $\frac{1}{16}$

Problem Solving Strategy: Using Estimation

Is $250 enough for a ski parka and 2 pairs of gloves? You can use estimation to find out.

- Estimate the total cost of the items at the regular price.
- How do you find out how much less the sale price is?
- How can the total cost be rounded to make the estimation easier?
- What is the estimated sale price?
- Is $250 enough money?

GUIDED PRACTICE

Estimate. Use the items pictured and the sale information.

1. About how much is the sale price of a pair of gloves?

2. About how much do you save when you buy a ski parka on sale?

3. Sue thinks she can buy a pair of skis and ski boots for under $150. Is she right? Explain.

4. Would $50 be enough to buy 4 pairs of the socks on sale?

5. You want to buy all the items shown in the chart.

    a. How many items are listed?
    b. What does it mean to say that the prices cluster around $16?
    c. How can you use multiplication to estimate the total cost of the items?

scarf	$16.25
mittens	$14.95
ear muffs	$15.50
day pack	$17.50

The prices **cluster** around $16.

PRACTICE

Use estimation to answer the questions about the sale.

6. The cost to the store for skates is $56.50 per pair. Will a profit still be made when two pairs are sold at the special price?

7. Jose bought a sweater, goggles, and gloves. Jon bought a parka and 2 pairs of socks. About how much more did Jon spend?

8. Is $200 a good estimate for the regular price of ski boots, ski poles, and gloves? Explain.

9. Gina bought a ski sweater, skis, and boots. Did she spend more than or less than $200?

CHOOSE Choose any strategy to solve. Use the items on pages 256 as needed.

10. The sale begins January 10 and ends March 1. Geraldo says he has 3 weeks left to get to the sale. If today is February 10, is he right? Explain.

11. Robert bought ski gloves on sale for $14.97. He spent $7.50 for a movie ticket and refreshments. If he ended the day with $1.38, how much money did he have to start?

12. Anna chose items totalling exactly $409 before the sale. What are the two possibilities for the 3 items she bought?

13. Harris High School ski team bought 36 ski hats on sale. How many hats could they have bought at the regular price for the same amount of money?

MIDCHAPTER CHECKUP

LANGUAGE & VOCABULARY

Write a problem that could be represented by this array. Then write the multiplication shown by the array.

QUICK QUIZ

Multiply. Write the product in lowest terms. *(pages 250–254)*

1. $\frac{2}{5} \times \frac{3}{8}$
2. $\frac{3}{4} \times 10$
3. $\frac{1}{6} \times \frac{2}{3} \times \frac{3}{10}$
4. $3\frac{1}{2} \times 2\frac{3}{7}$
5. $4\frac{3}{5} \times \frac{2}{9}$
6. $\frac{5}{8} \times 2\frac{1}{3}$

7. Show how to use the short cut to find $\frac{5}{12} \times \frac{9}{20}$.

Write the reciprocal. *(page 255)*

8. 15
9. $\frac{3}{5}$

Solve. *(pages 256–257)*

10. Each morning André jogs $4\frac{2}{5}$ mi. Last month he jogged on 27 mornings.
    a. How far does André jog each morning?
    b. What numbers could you use to estimate how many miles André jogged last month?
    c. About how many miles did André jog last month?

258 MIDCHAPTER CHECKUP

LEARNING LOG

Write the answers in your learning log

1. When you multiply two numbers that are greater than one, your answer gets bigger. Your friend thinks the same thing is true when you multiply two fractions that are less than one. Explain what is wrong with this thinking.

2. Describe what happens when you are multiplying fractions and you forget to use the short cut method.

MATH AMERICA

About $\frac{3}{4}$ of the over 20 million vehicles registered in California are automobiles. Find out the total number of vehicles and the number of automobiles in your state. Estimate the fractional part of the total number of vehicles that is represented by automobiles.

Write in standard form.

1. $\left(\frac{4}{5}\right)^2$
2. $\left(\frac{3}{8}\right)^3$
3. $\left(\frac{7}{10}\right)^1$

Exploring Dividing Fractions

If 1 batch of scones uses $\frac{2}{3}$ of a stick of margarine, how many batches can you make with 2 sticks of margarine? There are at least two ways to find out.

The Division Method

You will need 2 whole fraction bars and one $\frac{2}{3}$ fraction bar.

1. **IN YOUR WORDS** How does the diagram below show how many $\frac{2}{3}$ are in 2 wholes?

   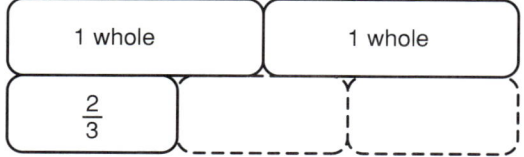

2. How many batches of the recipe can you make with 2 sticks of margarine? Write a division equation to show this.

The Multiplication Method

You will need 1 whole fraction bar and one $\frac{2}{3}$ fraction bar.

3. How many *complete* $\frac{2}{3}$ bars fit into 1 whole?

   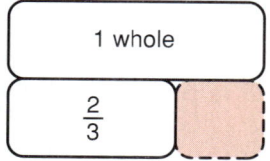

   There is at least one $\frac{2}{3}$ in 1, so you can make at least 1 batch of the recipe with 1 stick of margarine.

4. How much more than 1 batch can you make? (*Hint*: If you could fold the shaded section over the $\frac{2}{3}$, it wouldn't cover it all. What fraction of the $\frac{2}{3}$ would it cover?)

5. You can make $1\frac{1}{2}$, or $\frac{3}{2}$, batches with 1 stick. Write a multiplication equation to show how many batches you can make with 2 sticks.

260   LESSON 8–7

Look at the two equations you wrote:

$2 \div \frac{2}{3} = 3$ → the division method

$2 \times \frac{3}{2} = 3$ → the multiplication method

Since both answers are 3, you can write this equation:

$2 \div \frac{2}{3} = 2 \times \frac{3}{2}$

6. How are the two sides of the equation alike? How are the two sides different?

7. How are $\frac{2}{3}$ and $\frac{3}{2}$ related?

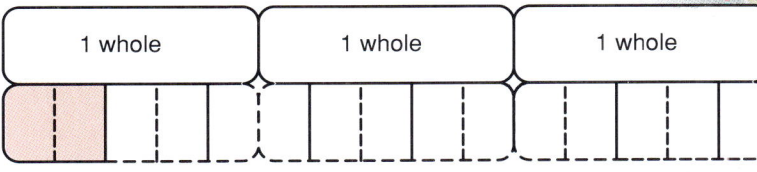

You can use fraction bars to find $3 \div \frac{2}{5}$.

8. How many *complete* $\frac{2}{5}$ bars fit into 3 wholes?

9. What extra fraction of a $\frac{2}{5}$ bar fits in into 3 wholes?

10. You found that $3 \div \frac{2}{5} = 7\frac{1}{2}$. Now find $3 \times \frac{5}{2}$.

11. What do you notice?

12. CRITICAL THINKING Describe how you can make a **MATH CONNECTION** to divide fractions.

> **To divide by a fraction, multiply by its reciprocal.**

13. The diagram shows that $\frac{1}{3} \div \frac{7}{8}$ is

  a. more than 1
  b. less than 1
  c. more than 2

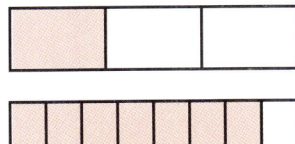

14. Now find $\frac{1}{3} \div \frac{7}{8}$ by multiplying $\frac{1}{3}$ by the reciprocal of $\frac{7}{8}$.

$\frac{1}{3} \div \frac{7}{8} = \frac{\boxed{}}{\boxed{}} \times \frac{\boxed{}}{\boxed{}} = \boxed{}$

CHAPTER 8

Dividing Fractions

If it takes $\frac{3}{8}$ lb of clay to make 1 teacup, how many teacups can the potter make from $\frac{3}{4}$ lb of clay?

To find out, divide $\frac{3}{4}$ by $\frac{3}{8}$.

Rewrite as multiplication. Use the reciprocal of the divisor.	Multiply. Use the short cut when possible.
$\frac{3}{4} \div \frac{3}{8} = \frac{3}{4} \times \frac{8}{3}$	$\frac{\overset{1}{\cancel{3}}}{\underset{1}{\cancel{4}}} \times \frac{\overset{2}{\cancel{8}}}{\underset{1}{\cancel{3}}} = \frac{2}{1} = 2$

The potter can make 2 teacups from $\frac{3}{4}$ lb of clay.

Other examples:

$$\frac{2}{3} \div \frac{1}{5} = \frac{2}{3} \times \frac{5}{1} = \frac{10}{3} = 3\frac{1}{3}$$

$$\frac{1}{5} \div \frac{4}{5} = \frac{1}{\underset{1}{\cancel{5}}} \times \frac{\overset{1}{\cancel{5}}}{4} = \frac{1}{4}$$

GUIDED PRACTICE

Write the reciprocal.

1. $\frac{1}{10}$
2. $\frac{3}{5}$
3. $\frac{5}{8}$
4. $\frac{1}{2}$

Complete. Use the short cut when possible.

5. $\frac{3}{5} \div \frac{1}{10} = \frac{3}{5} \times \blacksquare = \blacksquare$
6. $\frac{2}{3} \div \frac{3}{5} = \frac{2}{3} \times \blacksquare = \blacksquare$
7. $\frac{1}{2} \div \frac{5}{8} = \blacksquare$

PRACTICE

Divide. Use the short cut when possible.

8. $\frac{1}{2} \div \frac{1}{6}$
9. $\frac{9}{10} \div \frac{1}{5}$
10. $\frac{3}{4} \div \frac{2}{5}$
11. $\frac{4}{5} \div \frac{3}{10}$
12. $\frac{2}{3} \div \frac{8}{9}$
13. $\frac{1}{7} \div \frac{1}{3}$
14. $\frac{5}{8} \div \frac{2}{3}$
15. $\frac{1}{5} \div \frac{4}{5}$
16. $\frac{2}{9} \div \frac{1}{6}$
17. $\frac{3}{5} \div \frac{1}{3}$

18. $\frac{5}{8} \div \frac{5}{6}$ **19.** $\frac{8}{9} \div \frac{2}{3}$ **20.** $\frac{7}{10} \div \frac{2}{5}$ **21.** $\frac{1}{3} \div \frac{3}{6}$ **22.** $\frac{1}{2} \div \frac{7}{12}$

23. $\frac{8}{6} \div \frac{2}{3}$ **24.** $\frac{7}{8} \div \frac{3}{4}$ **25.** $\frac{1}{8} \div \frac{1}{6}$ **26.** $\frac{5}{12} \div \frac{1}{4}$ **27.** $\frac{2}{3} \div n = \frac{4}{9}$

MIXED REVIEW Find the answer.

28. $2.5 + 6$ **29.** $5.3 - 0.57$ **30.** $\frac{1}{2} \times \frac{4}{5}$ **31.** 0.5×0.8 **32.** $0.42 \div 0.6$

NUMBER SENSE Is the quotient greater than 1, less than 1, or equal to 1? Justify your answer. Do not compute.

33. $\frac{3}{4} \div \frac{3}{4}$ **34.** $8 \div \frac{1}{8}$ **35.** $\frac{1}{12} \div 2$ **36.** $\frac{3}{4} \div \frac{1}{12}$ **37.** $\frac{1}{5} \div \frac{1}{4}$

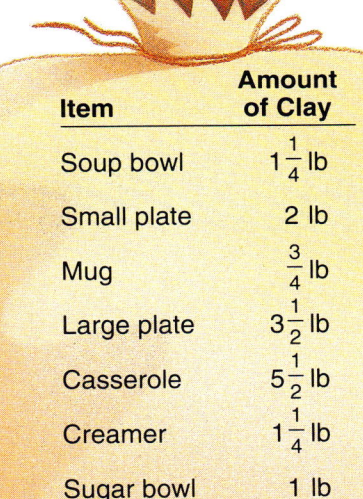

Item	Amount of Clay
Soup bowl	$1\frac{1}{4}$ lb
Small plate	2 lb
Mug	$\frac{3}{4}$ lb
Large plate	$3\frac{1}{2}$ lb
Casserole	$5\frac{1}{2}$ lb
Creamer	$1\frac{1}{4}$ lb
Sugar bowl	1 lb

PROBLEM SOLVING

Solve. Use the chart.

38. How many mugs can be made from a 21-lb container of clay?

39. What combinations of different items can you make using exactly 10 lb of clay? Give two answers.

40. To make a glaze for coating pottery, $\frac{1}{2}$ c of borate is mixed with $\frac{1}{3}$ c of flint. The glaze uses how many times as much borate as flint?

Estimate

You can use compatible numbers to estimate the product of a whole number and a fraction. Choose the nearest whole number that can be divided evenly by the denominator.

$\frac{1}{3} \times 14 = n$ $\frac{1}{3} \times 15 = 15 \div 3 = 5$ So, $\frac{1}{3} \times 14$ is about 5.

Estimate the product.

1. $\frac{1}{4} \times 17$ **2.** $\frac{1}{3} \times 28$ **3.** $23 \times \frac{1}{6}$ **4.** $\frac{1}{5} \times 42$ **5.** $13 \times \frac{3}{4}$

Dividing Fractions and Whole Numbers

One way of storing computer information is a disk. A $3\frac{1}{2}$-in. "floppy disk" is about $\frac{1}{7}$ in. thick. How many disks can you fit into a case that is 2 in. deep?

To find out, divide 2 by $\frac{1}{7}$.

Write the whole number as a fraction.	Rewrite as multiplication. Use the reciprocal of the divisor.	Multiply.
$2 \div \frac{1}{7} = \frac{2}{1} \div \frac{1}{7}$	$\frac{2}{1} \div \frac{1}{7} = \frac{2}{1} \times \frac{7}{1}$	$\frac{2}{1} \times \frac{7}{1} = \frac{14}{1} = 14$

You can fit 14 disks into a 2-in.-deep case.

Think about why dividing by $\frac{1}{7}$ is like multiplying by 7. Since there are 7 sevenths in 1 whole, there are 2×7, or 14, sevenths in 2 wholes.

THINK ALOUD How many eighths are in 2 wholes?

Other examples:

$\frac{1}{2} \div 12 = \frac{1}{2} \div \frac{12}{1} = \frac{1}{2} \times \frac{1}{12} = \frac{1}{24}$ $5 \div \frac{10}{11} = \frac{5}{1} \div \frac{10}{11} = \frac{\cancel{5}^1}{1} \times \frac{11}{\cancel{10}_2} = \frac{11}{2} = 5\frac{1}{2}$

=== GUIDED PRACTICE ===

Complete. Find the quotient.

1. $6 \div \frac{1}{2} = \frac{6}{\_\_} \times \frac{\_\_}{\_\_}$ 2. $\frac{1}{4} \div 8 = \frac{1}{4} \times \frac{\_\_}{\_\_}$ 3. $\frac{3}{5} \div 5 = \frac{3}{5} \times \frac{\_\_}{\_\_}$

4. $6 \div \frac{5}{6}$ 5. $\frac{2}{7} \div 3$ 6. $4 \div \frac{3}{5}$

=== PRACTICE ===

Divide.

7. $2 \div \frac{2}{3}$ 8. $9 \div \frac{1}{4}$ 9. $9 \div \frac{9}{10}$ 10. $6 \div \frac{3}{8}$ 11. $7 \div \frac{2}{5}$

12. $\frac{1}{4} \div 4$ 13. $\frac{4}{7} \div 2$ 14. $\frac{3}{8} \div 3$ 15. $\frac{5}{6} \div 4$ 16. $\frac{2}{3} \div 5$

264   LESSON 8-9

17. $5 \div \frac{2}{3}$ **18.** $\frac{1}{2} \div 10$ **19.** $\frac{7}{12} \div 7$ **20.** $\frac{3}{4} \div \frac{4}{5}$ **21.** $10 \div \frac{3}{4}$

MENTAL MATH Find the answer.

22. $7 \div \frac{1}{3}$ **23.** $5 \div \frac{1}{4}$ **24.** $6 \div \frac{1}{5}$ **25.** $3 \div \frac{1}{8}$

MIXED REVIEW Find the answer.

26. $3 \times \frac{3}{7}$ **27.** 0.8×9 **28.** $6 \div 0.3$ **29.** $0.324 \div 6$ **30.** $2.4 \div 1.5$

Choose a number from the box that makes the equation true.

| 8 | 6 | 2 | $\frac{1}{3}$ | $\frac{1}{2}$ | $\frac{3}{4}$ |

31. $n \div 2 = 3$ **32.** $24 \div n = 72$ **33.** $\frac{1}{4} \div n = \frac{1}{32}$ **34.** $n \div 6 = 1\frac{1}{3}$

35. $6 \div n = 8$ **36.** $n \div n = 1$ **37.** $n \div 1 = n$ **38.** $n \div \frac{2}{3} = \frac{3}{4}$

PROBLEM SOLVING

CHOOSE Choose mental math, calculator, or pencil and paper to solve.

39. A 1981 microprocessor chip has 68,000 transistors on a $\frac{1}{4}$-in. square. A 1991 chip with 1,000,000 transistors contains about how many times as many transistors as the earlier chip?

40. In 1990, large mainframe computers carried out about 10 million instructions in $\frac{1}{2}$ s. How long did it take them to carry out 1 million instructions?

Critical Thinking

Complete each table.

1. What is the result of multiplying by $\frac{1}{2}$?
2. What is the result of dividing by $\frac{1}{2}$?
3. Predict the product and the quotient if $n = 50$ and if $n = 100$.
4. What would be the result if you used $\frac{1}{4}$ instead of $\frac{1}{2}$ in each table?

n	$n \times \frac{1}{2}$
1	$1 \times \frac{1}{2} = \frac{1}{2}$
2	$2 \times \frac{1}{2} = 1$
3	$1\frac{1}{2}$
4	2
5	?
10	?
20	?

n	$n \div \frac{1}{2}$
1	$1 \div \frac{1}{2} = 2$
2	$2 \div \frac{1}{2} = 4$
3	6
4	8
5	?
10	?
20	?

Dividing Mixed Numbers

The 1890s bicycling outfit required $3\frac{5}{8}$ yd of 54-in.-wide fabric. How many similar outfits can be made from $7\frac{1}{2}$ yd of 54-in.-wide material?

> Divided-skirt trousers became popular for bicycle riding in the late 1800s. Some could be let down to look like an ordinary walking skirt.

To find out, divide $7\frac{1}{2}$ by $3\frac{5}{8}$.

Write the mixed numbers as fractions.	Multiply by the reciprocal of the divisor.
$7\frac{1}{2} \div 3\frac{5}{8} = \frac{15}{2} \div \frac{29}{8}$	$\frac{15}{\underset{1}{\cancel{2}}} \times \frac{\overset{4}{\cancel{8}}}{29} = \frac{60}{29} = 2\frac{2}{29}$

The whole number 2 tells us that 2 outfits can be made. What does the fraction $\frac{2}{29}$ tell us?

CRITICAL THINKING If you use fabric half as wide, you need more than twice as long a piece. Why might that be?

Another example: $6 \div 1\frac{3}{5} = \frac{6}{1} \div \frac{8}{5} = \frac{\overset{3}{\cancel{6}}}{1} \times \frac{5}{\underset{4}{\cancel{8}}} = \frac{15}{4} = 3\frac{3}{4}$

GUIDED PRACTICE

Write the reciprocal.

1. $4\frac{1}{3}$
2. $2\frac{3}{4}$
3. $1\frac{5}{6}$
4. $9\frac{7}{8}$

Find the quotient. Explain your steps.

5. $\frac{1}{3} \div 2\frac{2}{3}$
6. $4\frac{2}{5} \div 3\frac{1}{2}$
7. $4 \div 1\frac{1}{5}$
8. $3\frac{3}{4} \div \frac{2}{3}$

LESSON 8–10

PRACTICE

Find the quotient.

9. $2\frac{1}{2} \div \frac{1}{8}$
10. $3\frac{1}{4} \div 4\frac{1}{3}$
11. $\frac{1}{2} \div 4\frac{1}{2}$
12. $3\frac{3}{8} \div 3$
13. $3\frac{3}{5} \div 3\frac{1}{3}$

14. $\frac{1}{3} \div 6\frac{1}{3}$
15. $4\frac{4}{5} \div 5$
16. $\frac{5}{6} \div 7\frac{1}{2}$
17. $3 \div 2\frac{2}{3}$
18. $5\frac{1}{2} \div 3\frac{1}{6}$

19. $\frac{2}{3} \div \frac{1}{4}$
20. $2\frac{1}{10} \div 1\frac{4}{5}$
21. $1\frac{7}{8} \div 2\frac{1}{12}$
22. $5\frac{1}{4} \div 1\frac{2}{5}$
23. $2\frac{1}{6} \div 2\frac{1}{3}$

MIXED REVIEW Find the answer.

24. $\frac{1}{3} \times 3\frac{1}{2}$
25. 3.2×4.6
26. $2\frac{1}{4} \times 1\frac{1}{2}$
27. $1.2 \div 2.5$
28. $4.08 \div 0.2$

IN YOUR WORDS Explain the mistake.

29. $6 \div \frac{1}{3} = \frac{1}{3}$ of $6 = 2$.
30. $\frac{1}{2} \div 5 = 2 \times 5 = 10$
31. $1\frac{1}{3} \div 2\frac{1}{2} = \frac{4}{3} \times \frac{5}{2} = \frac{20}{6} = 3\frac{1}{3}$
32. $6 \div \frac{1}{5} = 30$, so $6 \div \frac{2}{5} = 60$.

PROBLEM SOLVING

Use the information at right.

33. How many blouses for a 4-yr-old could be made with a $6\frac{1}{2}$-yd bolt of 36-in.-wide fabric?

34. Is 6 yd of 36-in.-wide fabric enough to make 2 blouses for an 8-yr-old and 1 blouse for a 4-yr-old?

Girl's "Guimpe" (1894)
To make with 36-in.-wide fabric:
Use $1\frac{5}{8}$ yd for a child of 4.
Use $1\frac{7}{8}$ yd for a girl of 8.

This type of blouse was often worn with a jumper.

Critical Thinking

1. Work with a partner to find at least one value of n.
   a. $6 \div n = 6$
   b. $6 \div n < 6$
   c. $6 \div n > 6$
   d. $0 \div n = 0$

2. Compare your results with another group's. Write a description of the kind of numbers that will work for each case.

Problem Solving Strategy: Using Equations

Víctor is making *empanadillas*, an appetizer of meat-filled pastry, for his family's reunion in Havana.

Look at his recipe for the dough for 1 batch of empanadillas. How many batches can he make with 8 c flour?

- What do you need to know to solve the problem?

You can use an equation to help you solve the problem. The variable n can represent the number of batches Víctor can make.

Empanadillas Dough

2 c flour	1 large egg
$\frac{1}{2}$ c lard	1 tablespoon sugar
$\frac{1}{2}$ c cold water	$\frac{1}{2}$ tablespoon salt

number of cups for each batch		number of batches he can make		number of cups Víctor has
2	×	n	=	8

Solve the equation for n.
Then look back and check your answer.

LESSON 8-11

GUIDED PRACTICE

Read the passage. Answer the questions.

Víctor used $\frac{1}{4}$ lb ground pork for the empanadillas filling. He had $\frac{3}{4}$ lb of ground pork left. How much did he have to begin with?

1. What amount of pork did Víctor use?

2. How would you find the amount of pork Víctor has left?

3. Choose the equation that represents the problem.

   a. $n + \frac{1}{4} = \frac{3}{4}$ b. $n - \frac{1}{4} = \frac{3}{4}$ c. $n \times \frac{1}{4} = \frac{3}{4}$

PRACTICE

Choose the equation that best represents the problem. Solve.

4. Berto is making his family recipe for *arroz con pollo*. He spent $6.75 for some boxes of rice. If each box cost $1.35, how many did he buy?

   a. $\$6.75 \times \$1.35 = n$

   b. $n \times \$6.75 = 5$

   c. $\$6.75 \div n = \$1.35$

5. Berto uses $\frac{1}{2}$ lb more chicken than Carlos does in his *arroz con pollo* recipe. If Berto uses 6 lb of chicken, how much does Carlos use?

   a. $n \times \frac{1}{2} = 6$

   b. $n + \frac{1}{2} = 6$

   c. $6 + \frac{1}{2} = n$

6. To make black beans for 8 people, Lou buys $\frac{3}{4}$ lb of dried beans. If he doubles the recipe, how many pounds should he buy?

   a. $n = 2 \times 8$

   b. $n = 2 \times \frac{3}{4}$

   c. $8 \times \frac{3}{4} = n$

7. Rosa is making tostadas. She puts $2\frac{1}{2}$ teaspoons of cooked beans on 1 tortilla. She used 30 teaspoons. How many tostadas did she make?

   a. $n \div 2\frac{1}{2} = 30$

   b. $n \times 2\frac{1}{2} = 30$

   c. $n = 2\frac{1}{2} \times 30$

 Choose any strategy to solve.

8. Marco is bringing 36 tortillas to the reunion. His recipe calls for 2 c of flour for 4 tortillas. How much flour will he need?

9. Louisa is arranging the seating for the reunion. Each square table seats 4 people, one on each side. If she puts 4 square tables together to make 1 large rectangular table, how many people can be seated?

CHAPTER CHECKUP

LANGUAGE & VOCABULARY

Explain how to find $4 \div \frac{3}{8}$ using each of the following methods.
- division method and fraction bars
- multiplication method and fraction bars
- reciprocal method

TEST

CONCEPTS

1. Write the multiplication shown by the array. *(pages 248–249)*

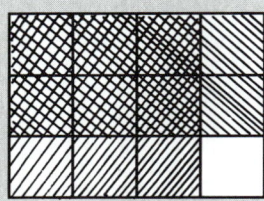

2. Multiply $\frac{4}{25} \times \frac{5}{6}$. Show how to use the short cut. *(page 254)*

Write the reciprocal. *(page 255)*

3. $\frac{3}{8}$
4. 5
5. $4\frac{1}{2}$

SKILLS

Multiply. Write the product in lowest terms. *(pages 250–253)*

6. $\frac{5}{8} \times \frac{8}{9}$
7. $3 \times \frac{3}{4}$
8. $\frac{3}{5} \times \frac{7}{12}$
9. $2\frac{5}{8} \times \frac{4}{9}$
10. $1\frac{1}{4} \times 1\frac{1}{6}$
11. $3\frac{5}{6} \times 2$
12. $\frac{8}{15} \times \frac{9}{10}$
13. $2\frac{2}{7} \times 1\frac{3}{8}$
14. $\frac{2}{3} \times \frac{1}{9} \times \frac{7}{8}$

Divide. *(pages 262–267)*

15. $\frac{1}{6} \div \frac{5}{6}$
16. $\frac{3}{8} \div \frac{2}{5}$
17. $\frac{8}{9} \div \frac{2}{3}$
18. $5 \div \frac{1}{4}$
19. $7 \div \frac{3}{4}$
20. $\frac{3}{10} \div 6$
21. $1\frac{3}{4} \div 2\frac{5}{8}$
22. $6 \div \frac{1}{3}$

PROBLEM SOLVING

Choose the equation that represents the problem. *(pages 268–269)*

23. The trip from her home to her friend's house usually takes Carol $4\frac{3}{4}$ h. If she has been traveling for $1\frac{1}{2}$ h, how much longer does she need to travel?

    a. $4\frac{3}{4} + 1\frac{1}{2} = n$
    b. $1\frac{1}{2} + n = 4\frac{3}{4}$
    c. $n - 1\frac{1}{2} = 4\frac{3}{4}$

Solve. *(pages 256–257, 268–269)*

24. The Herb Shop buys herbs in 92-oz packages from growers. The shop repackages the herbs for sale in $2\frac{3}{4}$-oz bags.
    a. How many ounces are in each of the growers' bags?
    b. How would you estimate the number of $2\frac{3}{4}$ oz bags in a 92 oz bag?
    c. About how many $2\frac{3}{4}$ oz bags can be filled from one 92-oz package?

25. A recipe calls for $\frac{1}{3}$ c milk per batch. How many batches can you make with 6 c milk?

LEARNING LOG

Write the answers in your learning log.

1. Your friend was absent yesterday. Explain what is meant by the reciprocal of a fraction.

2. Dividing with whole numbers can be checked by multiplying. Explain how to use this method with fractions.

CHAPTER 8   271

EXTRA PRACTICE

Write the multiplication shown by the array. *(pages 248–249)*

1.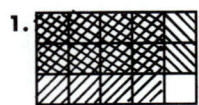
2.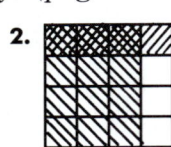

Multiply. Write the product in lowest terms. *(pages 250–254)*

3. $\frac{3}{8} \times \frac{2}{3}$
4. $4 \times \frac{5}{8}$
5. $\frac{1}{9} \times \frac{1}{8}$
6. $\frac{7}{12} \times \frac{4}{5}$
7. $\frac{8}{15} \times \frac{9}{16}$

8. $\frac{4}{5} \times 1\frac{2}{3}$
9. $2\frac{1}{2} \times 3\frac{3}{4}$
10. $1\frac{5}{9} \times \frac{3}{10}$
11. $4\frac{2}{5} \times 10$
12. $3\frac{1}{3} \times 2\frac{1}{4}$

13. $\frac{1}{3} \times \frac{2}{5} \times \frac{3}{7}$
14. $4 \times \frac{3}{8} \times \frac{7}{10}$
15. $\frac{5}{9} \times \frac{6}{7} \times \frac{14}{15}$

Write the reciprocal. *(page 255)*

16. 8
17. $\frac{2}{5}$
18. $1\frac{3}{4}$
19. $\frac{1}{9}$
20. 16
21. $2\frac{1}{3}$

Divide. *(pages 262–267)*

22. $\frac{1}{10} \div \frac{1}{2}$
23. $\frac{2}{3} \div \frac{5}{9}$
24. $\frac{1}{4} \div \frac{1}{5}$
25. $\frac{1}{8} \div \frac{7}{10}$
26. $\frac{5}{12} \div \frac{3}{4}$

27. $5 \div \frac{5}{6}$
28. $\frac{6}{7} \div 2$
29. $\frac{5}{8} \div 4$
30. $6 \div \frac{7}{12}$
31. $3 \div \frac{9}{10}$

32. $\frac{4}{5} \div 1\frac{1}{2}$
33. $4\frac{3}{4} \div 6$
34. $5\frac{1}{3} \div 2\frac{1}{6}$
35. $4 \div 1\frac{1}{3}$
36. $2\frac{5}{8} \div \frac{3}{4}$

Solve. *(pages 256–257)*

37. A cafeteria plans to use $3\frac{3}{4}$ oz of fish per serving. About how many ounces of fish will the cafeteria use if 58 people order fish?

Choose the equation that represents the problem. *(pages 268–269)*

38. Yonah used $8\frac{1}{2}$ c flour to make 4 loaves of bread. How many cups of flour did she use per loaf?

    a. $4 \times 8\frac{1}{2} = n$
    b. $4 \div 8\frac{1}{2} = n$
    c. $4 \times n = 8\frac{1}{2}$

ENRICHMENT

Class Party

Work with a partner to find the cost of food for a class party.

- Find a recipe for fruit punch and a snack food you would like to serve at the party.
- Rewrite the ingredients list of each recipe so that your entire class could be served.
- Visit a grocery store or use a newspaper. Complete a table like the one below for all the ingredients on your list. Then find the total cost of preparing the recipes.

TIME SURVEY

- Select three activities you do on a regular basis, such as going to school, riding your bike, or sleeping.
- Keep track of the time you spend on each activity during one week. Record your times to the nearest quarter hour.
- Find the average time you spend on each activity per day, per month (4 wk), and per year (52 wk).

Ingredient	Amount to purchase	Cost
orange juice	2 half gallon bottles	$3.78

GRID MULTIPLICATION

Here's another way to multiply $4\frac{1}{2}$ by $2\frac{1}{4}$.

Write each mixed number on the side of a grid.	Fill in the grid as you would a multiplication table.	Add all the numbers inside the grid.

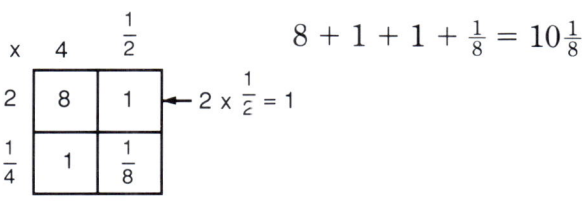

Use grid multiplication to find each product.

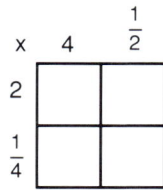

1. $5\frac{4}{7} \times 7\frac{2}{5}$
2. $8\frac{1}{3} \times 9\frac{3}{4}$
3. $10\frac{2}{3} \times 6\frac{4}{5}$
4. $12\frac{1}{2} \times 4\frac{5}{6}$

CUMULATIVE REVIEW

Find the perimeter of each.

1. a square with a side of 10 in.
   a. 20 in.
   b. 100 in.
   c. 14 in.
   d. none of these

2. (rectangle: 4.5 mi by 1.2 mi)
   a. 5.7 mi
   b. 11.4 mi
   c. 6.9 mi
   d. none of these

3. (right triangle: 6 ft 1 in., 7 ft 4 in., 4 ft 2 in.)
   a. 17 ft 7 in.
   b. 13 ft 5 in.
   c. 18 ft 3 in.
   d. none of these

Which number is divisible by the given numbers?

4. 3, 5, and 10
   a. 45
   b. 103
   c. 60
   d. none of these

5. 5 and 6
   a. 40
   b. 75
   c. 54
   d. none of these

6. 2, 4, and 5
   a. 80
   b. 30
   c. 56
   d. none of these

Find the fraction that is equivalent to each.

7. $\frac{4}{9}$
   a. $\frac{2}{3}$
   b. $\frac{16}{36}$
   c. $\frac{8}{27}$
   d. none of these

8. $\frac{15}{24}$
   a. $\frac{5}{6}$
   b. $\frac{3}{8}$
   c. $\frac{5}{8}$
   d. none of these

9. $\frac{5}{12}$
   a. $\frac{10}{24}$
   b. $\frac{10}{12}$
   c. $\frac{15}{24}$
   d. none of these

Add or subtract.

10. $1\frac{9}{10} + 3\frac{7}{10}$
    a. $4\frac{3}{5}$
    b. $4\frac{4}{5}$
    c. $5\frac{3}{5}$
    d. none of these

11. $5\frac{7}{8} - 3\frac{3}{8}$
    a. $1\frac{1}{2}$
    b. $2\frac{1}{2}$
    c. $2\frac{1}{4}$
    d. none of these

12. $6\frac{1}{6} - 2\frac{5}{6}$
    a. $4\frac{1}{3}$
    b. $4\frac{2}{3}$
    c. $3\frac{2}{3}$
    d. none of these

PROBLEM SOLVING REVIEW

Remember the strategies and types of problems you have had so far. Solve.

Problem Solving Check List
- Too much information
- Too little information
- Drawing a diagram
- Making a table
- Using estimation
- Using guess and check
- Multistep problems

1. Sally read 5 consecutive pages in *Moby Dick*. The sum of the page numbers was 450. What pages did Sally read?

2. The 392 students at Lincoln School are taking a field trip to a wildlife preserve. Each bus carries 60 people.
   a. How many people does one bus carry?
   b. How can you decide how many buses are needed?
   c. How many buses are needed?

3. The Sport mini-van is available in five colors: white, silver, red, black, and green; and with two types of transmission: manual and automatic. How many combinations of color and transmission are available?

4. Marc bought 2 tacos and lemonade for $2.34. He gave the clerk a $10 bill. What are the fewest coins and bills he could receive as change?

5. Mrs. Pease is putting a fence around her garden. The garden is a rectangle 16 ft by 30 ft. How much fencing does Mrs. Pease need?

6. This year 8,343 senior citizens, 22,925 adults, and 7,115 students attended City Orchestra concerts. About how many people attended City Orchestra concerts this year?

7. While fishing, Mr. Miles and his daughter caught 42 fish. Mr. Miles caught 8 fewer fish than his daughter. How many fish did his daughter catch?

8. How much is saved with a single payment?

   STEREO—$329
   Or only 6 payments of $60 each!

Is the answer *reasonable* or *unreasonable*? Explain.

9. The Lees drove 225 mi from their house to the lake in $4\frac{1}{2}$ h. They left at 1:30 P.M. On average, how many miles did they drive per hour?

   Answer: 6:00 P.M.

10. The Lees can rent a canoe for $4 for the first hour and $1 for each additional hour. How much will it cost to rent a canoe for 5 h?

    Answer: $20

TECHNOLOGY

DIGIT DETECTIVES

In the computer game "Fraction Challenge," players solve missing-digit fraction problems. Sharpen your skills with this paper-and-pencil version of the computer game. Place digits correctly. You may use each digit only once.

1. Digits to place:
   3 3 4 7 8 8

   a. $\dfrac{2}{3} \times \dfrac{\square}{5} = \dfrac{\square}{15}$
   b. $\dfrac{\square}{\square} \times \dfrac{6}{5} = \dfrac{14}{5}$
   c. $\dfrac{4}{5} \times \dfrac{10}{3} = \dfrac{\square}{\square}$

2. Digits to place:
   2 2 3 5 7 9

   a. $\dfrac{4}{\square} \div \dfrac{\square}{7} = \dfrac{14}{5}$
   b. $\dfrac{\square}{5} \div \dfrac{3}{\square} = \dfrac{14}{15}$
   c. $\dfrac{\square}{5} \div \dfrac{\square}{10} = \dfrac{6}{1}$

3. Digits to place:
   1 2 4 5 6 7 9

   a. $\dfrac{\square}{2} \times \dfrac{4}{\square} = \dfrac{2}{5}$
   b. $\dfrac{2}{3} \div \dfrac{5}{\square} = \dfrac{\square}{5}$
   c. $\dfrac{\square}{7} \times \dfrac{\square}{8} = \dfrac{1}{\square}$

SAVING BY FRACTIONS

Danny started by saving $10 in January. He planned to save $\frac{1}{10}$ more each month than he saved the previous month. How many months will it take for him to be saving more than $100 per month?

TWO WAYS

You can add mixed numbers on the calculator in two different ways—with or without using memory keys. First experiment with your calcualtor. Then show the keystrokes you would use to add $2\frac{1}{2} + 1\frac{3}{4}$ using both methods.

Geometry and Measurement 9

DID YOU KNOW...?
The smallest rideable bicycle in the world has wheels that are less than 1 in. high.

USING DATA
Collect
Organize
Describe
Predict

Do you think the smaller wheel makes twice as many turns or half as many turns as the larger wheel over the same distance?

Work with a partner. Organize the information from the timeline into a newspaper article about the history of bicycles.

2000
F. Reese (England) sets record, bikes around the world in 143 days
← Dr. Abbott (USA) bikes 140.5 mi/h

1950
A. Letourer (USA) → bikes 108.92 mi/h

1900
First air filled tire →
First chain driven rear wheel →
Word bicycle patented →
← First high wheeler

1850
Pedal powered → bicycle
Handle steers front → wheel

1800
← Foot powered "Hobby Horse"

CHAPTER 9

 **Exploring Basic Figures**

Place a piece of dot paper on your desk. The flat paper suggests part of a plane that goes on and on without end in all directions. It contains an infinite number of points.

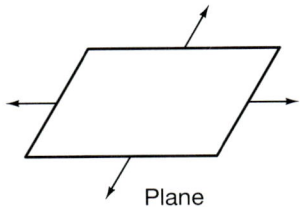
Plane

1. How many points are represented by dots on your paper? How many more points are there that are not represented by dots?

Label two points A and B as shown. A **line** is a straight path of points that goes on infinitely in both directions. Use a ruler to draw line AB.

 We write $\overleftrightarrow{AB}$ or $\overleftrightarrow{BA}$.

Mark a point C on $\overleftrightarrow{AB}$ as shown below.

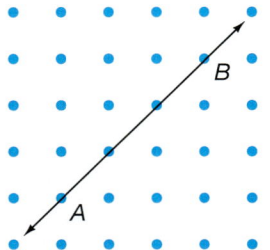

2. Give three different names for the line.

A **segment** is a part of a line with two endpoints. Look at the part of the line with endpoints A and B.

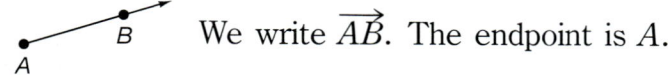 We write $\overline{AB}$ or $\overline{BA}$.

Two segments that have the same length are **congruent segments**.

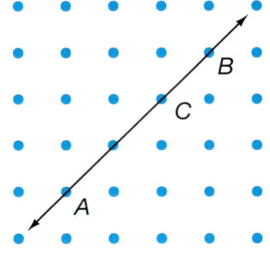

3. Name two segments on your line that have C as an endpoint. Are they congruent? Where could you place a point E to form three congruent segments?

A **ray** is a part of a line with one endpoint.

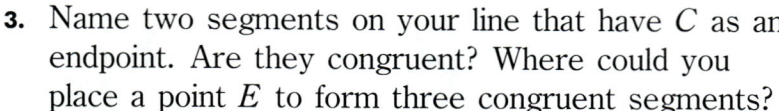 We write $\overrightarrow{AB}$. The endpoint is A.

4. Name three other rays that are a part of the line on your paper.

Draw line CD as shown. Label points D and G.

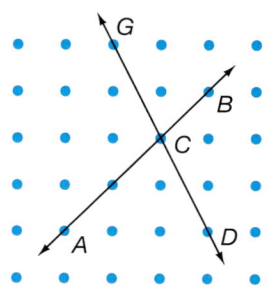

5. At what point do $\overleftrightarrow{AB}$ and $\overleftrightarrow{CD}$ intersect?

On your paper, darken $\overrightarrow{CB}$ and $\overrightarrow{CD}$. These two rays form angle *BCD*. Two rays with the same endpoint form an **angle**. The common point *C* is called the **vertex**. We write ∠*BCD*.

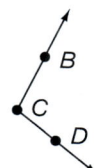

6. Where, in the angle's name, is the letter for the vertex?

7. Name ∠*BCD* another way. Then name as many other angles as you can find on your paper.

On another part of your paper, draw $\overleftrightarrow{PQ}$ and $\overleftrightarrow{MN}$ as shown. **Parallel lines** are lines in the same plane that never intersect. $\overleftrightarrow{MN}$ and $\overleftrightarrow{PQ}$ are parallel lines.

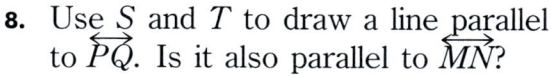

8. Use *S* and *T* to draw a line parallel to $\overleftrightarrow{PQ}$. Is it also parallel to $\overleftrightarrow{MN}$?

9. Draw two other lines parallel to these lines.

You can make interesting designs by using basic figures. Work with a partner.

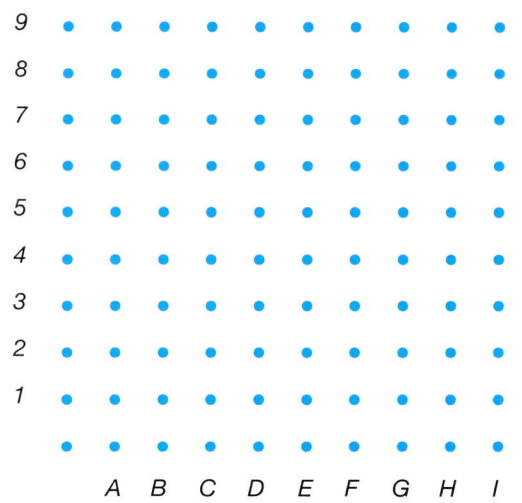
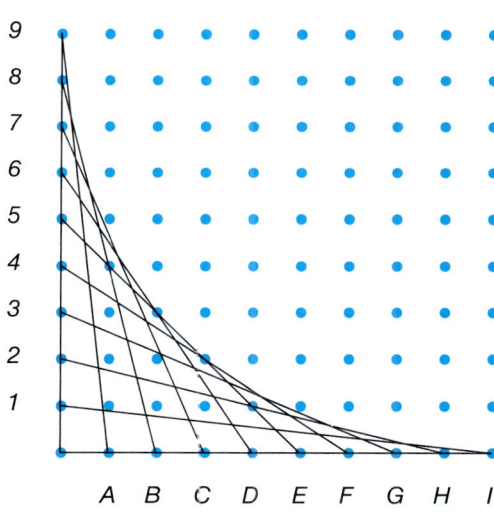

10. Use dot paper to set up a grid like the one shown on the left above. Use the grid to make the design shown on the right above.

11. Find and name as many basic figures as you can in the design.

12. **CREATE YOUR OWN** Set up a different grid. Use it to make a design of your own.

CHAPTER 9   279

Angles

When hit by a bowling ball, a bowling pin will fall when its base tips beyond 7° or more. In this lesson, we will learn how to measure and draw this and other angles.

To measure ∠CAB, use the outer scale of the protractor.
- Put the center of the protractor at the vertex, A.
- Line up one side of the angle with 0 degrees.
- Read the angle along the other side. The measure of ∠CAB is between 5° and 10° or about 7°.

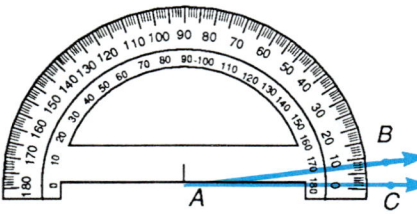

To draw an angle with a measure of 120°, use the outer scale of the protractor.
- Draw one ray, $\overrightarrow{YZ}$.
- Place the center of the protractor at the vertex, Y.
- Line up 0 degrees with $\overrightarrow{YZ}$.
- Mark point X at 120° and draw $\overrightarrow{YX}$.

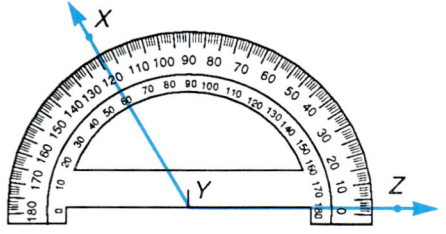

The drawing at right shows another angle with a measure of 120°. Two angles with the same measure are called **congruent angles.**

A **right angle** measures 90°.

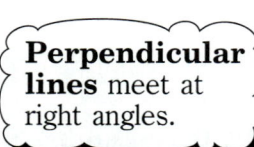

Perpendicular lines meet at right angles.

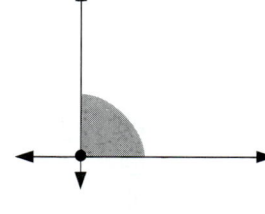

An **acute angle** measures less than 90°.

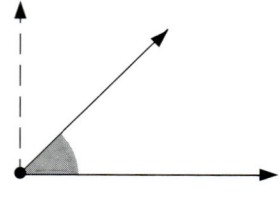

An **obtuse angle** measures more than 90°.

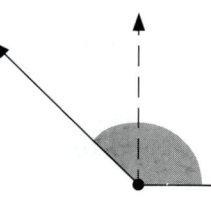

A **straight angle** measures 180°.

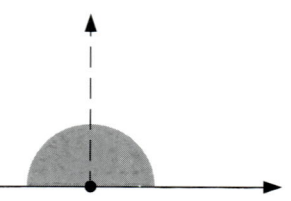

280 LESSON 9–2

GUIDED PRACTICE

Use a protractor to draw an angle with the given measure.

1. 90°   2. 30°   3. 140°   4. 180°

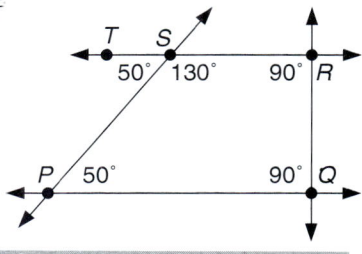

Use the figure at the right. Name the angle or angles.

5. acute angle
6. congruent angles
7. obtuse angle
8. right angle

PRACTICE

Measure the angle. Use the figure at right.

9. $\angle ABC$ 10. $\angle CBD$
11. $\angle DBE$ 12. $\angle ABD$

13. Which angles are congruent angles?

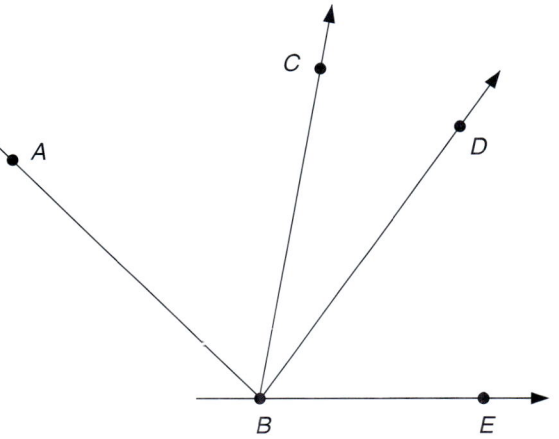

Draw an angle with the given measure.
Is the angle acute, right, or obtuse?

14. 160°   15. 65°   16. 140°

17. Draw a right angle. Use the right angle to draw perpendicular lines.

PROBLEM SOLVING

18. What is the angle made by the bowling pins?

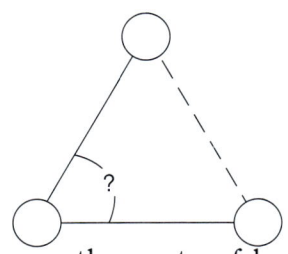

19. Name other sets of bowling pins that form this pattern.

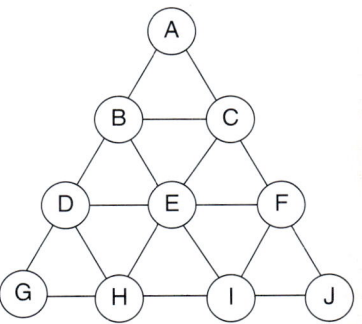

CHAPTER 9   281

Triangles

Triangles are used in the construction of many bridges. A triangle is a rigid figure. Its three sides will not move.

Triangles can be named by the lengths of their sides.

An **equilateral triangle** has all sides congruent. What is the length of each side?

An **isosceles triangle** has two sides congruent. Which two sides are congruent?

A **scalene triangle** has no sides congruent.

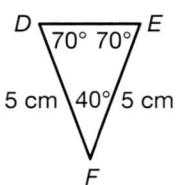
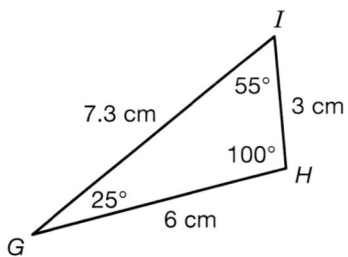

THINK ALOUD What do you notice about the sum of the measures of the angles in triangle *ABC*? *DEF*? *GHI*?

The sum of the measures of the angles of a triangle is 180°.

Triangles can also be named by the sizes of their angles.

A **right triangle** has one right angle. How can you find the measure of ∠*RST*?

An **acute triangle** has all acute angles.

An **obtuse triangle** has one obtuse angle.

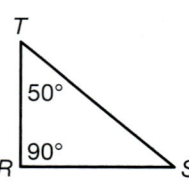
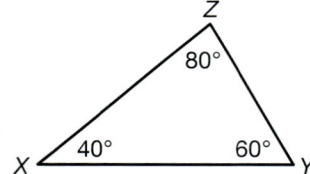
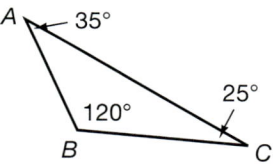

━━━━━━━━━━━━━━━━━━━━ **GUIDED PRACTICE** ━━━━━━━━━━━━━━━━━━━━

Match the figure with its description.

1. isosceles triangle
2. equilateral triangle
3. right triangle
4. scalene triangle
5. acute triangle
6. obtuse triangle

a. no congruent sides
b. all acute angles
c. all congruent sides
d. exactly two congruent sides
e. one obtuse angle
f. one right angle

LESSON 9-3

7. Explain how to find the measure of ∠KJL. Then find it.

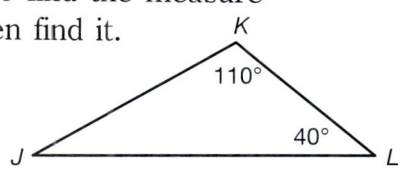

PRACTICE

Write *equilateral*, *isosceles*, or *scalene*.

8.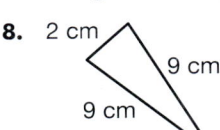

9.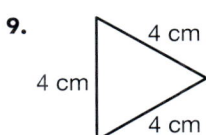

10.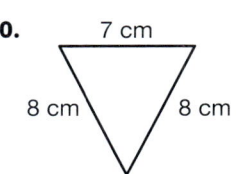

11.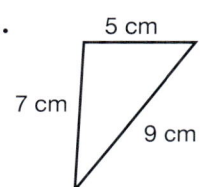

Write *acute*, *obtuse*, or *right*.

12.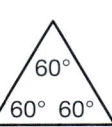

13.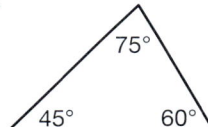

14.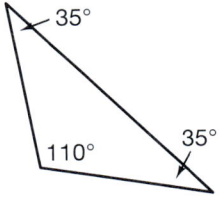

15.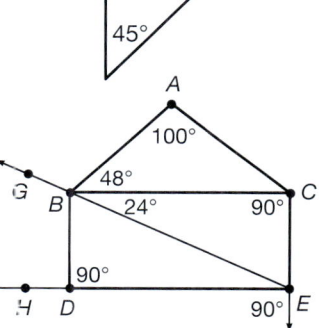

Find the measure.

16. ∠ACB
17. ∠BDH
18. ∠CEB
19. ∠GBA
20. ∠BED
21. ∠GBD

CRITICAL THINKING Can you make a triangle that can have both names? Explain how or why not.

22. right, isosceles
23. right, scalene
24. right, equilateral

PROBLEM SOLVING

Can you make a triangle with sides of these lengths? Try it using measured strips of paper.

25. 2 cm, 5 cm, 8 cm
26. 3 cm, 4 cm, 5 cm
27. 8 cm, 9 cm, 7 cm
28. 16 cm, 4 cm, 7 cm

29. Try some other examples. Write a statement about what you can conclude.

CHAPTER 9 283

Problem Solving: Venn Diagrams

A music teacher surveyed classrooms A and B to determine who is interested in taking music lessons.

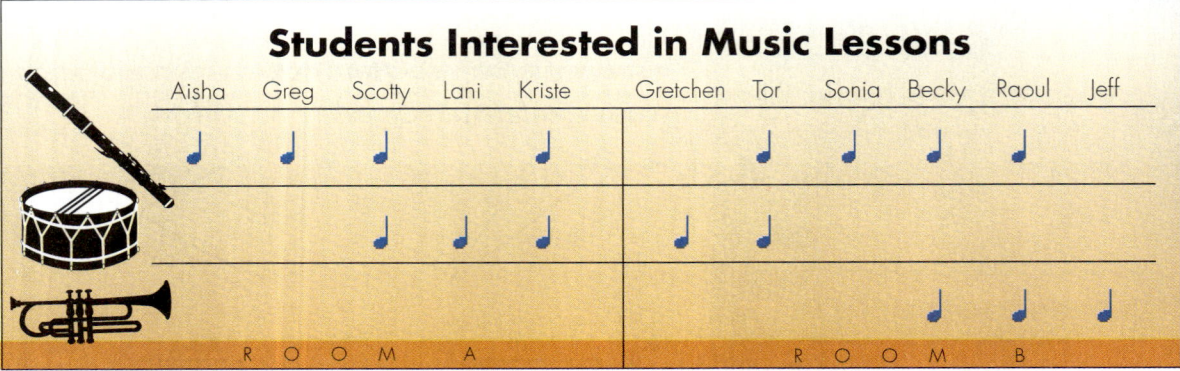

You can use **Venn diagrams** to represent collections of people or objects and their relationships. When two collections have no people or objects in common, we draw the diagram like this.

Venn diagram

Room A: Scotty, Greg, Kristie, Lani, Aisha

Room B: Gretchen, Tor, Sonia, Becky, Raoul, Jeff

When one collection is contained entirely within the other, we draw the diagram like this.

Room B: Gretchen, Tor, Sonia
Trumpet: Raoul, Becky, Jeff

THINK ALOUD Why is Becky inside the inner region, but Gretchen is not? Why isn't Greg in either region?

Now let's make a Venn diagram to show collections that have some things in common.

        Collection 1   Collection 2
Students interested in trumpet   Students interested in flute

(*Hints:* How many figures should you draw? Should the figures overlap? Which students will be within each region?)

Does your Venn diagram look like this?

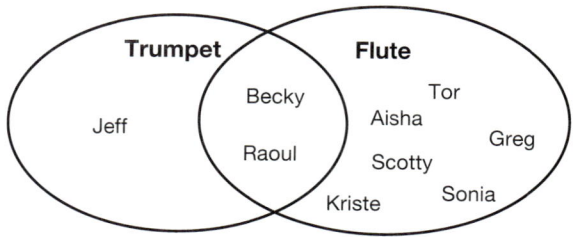

284   LESSON 9-4

GUIDED PRACTICE

1. Use the Venn diagram to answer.

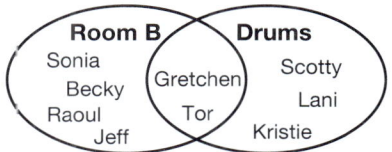

a. What are the two collections?
b. Why are Gretchen and Tor in the overlapping region?
c. Where would you place a new student in room A who is interested in drums?

PRACTICE

Draw a Venn diagram. Use the information on page 284.

2. students in Room A
   students interested in trumpet

3. students interested in flute
   students interested in drums

4. students interested in trumpet
   students interested in drums

5. all students surveyed
   students interested in flute

6. **CRITICAL THINKING** Describe the three ways you could picture two collections of objects using Venn diagrams.

CHOOSE Choose any strategy to solve.

7. Copy and complete the Venn diagram using the chart on page 284.
   a. In which sections are there no names? What does this mean?
   b. Which sections represent only one instrument? Which students are interested in only one instrument?

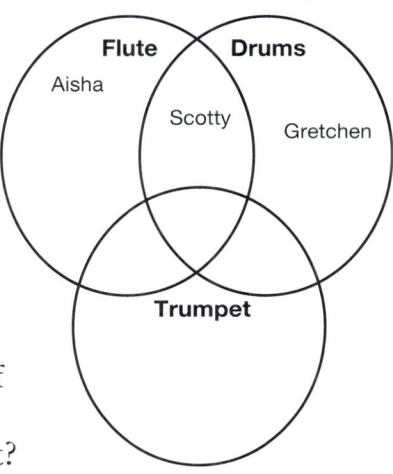

8. Vince, Connie, and Orji each play an instrument. One plays violin, one plays clarinet, and one plays oboe. The name of the instrument and the name of its player do not start with the same letter. Vince does not play the oboe. Who plays each instrument?

CHAPTER 9   285

Polygons

We can see many shapes in nature. The honeycomb is made from beeswax to hold honey and eggs. It contains many hexagons. A **hexagon** has six sides and six angles.

A **polygon** is a closed plane figure made from line segments. How many sides and angles does each polygon have?

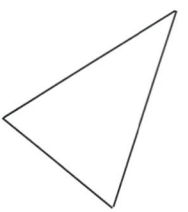
triangle

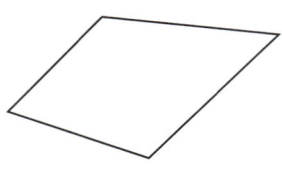
quadrilateral

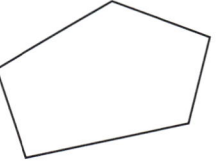
pentagon

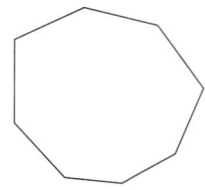
octagon

When all the sides of a polygon are the same length, and all angles have the same degree measure, the polygon is a **regular polygon**. Does a honeycomb form regular hexagons?

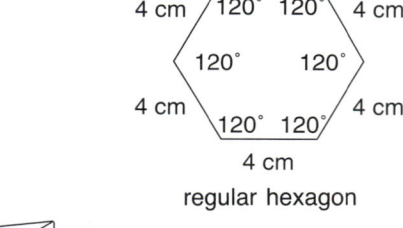
regular hexagon

A **diagonal** of a polygon is a line segment that joins two vertexes but is not itself a side.

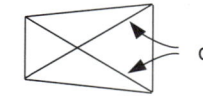
diagonals

──────── GUIDED PRACTICE ────────

Draw an example of each figure.

1. pentagon
2. quadrilateral
3. octagon
4. hexagon
5. triangle
6. regular quadrilateral

──────── PRACTICE ────────

Name the polygon that has

7. eight sides.
8. six angles.
9. three sides.
10. four sides.
11. five angles.
12. eight angles.

286   LESSON 9–5

Is the figure a polygon? Write *yes* or *no*.
If so, is it a regular polygon? Write *yes* or *no*.

13. 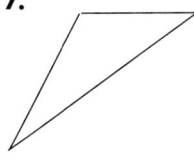 **14.** 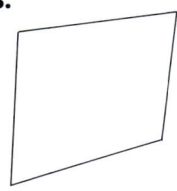 **15.** 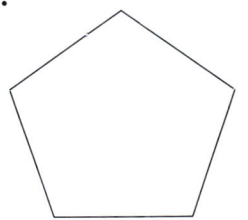 **16.**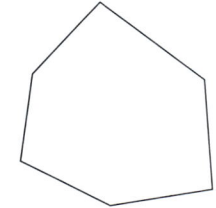

Trace each figure. Draw all possible diagonals.
How many are there?

17. **18.** **19.** **20.**

21. CRITICAL THINKING Use Exercises 17–20 to predict the number of diagonals for a 7-sided polygon. Try it out to see if your answer was correct.

PROBLEM SOLVING

In the polygons at right, congruent sides are marked with —+— or —#— . Congruent angles are marked with ⌒ or ⌒⌒

Draw a Venn diagram to show how the two collections are related. Write the letters of the figures in the correct regions.

22. regular polygons and polygons

23. triangles and quadrilaterals

24. polygons and triangles

25. polygons with all sides congruent and polygons with all angles congruent.

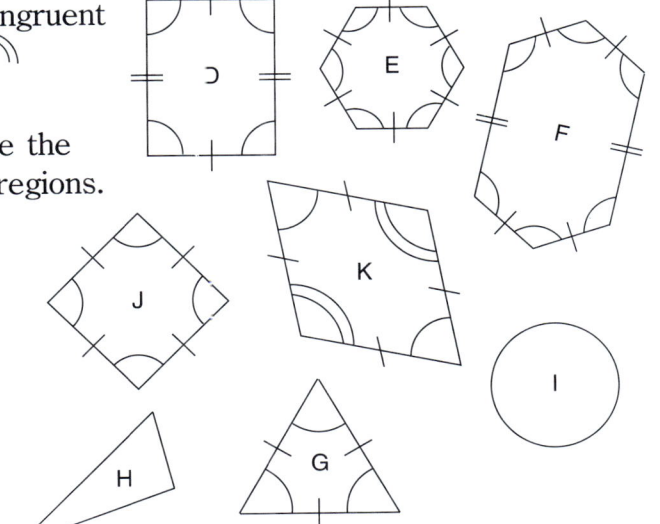

CHAPTER 9 287

Exploring Quadrilaterals

1. Describe some quadrilaterals in the picture. How are they alike? How are they different?

Work with a partner. Use the quadrilaterals workmaster, a protractor, and a ruler to explore the properties of some special quadrilaterals. Cut out the figures.

2. Figures A, B, C, and D are special quadrilaterals called parallelograms. Discover which of these properties are true for all parallelograms. Remember, all figures must have the property.

Scharfruhiges Rosa (Shrill Peaceful Pink), by Wassily Kandinsky, 1924.

   a. Are all sides congruent?
   b. Are opposite sides congruent?
   c. Are opposite sides parallel?
   d. Are all angles congruent?
   e. Are opposite angles congruent?
   f. Are diagonals congruent?

3. **IN YOUR WORDS** Write a sentence about the properties of all parallelograms.

> A **parallelogram** is a quadrilateral with opposite sides parallel.

4. Which other quadrilaterals from the workmaster are parallelograms?

Place these figures in two rows.
   Row 1: B, D, L, O ↩ These are rectangles.
   Row 2: A, C, H, I, K ↩ These are not rectangles.

5. Do rectangles have all the properties of a parallelogram? Are rectangles parallelograms?

6. What other properties do rectangles have? Use the list from Exercise 2.

7. **IN YOUR WORDS** Write a sentence about the properties of all rectangles.

> A **rectangle** is a special parallelogram with all angles right angles.

Place these figures in two rows.
   Row 1: C, D, H, O ↩ These are rhombuses.
   Row 2: A, B, I, J, K ↩ These are not rhombuses.

8. Do rhombuses have all the properties of a parallelogram? Are rhombuses parallelograms?

9. What other properties do rhombuses have? Use the list from Exercise 2.

10. **IN YOUR WORDS** Write a sentence about the properties of all rhombuses.

> A **rhombus** is a parallelogram with all sides congruent.

11. Can a figure be both a rectangle and a rhombus? To decide, look for a parallelogram with all sides congruent and all angles right angles.

12. What is this figure called? Write the letters of the figures that fit this description.

> A **square** is a special rectangle with all sides congruent. It is also a rhombus with all angles right angles.

Figures J and M are **trapezoids**.

13. Are opposite sides parallel?

14. Are opposite sides congruent?

15. **IN YOUR WORDS** Write a sentence about the properties of all trapezoids.

> A **trapezoid** is a quadrilateral with exactly one pair of parallel sides.

Give as many names as you can for the figure.

16. 17. 18. 19.

CHAPTER 9   289

Problem Solving Strategy: Using Generalizations

Ms. Skolnick uses a scissors jack to reach the letters on this movie sign. When the jack starts to rise, the sides look like rhombuses. When it reaches the top, the sides look like squares.

THINK ALOUD When is a rhombus a square?

When the jack is closed, the ticket office looks like a rectangle. As the jack opens, your view of the ticket office is blocked so you see only a square. When is a rectangle a square?

When we generalize, we tell how ideas or things are related.

You can use what you know about the properties of squares to make these generalizations:

All squares are rhombuses.   All squares are rectangles.

A Venn diagram can help you understand generalizations.

Squares are inside the region representing rectangles. All squares are rectangles. Some rectangles are squares.

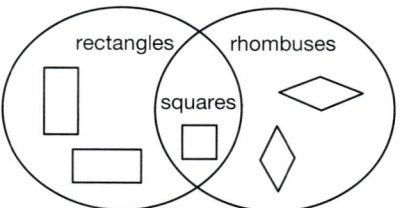

Squares are inside the region representing rhombuses. All squares are rhombuses. Some rhombuses are squares.

GUIDED PRACTICE

1. Look at the Venn diagram and answer the question.

   a. Are there any trapezoids in the region representing parallelograms?

   b. What are some words we often use in generalizations?

   c. Use the Venn diagram to write a generalization about trapezoids and parallelograms.

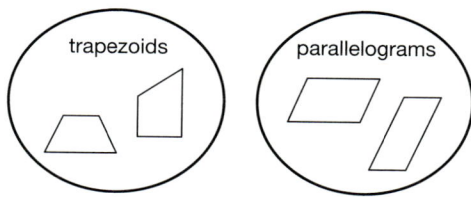

PRACTICE

Name the figures in the Venn diagram that show that each statement is true.

2. Some rectangles are squares.
3. All squares are rhombuses.
4. Some rectangles are not squares.
5. All squares are rectangles.
6. Some rhombuses are not squares.
7. All squares are rectangles and rhombuses.

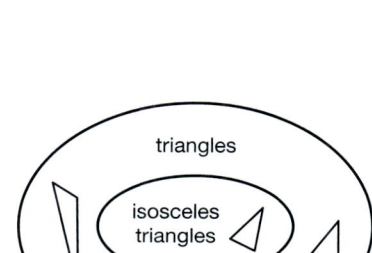

Use the diagram to tell if the generalization is true or false.

8. All triangles are isosceles triangles.
9. All isosceles triangles are triangles.
10. Some triangles are not isosceles triangles.
11. No isosceles triangle is a triangle.
12. Make a diagram to show the relationship among polygons, triangles, and equilateral triangles. Write three generalizations.

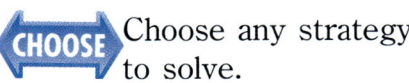

CHOOSE Choose any strategy to solve.

13. Figure A shows the top of a drying rack. How many rectangles do you see?

14. Marjory's drying rack has 3 fewer shelves than David's. Together they have 17 shelves. How many shelves are on David's rack?

15. How can you find the sum of the measures of the angles in the quadrilateral by using triangles? Try it with other quadrilaterals. What generalizations can you make?

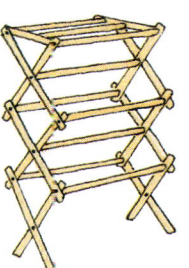

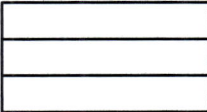

Figure A

A drying rack opens and closes like a scissors jack.

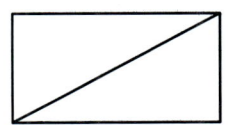

MIDCHAPTER CHECKUP

LANGUAGE & VOCABULARY

Organize the following terms into several sets. The terms within each set must be related. Use each term once.

acute, equilateral, hexagon, isosceles, line, obtuse, parallelogram, ray, rhombus, right, scalene, segment, square, trapezoid, triangle

QUICK QUIZ

Draw each figure. *(pages 278–281)*

1. $\overleftrightarrow{RS}$
2. $\overrightarrow{YZ}$
3. obtuse angle *BCD*

Use two of these words to describe each triangle: equilateral, isosceles, scalene, acute, obtuse, right *(pages 282–283)*

4.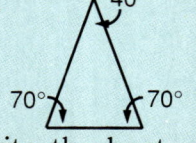
5.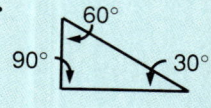

Write the best name for the figure. *(pages 286–287)*

6.
7.
8.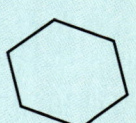
9.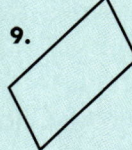

Solve. *(pages 284–285)*

10. The survey shows which sports 8 students participate in.
    a. In which sports does Darryl participate?
    b. How does a Venn diagram show who participates in softball and tennis?
    c. Draw a Venn diagram to show how many students participate in softball or tennis, but not both.

Name	Softball	Basketball	Tennis
Brad	X		
Amy	X		X
Kara		X	
Luis	X		X
Marie		X	
Dawn	X		
Darryl		X	X
Adigun	X	X	

Write the answers in your learning log.

1. Describe how $\overrightarrow{CD}$ and $\overrightarrow{DC}$ are different.

2. Describe the sides and angles of a right triangle.

On July 4, 1976, bicentennial celebrations were held across America. They marked the 200th anniversary of the signing of the Declaration of Independence. Find out when your city or town was founded. What are or what will be the dates of its centennial, bicentennial, and tricentennial celebrations.

How many triangles are in this figure?

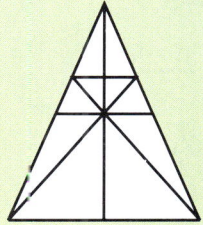

Exploring Slides, Flips, and Turns

Slides, flips, and turns are motions that change the position of figures.

To make a cartoon character appear to **slide** along a line, an artist draws the character again and again along a line on separate sheets of celluloid.

1. Trace and cut out a copy of the triangle below. Slide it along line AK to each of the positions shown.

2. Use the figure below. Name the segments congruent to $\overline{AB}$. Name the segments parallel to $\overline{AC}$.

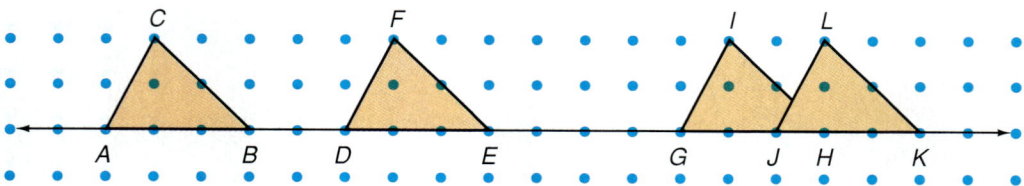

To make the cartoon character **turn**, the artist draws the character in different positions around point P on separate sheets of celluloid.

3. Copy the triangle shown below. Turn it to the three positions shown. How many degrees are in turn 1? turn 2? turn 3?

4. What can you say about $\overline{MN}$ and $\overline{NP}$?

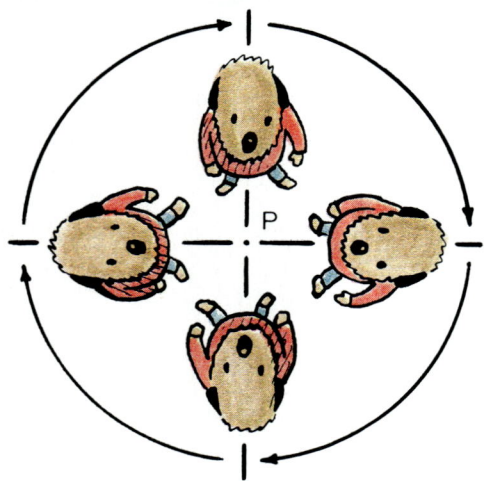

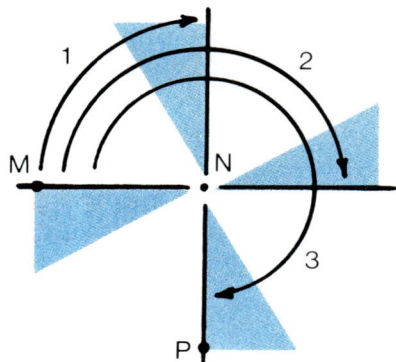

294 LESSON 9-8

To make the cartoon character **flip** over, the artist draws it on one side of the line and then flips it over and draws it in the new position.

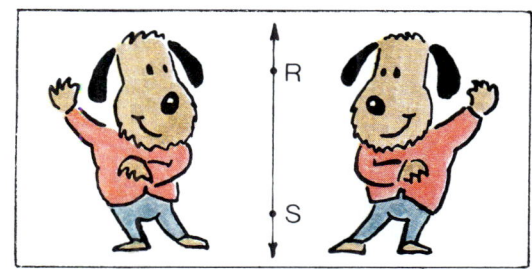

5. Draw quadrilateral *ABCD* on grid paper. Flip it over $\overleftrightarrow{XY}$. Label it *EFGH*. Which vertex is the image of *A*? of *C*?

6. Which segment is the image of $\overline{AD}$? Is the segment congruent to $\overline{AD}$?

7. What happens if you flip a figure and then flip it again, using the same line?

8. **CRITICAL THINKING** Complete by drawing the correct figure.

   a.

   b.

A **tessellation** uses figures that touch but do not overlap. The tessellation at the left was made by sliding parallelograms. Draw and cut out each figure below. Use it to make a tessellation if possible.

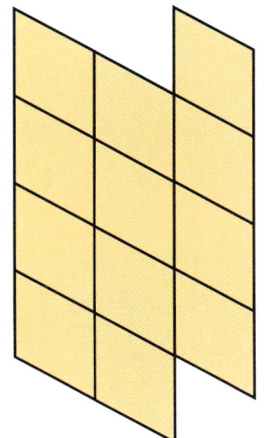

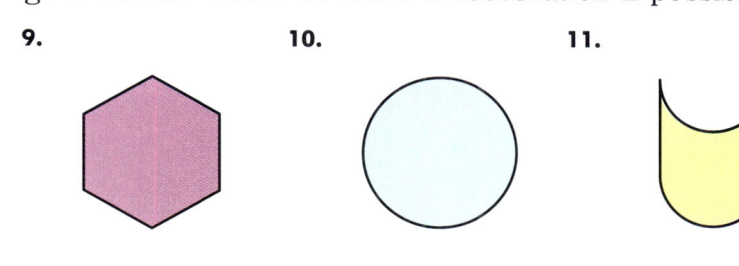

9.   10.   11.

12. Create your own tessellation by drawing any figure and tracing it many times.

CHAPTER 9   295

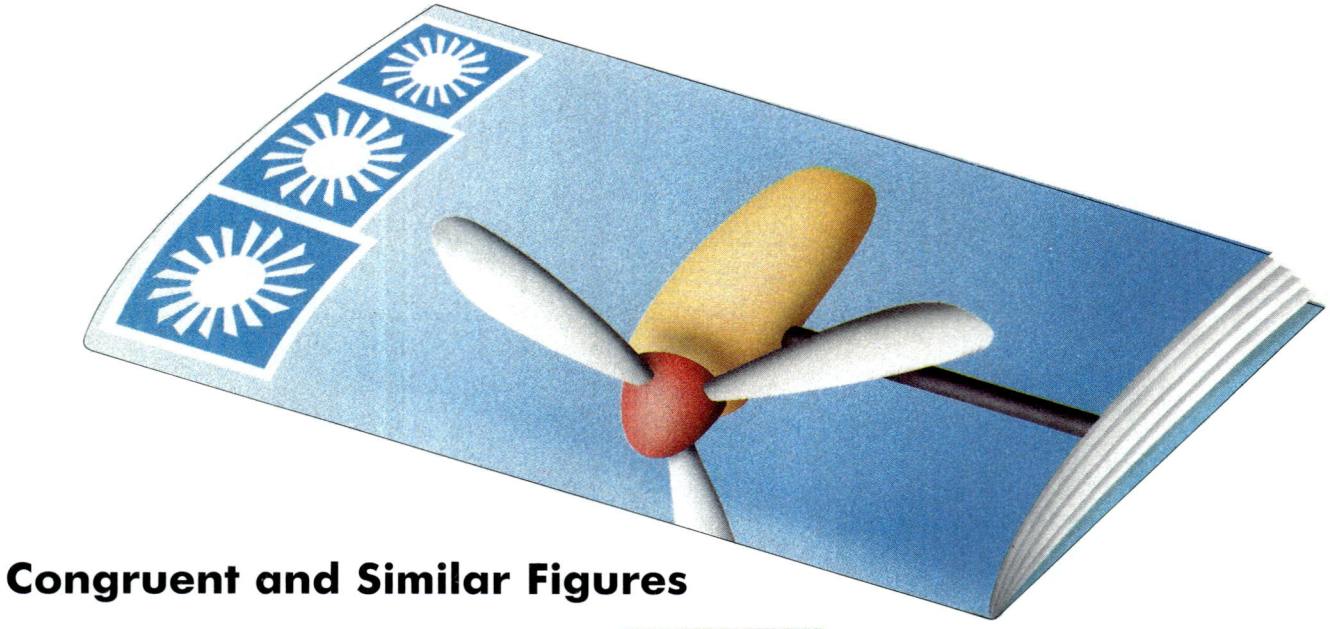

Congruent and Similar Figures

This logo appears in different sizes in a company's ads and catalogs.

Similar figures have the same shape. They do not need to be the same size.

The windmill is exactly the same size and shape on each seal.

Congruent figures have the same size and shape. Slides, flips, and turns give congruent figures. Are the windmills slides, flips, or turns of one another?

THINK ALOUD Are congruent figures also similar figures?

Similar Figures

Congruent Figures

The triangles below are congruent. We can place one triangle on top of the other to make parts match. The parts that match are called **corresponding parts**. The symbol ≅ means "is congruent to."

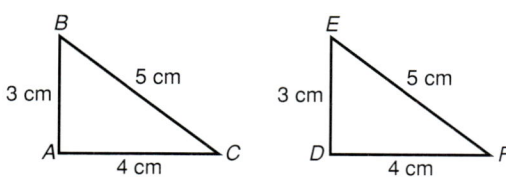

Triangle $ABC \cong$ Triangle DEF

Corresponding parts of congruent figures are congruent.

$\overline{AC} \cong \overline{DF}$ $\overline{BC} \cong \overline{EF}$ $\angle CAB \cong \angle FDE$

Name other corresponding parts.

GUIDED PRACTICE

1. Triangle *LMN* and triangle *PGR* are congruent. Trace triangle *PGR*. Place it on top of triangle *LMN*.
   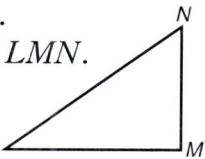 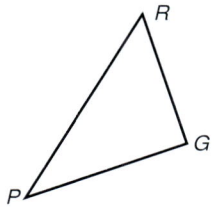
   a. Name the corresponding sides.
   b. Name the corresponding angles.

PRACTICE

Are the figures similar, congruent, or both?

2. 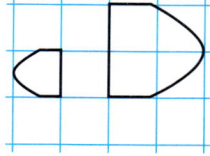   3. 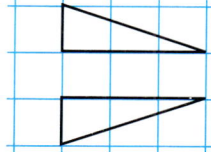   4.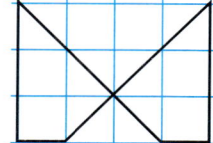

The pairs of triangles below are congruent. Answer the questions about them.

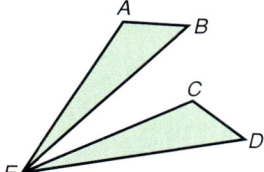   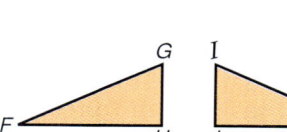   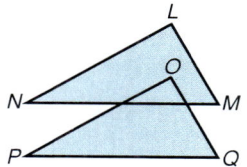

5. Name the corresponding sides and angles.
   a. for the green pair   b. for the orange pair   c. for the blue pair

6. For each pair, tell whether the triangles are slides, flips, or turns of each other.

PROBLEM SOLVING

This logo is on a company's stationery. To put it on a book, the company wants to double the 4-cm side. The logo is to have the same shape as the original one.

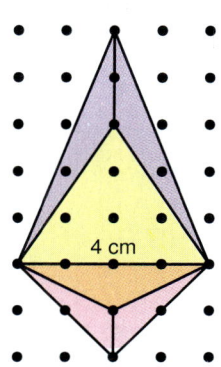

7. How many triangles are in the figure?

8. Which triangles are flips of each other?

9. Draw the new logo.

10. What can you say about all sides of the triangles in the enlarged logo?

11. What can you say about the angles in the enlarged logo?

CHAPTER 9   297

Symmetry

Many famous buildings have floor plans that show symmetry.

A figure has **symmetry** when it can be folded so that both parts match. A **line of symmetry** is a line that separates the figure into two congruent parts.

The floor plan of the Colosseum has two lines of symmetry, $\overleftrightarrow{MN}$ and $\overleftrightarrow{PQ}$.

THINK ALOUD With $\overleftrightarrow{PQ}$ as a line of symmetry, describe the congruent parts that you see.

What segment is congruent to $\overline{PO}$? to $\overline{MO}$?

Colosseum, Italy
70 – 224

GUIDED PRACTICE

Trace the outline of each floor plan. Fold the floor plan to find the lines of symmetry. Draw the lines of symmetry.

1.

Pyramid of Khufu, Egypt, 2600 B.C.

2.

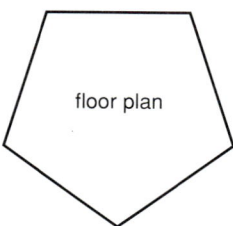

Pentagon, United States, 1941–43

PRACTICE

Draw each figure. Then cut and fold the figure to find lines of symmetry. Draw the lines of symmetry.

3.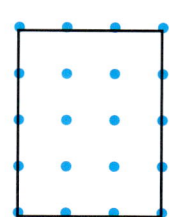
4.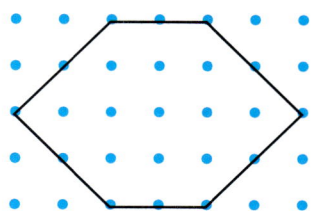
5.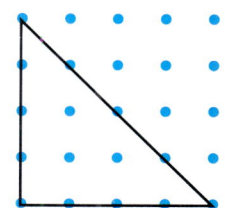
6.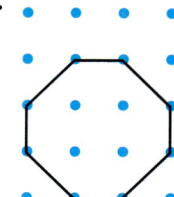

The design shows two lines of symmetry. Name all figures congruent to each figure.

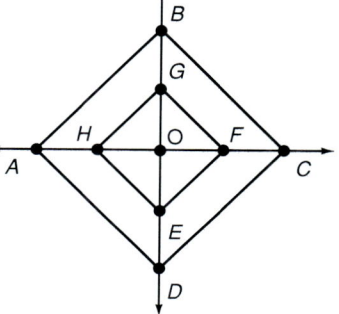

7. triangle *OFG*
8. triangle *OBC*
9. trapezoid *AHED*
10. triangle *ABC*
11. How many other lines of symmetry does the design have? Describe them.
12. Can you draw a figure with no lines of symmetry?

PROBLEM SOLVING

13. This plan has one line of symmetry. Draw the part of the plan that has been torn off.

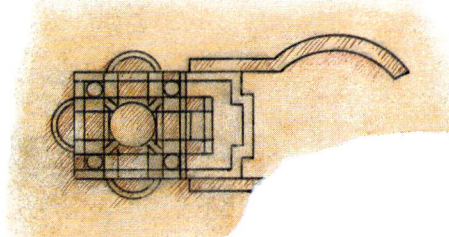

St. Peter's, Vatican
1506 – 1626

14. The Taj Mahal has four lines of symmetry. Draw the part of the plan that is missing.

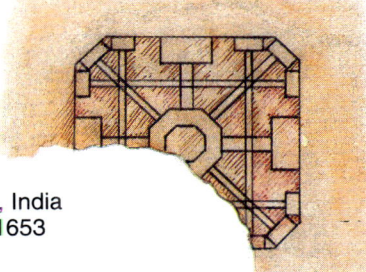

Taj Mahal, India
1636 – 1653

15. **CREATE YOUR OWN** Draw a floor plan that has two lines of symmetry.

Critical Thinking

Draw the complete figure.

1.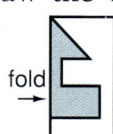
2.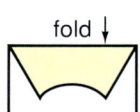
3.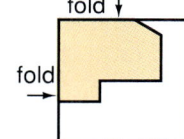
4.

CHAPTER 9

Circle and Circumference

Will's bike wheels have a diameter of 20 in. The **diameter** (d) is the length of a segment through the center, with endpoints on the circle.

How far does Will's wheel travel when it revolves once? To find out, you need to find the distance around the circle or the **circumference** (C).

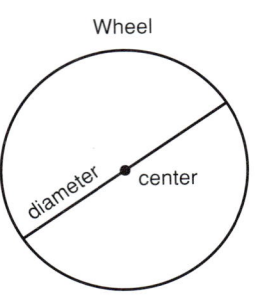

Here is an experiment that will help you discover how to find the circumference.

Take a narrow strip of paper. Mark off the diameter of the above circle.

Measure the circumference by bending the strip of paper around the wheel.

The circumference is about how many times as long as the diameter?

The circumference is about three times as long as the diameter.

So, Will's bike wheel travels 3 × 20 in., or about 60 in. when it revolves once.

We can get a better estimate for the circumference of the circle by using the number π (pi), which is approximately equal to 3.14.

$C = \pi \times d$
$\approx 3.14 \times d$
$\approx 3.14 \times 20$ in., or 62.8 in.

$\approx$ means "is approximately equal to."

The **radius** (r) of a circle is one half the diameter. We can use the radius to find the circumference.

$C = \pi \times d$
$= \pi \times 2 \times r$
$\approx 3.14 \times 2 \times 10$ in., or 62.8 in.

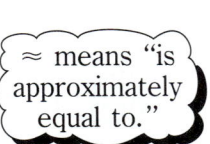

Unicycle champion Nick Kaufman, about 1888

300 LESSON 9-11

GUIDED PRACTICE

Find the circumference. Round to tenths.

1.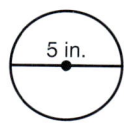
   Use $\pi \approx 3$.

2.
   Use $\pi \approx 3.14$.

3.
   Use $\pi \approx 3$.

4.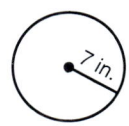
   Use $\pi \approx 3.14$.

PRACTICE

Find the circumference. Use $\pi \approx 3$.

5.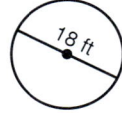

6.

7.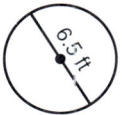

8.

Find the circumference. Use $\pi \approx 3.14$. Round to tenths.

9. $d = 30$ ft
10. $d = 2.8$ ft
11. $r = 9$ ft
12. $r = 1.2$ ft
13. $d = 1$ in.
14. $d = 4$ in.
15. $r = 4$ in.
16. $r = 6$ ft

CALCULATOR Find the diameter. Use $\pi \approx 3.14$.

17. $C \approx 50.24$ in.
18. $C \approx 188.4$ in.
19. $C \approx 47.1$ ft
20. $C \approx 72.22$ ft

ESTIMATE What is the perimeter of the figure?

21.

22.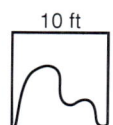

23.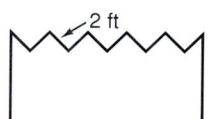

PROBLEM SOLVING

CHOOSE Choose calculator, estimation, or pencil and paper to solve.

24. Olivia's bike has wheels with a diameter of 26 in. The wheels on Will's bike are 20 in. in diameter. Whose bike wheels will revolve more often during a 1-km ride?

25. About how many times does a bike wheel with a diameter of 24 in. revolve when the bike travels 100 ft?

CHAPTER 9   **301**

EXPLORE

Exploring Area

A merchant ship has arrived in port with a torn sail. Can a different sail be cut and re-sewn to replace the torn one?

Trace and cut out the figures to help you decide.

| torn sail | replacement sail |

Area is a measure of the surface covered. The new sail can be used to replace the torn sail since it has a larger area.

1. Draw a sail that has the same area as the torn sail, but a different shape.

2. Do two figures that are congruent have the same area?

3. Can two figures that are not congruent have the same area? Where did you show this?

Use the figure at the right to help you decide how the area of triangle *ABC* compares with the area of the rectangle.

Draw and cut out the rectangle. Cut along the dotted segments and place the two pieces on top of triangle *ABC*.

4. How could you find the area of the triangle by using the area of the rectangle?

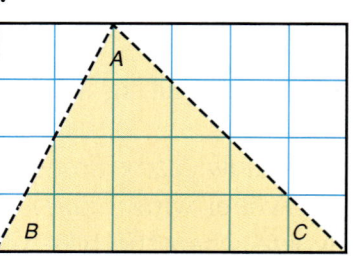

Cut out five rectangles congruent to rectangle A below. By making only one cut in a rectangle and piecing together the two parts, you can make many different figures. Here is an example.

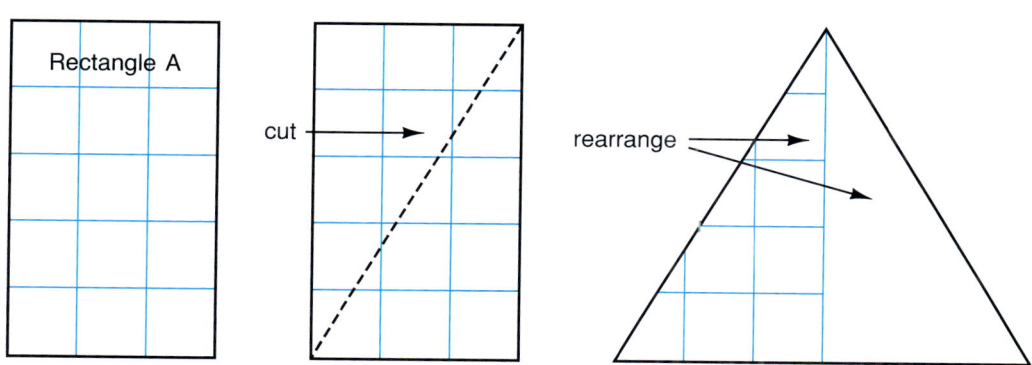

Make only one cut in each rectangle to make each figure. Draw the figure.

5. a different isosceles triangle
6. a parallelogram
7. a trapezoid
8. any figure you wish
9. What can you say about the areas of the figures you made in Exercises 5–8?

What is the area of each of these figures?

← 1 square centimeter

10.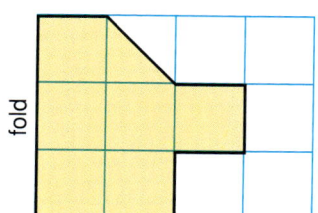

11.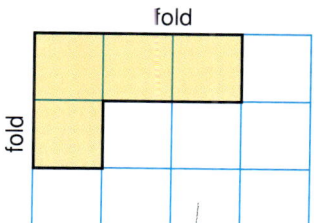

ESTIMATE Each square represents 1 square meter. Estimate the area of each figure.

12. 13. 14.

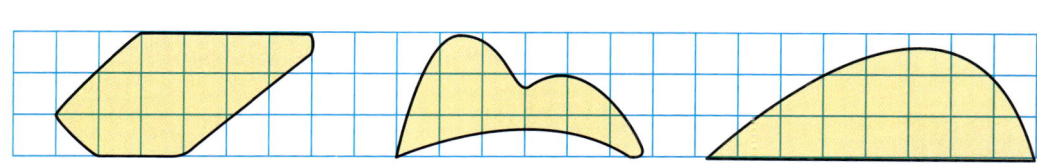

CHAPTER 9 303

Area of Rectangles and Triangles

This woven blanket measures 4 ft by 3 ft. What is its area?

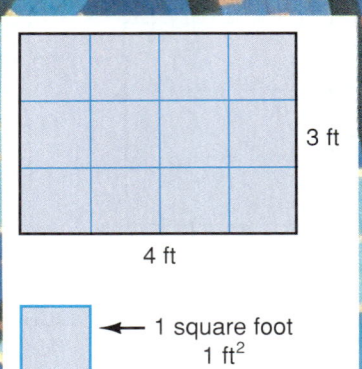

A unit for measuring area is the square foot (ft^2). You can add the number of square-foot units in each row to find the area of a rectangle.

$$4 \text{ ft}^2 + 4 \text{ ft}^2 + 4 \text{ ft}^2 = 12 \text{ ft}^2$$

Or you can multiply the length by the width.

$$\begin{aligned}\text{Area of rectangle} &= \text{length} \times \text{width} \\ &= l \times w \\ &= 4 \times 3, \text{ or } 12 \text{ ft}^2\end{aligned}$$

The area of the blanket is 12 ft^2.

THINK ALOUD Explain how you can use the formula for the area of a rectangle to find the area of a square.

The area (A) of triangle ABC is one half the area of the rectangle.

How could you cut the rectangle to show this? What is the area of triangle ABC?

The height (h) of a triangle is perpendicular to the base (b). You can use a formula to find the area.

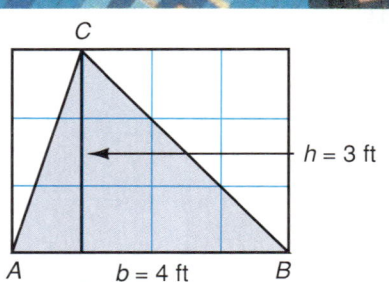

$$\begin{aligned}\text{Area of triangle} &= \tfrac{1}{2} \text{ of the area of the rectangle} \\ &= \tfrac{1}{2} \times (b \times h) \\ &= \tfrac{1}{2} \times (4 \times 3) \\ &= \tfrac{1}{2} \times 12, \text{ or } 6 \text{ ft}^2\end{aligned}$$

The length and width of the rectangle are also the base (b) and the height (h) of the triangle.

LESSON 9-13

GUIDED PRACTICE

Find the area.

1.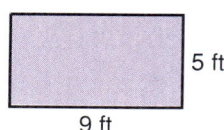
2.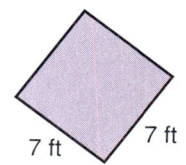
3.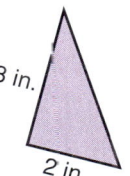
4.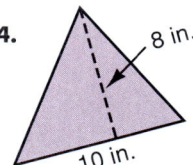

PRACTICE

Find the area. Use a calculator when needed.

5. rectangle
   $l = 6$ in., $w = 3$ in.

6. rectangle
   $l = 12$ ft, $w = 8$ ft

7. rectangle
   $l = 9$ yd, $w = 4.5$ yd

8. square
   side $s = 5$ in.

9. square
   side $s = 2.1$ ft

10. square
    $s = 17$ in.

11. triangle
    $b = 4$ in., $h = 6$ in.

12. triangle
    $b = 12$ yd, $h = 15$ yd

13. triangle
    $b = 24$ in., $h = 14$ in.

14. a. 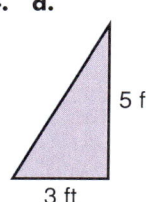 b.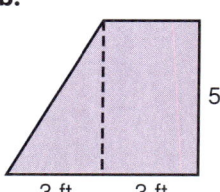
15. a. 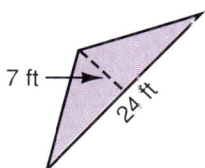 b.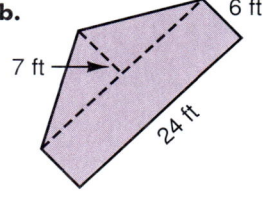

CALCULATOR Find the width of the rectangle.

16. area 134.9 ft², length 19 ft
17. area 6.6 ft², length 2.75 ft

18. Give the length and width for three different rectangles that each have an area of 64 in.²

PROBLEM SOLVING

19. It takes fleece from as many as 14 sheep to provide wool to make a 9 ft by 12 ft Navajo rug. The fleece from how many sheep is needed to make an 18 ft by 24 ft rug?

20. What would happen to the area of a rectangle if you doubled the length and width? Give examples.

21. A Germantown saddle blanket measures 48 in. by 34 in. There are 1,224 weft threads and 576 warp threads in the blanket. How many weft and warp threads are there per inch?

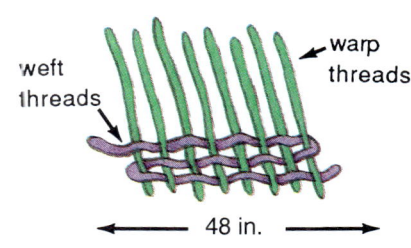

CHAPTER 9   305

Area of a Circle

The Pizzeria makes a 10-in. pizza. What is the area of the top of the pizza?

The area of a circle can be estimated by counting the square units and the parts of square units.

- There are about 68 square units in the regions shaded yellow.
- There are about 10 square units in the regions shaded red.
- There are about 78 square units in all.

The area of the pizza is about 78 in.²

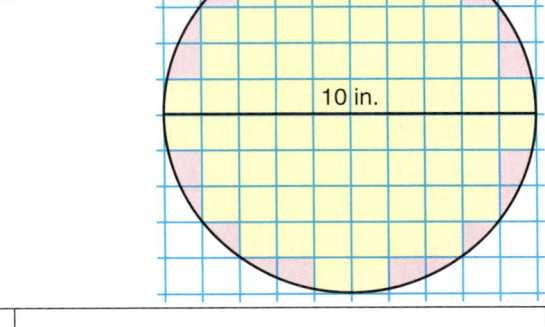

We can find a better approximation for the area of a circle by using this formula. Area of a circle = $\pi \times$ radius $\times$ radius $A = \pi \times r^2$	Since the diameter of the pizza is 10 in., the radius is 5 in. $A = \pi \times r^2$ $\approx 3.14 \times (5 \times 5)$ $\approx 3.14 \times 25$, or 78.5 in.² Remember: $\pi \approx 3.14$

GUIDED PRACTICE

1. Estimate the area of the circle by counting the units and the parts of units.

2. Find the area of the circle by using the formula. Explain your work. Compare your results with Exercise 1.

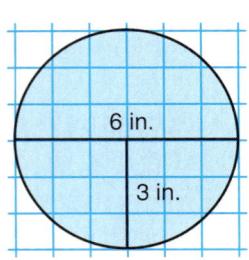

3. Find the area of a circle with a radius of 8 in.

PRACTICE

Find the area of the circle. Use $\pi \approx 3.14$.

4.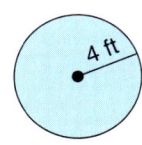
5.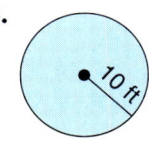
6.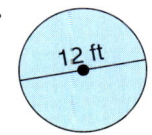
7.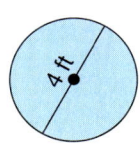

Find the area of the circle. Use $\pi \approx 3.14$.

8. $r = 12$ ft
9. $r = 20$ in.
10. $d = 2$ ft
11. $d = 18$ ft
12. $r = 7$ in.
13. $r = 13$ in.
14. $d = 8$ in.
15. $d = 11$ in.

Find the area. Use $\pi \approx 3.14$. Round to tenths.

16. a. the rectangle
    b. the semicircle
    c. the two figures combined

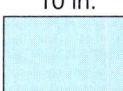
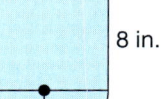

17. a. the triangle
    b. the shaded part of the circle
    c. the two figures combined

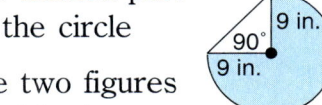
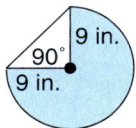

18. **NUMBER SENSE** Order the circles from the least to the greatest areas. Do not compute.

    a.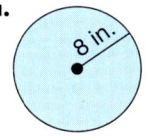
    b.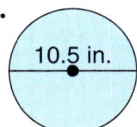
    c.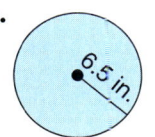
    d.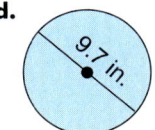

PROBLEM SOLVING

19. Jim cut a 20-in. pizza into 10 equal pieces. Estimate the area of each piece.

20. Dana cut a 9 in. by 12 in. pizza into 9 equal pieces. What is the area of each piece?

21. A pizza costs $6.00. Each topping costs $1.25. If you have $10.75, how many toppings can you get on your pizza?

Critical Thinking

Rosa said that a pizza with a 10-in. diameter is about twice as big as a pizza with a 7-in. diameter.

Is this true? Explain how you decided.

CHAPTER 9   307

Planning an Apartment

Imagine that you are an architect responsible for designing apartments in a new apartment building.

Remember: Area of a rectangle = length × width

Work with a partner. Use a calculator to help you find the answers.

1. If an apartment contains an area of 800 sq ft, it could be 40 ft long and 20 ft wide. What are some other possible combinations of length and width of a rectangular 800-sq-ft apartment?

2. A room 160 ft × 5 ft would have an area of 800 sq ft. Is this a reasonable design for an apartment? Explain your answer.

3. Suppose you were designing an 800-sq-ft one-bedroom apartment. What other rooms would you include in your apartment floor plan?

4. Draw a diagram like the one shown. Mark the length and the width of your total apartment space. Decide how to divide the space to include the rooms in your plan.

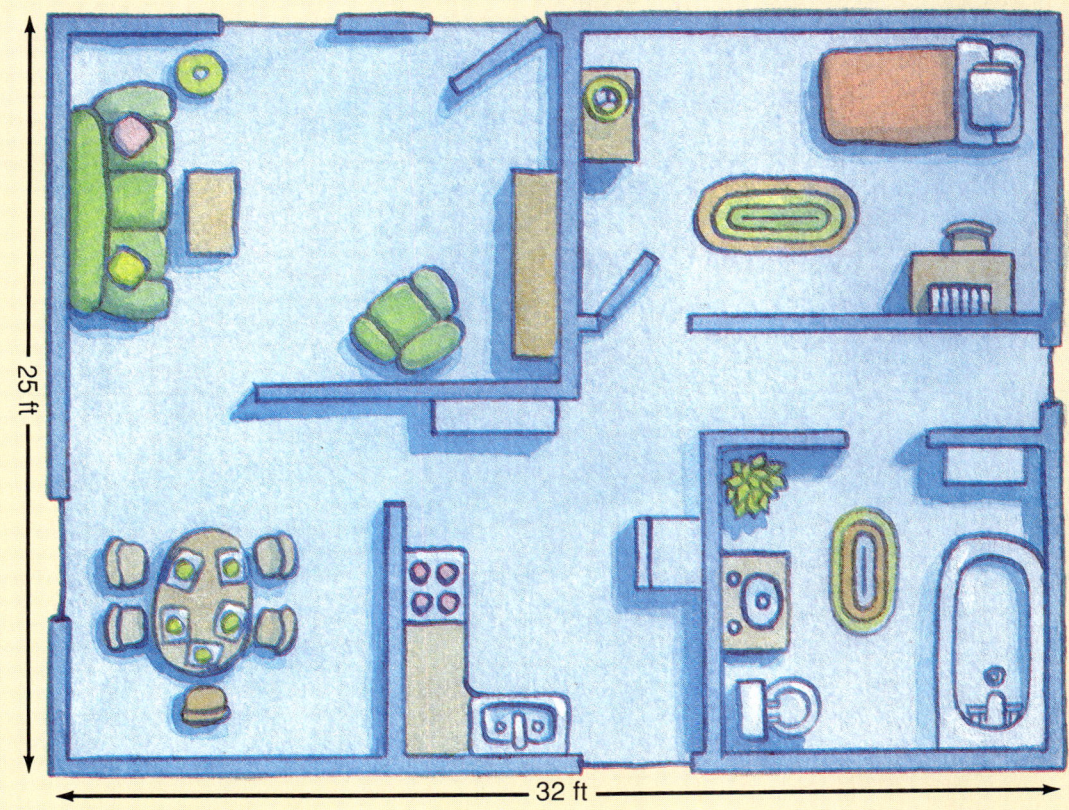

5. Show the rooms on your diagram. Give the area of each room. Be sure to include corridor space, if needed. Make sure the area of your apartment is 800 sq ft.

6. Draw another diagram for an 800 sq-ft apartment. This time plan a two-bedroom apartment. Would you still include the same rooms you included in your one-bedroom apartment? How else might you redesign your apartment?

7. Complete the floor plan for your two-bedroom design. Give the length, width, and area of each room.

8. Share your apartment designs with the class. Display and compare the different designs.

CHAPTER 9 309

Mixed Review

Write in word form.
1. 16,074
2. 0.052
3. 2.15
4. 0.101
5. 6.015

Round to the place of the underlined digit.
6. <u>7</u>9,842
7. 17,<u>4</u>31
8. 0.9<u>5</u>4
9. 3<u>4</u>.2
10. $7\frac{5}{9}$

Compare. Write >, <, =.
11. $\frac{7}{8}$ ▬ $\frac{7}{5}$
12. 8.075 ▬ 8.057
13. $\frac{3}{4}$ ▬ 0.75
14. 2.4 ▬ $2\frac{1}{2}$
15. 0.98 ▬ $\frac{98}{100}$
16. 38,541 ▬ 38,581

Find the answer.
17. 21,824 + 7,049
18. 561 ÷ 33
19. $40.30 − $17.65
20. 784 × 65
21. 7.06 − 3.244
22. 16.7 + 7.85
23. 0.0364 ÷ 0.07
24. 24.2 × 0.57
25. 60 × 4.8
26. 24.68 + 3.8
27. 83.3 × 9.2
28. 4.072 − 0.897
29. $\frac{7}{8} \div 14$
30. $3 \times \frac{5}{6}$
31. $6\frac{3}{5} - 4\frac{1}{2}$
32. $\frac{1}{10} + \frac{7}{10}$
33. $\frac{9}{10} \div \frac{1}{5}$
34. $3\frac{2}{3} \div \frac{1}{2}$
35. $\frac{7}{12} + \frac{1}{3}$
36. $10\frac{1}{2} - 8\frac{2}{3}$
37. $6\frac{4}{5} \times 1\frac{3}{7}$
38. $3\frac{2}{3} + \frac{1}{3}$
39. $\frac{5}{6} - \frac{5}{12}$
40. $\frac{1}{2} \times \frac{4}{5}$
41. $\frac{7}{12} - \frac{3}{8}$
42. $8\frac{2}{5} \times 10$
43. $6\frac{1}{4} + 1\frac{4}{5}$
44. $5 \div \frac{1}{6}$

PROBLEM SOLVING

CHOOSE Choose estimation, mental math, or paper and pencil to solve.

45. Ali walked 3.5 mi in 1 h. How far can she walk in $\frac{1}{2}$ h?

46. Jesse walked $1\frac{1}{2}$ mi to the bank and then home again. How many miles did he walk in all?

47. Harriet bicycled $\frac{3}{4}$ mi to her friend's house. Then they both rode $1\frac{3}{5}$ mi to the library. How far did Harriet ride her bike?

48. Of the students at the Spellman School, 0.5 ride the bus to school. Of the school's students, 0.4 walk. What fraction of the students use other forms of transportation?

49. Mark lives $\frac{4}{5}$ mi from school and Eddie lives $\frac{2}{3}$ mi from school. Who lives farther from school? How much farther?

50. Anna brought $3\frac{1}{4}$ lb of cheese to the class picnic. About $\frac{1}{2}$ of the cheese was eaten. About how many pounds were left?

51. A certain number is multiplied by 8. The product is 848. What is the number?

52. You subtract $15\frac{3}{4}$ from a certain number. The difference is $11\frac{2}{3}$. What is the number?

Use the chart to solve Problems 53–55.

53. Mr. Sands is cooking dinner and he wants everything to be ready at the same time. How long after he puts the chicken into the oven should he put in the potatoes?

54. How much longer than the broccoli will the chicken take to cook?

55. What time should Mr. Sands start cooking if he wants to serve dinner at 6:45 P.M.?

Cooking Times	
Roast chicken	$1\frac{1}{2}$ h
Baked potatoes	$1\frac{1}{4}$ h
Broccoli	15 min

CHAPTER CHECKUP

LANGUAGE & VOCABULARY

Use each word in two sentences. In the first sentence show that you understand the mathematical meaning of the word. In the second sentence show that you understand another, nonmathematical, meaning of the word.

acute   area   corresponding   similar

TEST

CONCEPTS

Write the word or phrase that best describes each figure. *(pages 278–281, 286–289)*

1.
2.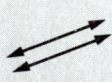
3.
4.
5.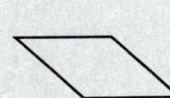
6.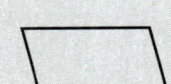

Use two of these words to describe each triangle: equilateral, isosceles, scalene, acute, obtuse, right. *(pages 282–283)*

7.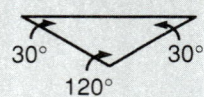
8.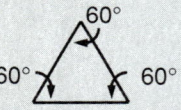
9.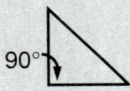

Are the figures similar or congruent? *(pages 296–297)*

10.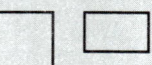

Triangles *MNO* and *PQR* are congruent. Find the corresponding angle or side in triangle *PQR*. *(pages 296–297)*

11. ∠ NOM

12. $\overline{MO}$

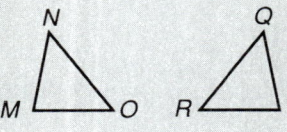

Trace the figure and draw the line(s) of symmetry. *(pages 298–299)*

13.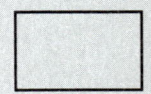

312   CHAPTER CHECKUP

SKILLS

Find the circumference. Use $\pi \approx 3.14$. Round to tenths. *(pages 300–301)*

14. $d = 8$ in.

15. $r = 3.2$ ft

Find the area. Use $\pi \approx 3.14$. Round to tenths. *(pages 304–307)*

16. Rectangle: $l = 7$ yd, $w = 4.2$ yd

17. Triangle: $b = 20$ in., $h = 5$ in.

18. Square: $s = 15$ ft

19. Circle: $r = 9$ mi

20. Circle: $d = 12$ ft

PROBLEM SOLVING

Use the Venn diagram. *(pages 284–285, 290–291)*

21. a. What two groups of figures are represented by this diagram?

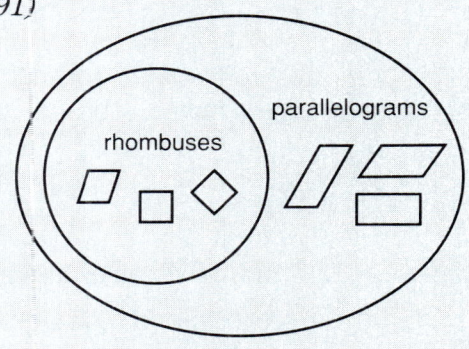

  b. How does a Venn diagram help you understand the relationship between objects in two groups?

  c. Write a generalization based on the relationship shown by the diagram.

Solve. *(pages 304–305)*

22. The Eagletons bought carpet for their 15-ft-square bedroom. How much carpet did they need to buy?

LEARNING LOG

Write the answers in your learning log.

1. What is true about the circumference of two circles if one has a radius of 12 in. and the other has a diameter of 24 in.? Explain.

2. Your friend said that of the upper case letters of the alphabet, the first letter without symmetry is B. Do you agree? Explain.

EXTRA PRACTICE

Write the best name for the figure. *(pages 278–281, 286–289)*

1.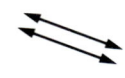
2.
3.
4.

5.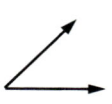
6.
7.
8.

9.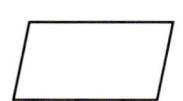
10.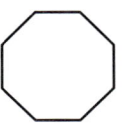
11.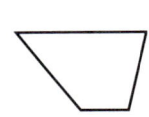
12.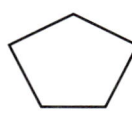

13.
14.
15.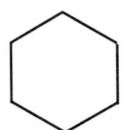
16.

Use two of these words to describe each triangle: equilateral, isosceles, scalene, acute, obtuse, right. *(pages 282–283)*

17.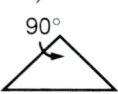
18.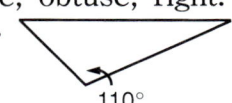
19.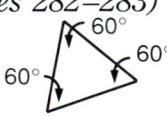
20.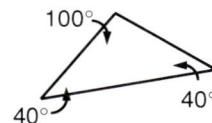

Triangles *JKL* and *MNO* are congruent. *(pages 296–297)*

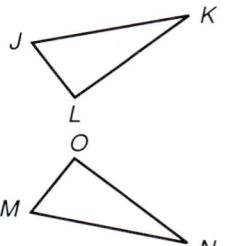

21. Name the corresponding sides.

22. Name the corresponding angles.

Trace each figure in Exercises 9–16. *(pages 298–299)*

23. Draw any lines of symmetry.

Find the circumference. Use $\pi \approx 3.14$. Round to tenths. *(pages 300–301)*

24. $d = 20$ ft
25. $d = 1.5$ in.
26. $r = 4$ ft
27. $r = 2.4$ yd

Find the area. Use $\pi \approx 3.14$. Round to tenths. *(pages 304–307)*

28. Square: $s = 10$ yd
29. Circle: $r = 4$ ft
30. Triangle: $b = 10$ in., $h = 3$ in.

Make a Venn diagram. *(pages 284–285, 290–29)*

31. Show the relationship between rectangles and parallelograms. Write a generalization.

ENRICHMENT

ENLARGEMENTS

The enlargement of this dog was made by using similar figures. What was in the small grids was copied onto larger grids.

Follow the steps to enlarge your favorite cartoon character.

a. Find a small picture of a character.
b. Draw horizontal and vertical lines $\frac{1}{4}$ in. apart to form a grid completely covering the character.
c. On a piece of poster board or construction paper, lightly draw horizontal and vertical lines 1 in. apart to form a grid.
d. Copy the part of the character found in each square of the smaller grid into the corresponding square on your poster board.
e. Erase the grid lines on your poster board.

Pennsylvania's capitol building in Harrisburg is considered one of the most beautiful capitols in the world.

Work with a partner. Locate a photograph or drawing of the front of a well-known building in your town, city, or state. Make a copy or tracing of it. Then do the following.

- Outline and identify geometric shapes on the building.
- Does the building front have symmetry? If so, draw the line(s) of symmetry. Explain how the symmetry or lack of symmetry affects the building's appearance.

ENRICHMENT 315

CUMULATIVE REVIEW

1. Which statement is true?
   a. $\frac{8}{15} > \frac{2}{5}$
   b. $\frac{2}{5} < \frac{6}{15}$
   c. $\frac{3}{5} = \frac{8}{15}$
   d. none of these

2. Which statement is true?
   a. $4\frac{1}{3} < 4\frac{1}{6}$
   b. $4\frac{2}{4} > 4\frac{1}{2}$
   c. $4\frac{2}{3} = 4\frac{4}{9}$
   d. none of these

3. Which is in order from least to greatest?
   a. $\frac{3}{4}, \frac{7}{8}, \frac{11}{16}$
   b. $\frac{11}{16}, \frac{7}{8}, \frac{3}{4}$
   c. $\frac{11}{16}, \frac{3}{4}, \frac{7}{8}$
   d. none of these

4. Write $\frac{7}{100}$ as a decimal
   a. 0.70
   b. 0.07
   c. 0.007
   d. none of these

5. Write $3\frac{9}{10}$ as a decimal.
   a. 0.39
   b. 3.09
   c. 9.3
   d. none of these

6. Write 0.121 as a fraction.
   a. $\frac{121}{1,000}$
   b. $\frac{121}{100}$
   c. $1\frac{21}{100}$
   d. none of these

Compute.

7. $\frac{5}{6} + \frac{1}{3}$
   a. $\frac{6}{9}$
   b. 1
   c. $1\frac{1}{6}$
   d. none of these

8. $\frac{4}{5} + \frac{3}{4}$
   a. $1\frac{3}{5}$
   b. $1\frac{11}{20}$
   c. $\frac{7}{9}$
   d. none of these

9. $\frac{7}{10} - \frac{2}{3}$
   a. $\frac{5}{7}$
   b. $\frac{1}{30}$
   c. $1\frac{11}{30}$
   d. none of these

10. $\frac{5}{9} \times \frac{3}{4}$
    a. $1\frac{1}{4}$
    b. $\frac{8}{13}$
    c. $\frac{1}{4}$
    d. none of these

11. $\frac{3}{8} \times 6$
    a. $\frac{1}{16}$
    b. $\frac{9}{24}$
    c. $2\frac{1}{4}$
    d. none of these

12. $\frac{2}{5} \times 3\frac{1}{2}$
    a. $1\frac{2}{5}$
    b. $6\frac{1}{5}$
    c. $\frac{4}{35}$
    d. none of these

316 CUMULATIVE REVIEW

PROBLEM SOLVING REVIEW

Problem Solving Check List
- Using operations
- Using estimation
- Using guess and check
- Too much information
- Too little information
- Using a simpler problem
- Multistep problems

Remember the strategies and types of problems you have done so far. Solve.

1. Mrs. Owasu drives $8\frac{3}{5}$ mi round-trip to and from work each day. Last month she worked 21 days.
   a. How far does Mrs. Owasu drive to and from work each day?
   b. What numbers would you use to estimate how many miles Mrs. Owasu drove to and from work last month?
   c. About how many miles did Mrs. Owasu drive to and from work last month?

2. Each day Mai-Lin practices skating from 6:40 A.M. to 8:10 A.M. She has been doing this for 3 yr. How long does Mai-Lin practice each day?

3. Andrew has 7 coins in his pocket. They total $.67. What coins does Andrew have?

4. A survey shows an average of 3 out of 5 families recycling cans and paper. Green River has 1,525 families. How many would you expect to recycle cans and paper?

5. This season the Jets won $\frac{3}{4}$ of their games, the Panthers won $\frac{5}{6}$ of their games, and the Bears won $\frac{7}{12}$ of their games. Which team had the best winning record?

6. The 1980 census showed that Boulder City had a population of 79,324. The 1990 census showed that Boulder City's population had decreased by 5,476. What was Boulder City's population in 1990?

7. Mr. Green needs to find the area of his backyard so he can buy enough grass seed. He measures and finds that the backyard is $92\frac{1}{2}$ ft long and $84\frac{1}{2}$ ft wide. What is the area of Mr. Green's backyard?

Write a simpler problem. Use it to solve the original problem.

8. In February, Plains City received $9\frac{3}{4}$ in. of snow during each of two storms and $8\frac{1}{2}$ in. of snow in another storm. How much snow fell on Plains City in February?

9. Miriam and Derrick found $2\frac{1}{4}$ gal of paint in their garage. They used all but $\frac{7}{8}$ gal to paint their tree house. How much paint did Miriam and Derrick use?

PROBLEM SOLVING REVIEW

TECHNOLOGY

FLIP SLIDING AWAY

In the computer game "Slide-Flip-Turn," partners predict combinations of flips, slides, and turns that will move a figure onto a congruent figure. Do this activity to sharpen your ability to visualize moves before actually making them.

Describe the slides, flips, and/or turns needed to move figure *A* onto figure *B*. Then try your moves by tracing and cutting out figure *A* and making the slides, flips, and/or turns on grid paper.

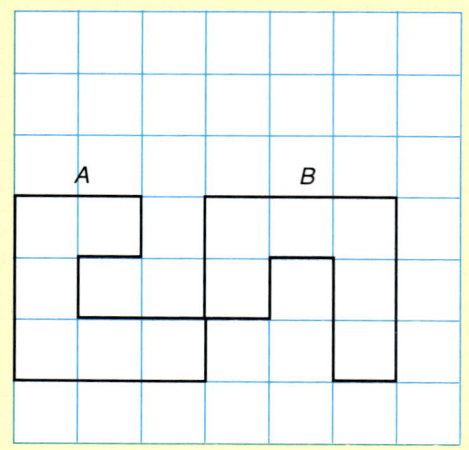

LINE PATTERNS

Make a chart of the number of points and the number of line segments in each drawing. Notice that each drawing is made by adding line segments to the previous drawing.

There is a relationship between the number of points in the new drawing and the number of line segments in the previous drawing. Find the pattern.

Use this pattern and your calculator to find the number of line segments there are between 25 points.

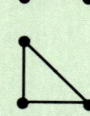

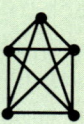

CIRCLE ESTIMATE

Choose the diameter or radius that will give the circumference shown. Use your calculator to find the circumference. Use 3.14 for π.

1. $d = 10$ $d = 14$ $C \approx 21.98$
   $d = 7$ $d = 6$

2. $r = 5$ $r = 6$ $C \approx 25.12$
   $r = 7$ $r = 4$

3. $r = 11$ $d = 18$ $C \approx 18.84$
   $d = 7$ $r = 3$

Ratio, Proportion, and Percent 10

DID YOU KNOW...?

Esperanto is a universal language that was created in 1887. It uses root words common to a large family of languages making it easier for many people to learn.

60% of the world's people speak one of the languages listed below.

LANGUAGE	PERCENT
Swahili (E. & Central Africa)	1%
French	2%
German	2%
Japanese	2%
Bengali (Bangladesh; India)	3%
Portuguese	3%
Malay-Indonesian	3%
Arabic (N. Africa; S.W. Asia)	4%
Russian	5%
Hindi (India)	6%
Spanish	6%
English	7%
Mandarin (China)	16%

USING DATA
- Collect
- Organize
- Describe
- Predict

Take a survey of 50 people in your community to find out which languages they speak. Record your results in a chart. Use the results of your survey to predict the responses of 100 people.

CHAPTER 10   319

Ratios

In 1927, Charles Lindbergh flew the *Spirit of St. Louis* from New York to Paris in about 33 h. Today it takes the Concorde about 3 h to fly the same distance.

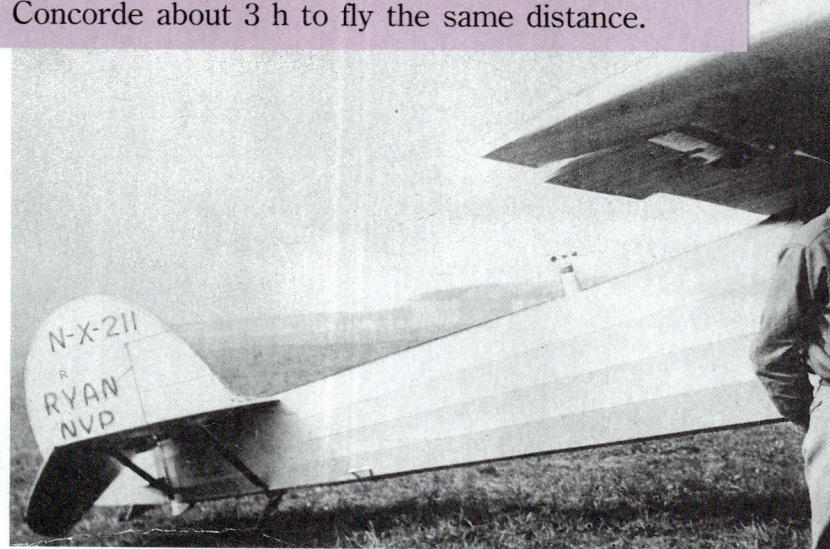

A **ratio** compares two quantities. The ratio of Lindbergh's flying time to the Concorde's flying time is 33 h to 3 h.

You can write a ratio in three ways.

    33 to 3   33:3 $\frac{33}{3}$

Read these ratios as *thirty-three to three*.

You can write a ratio in lowest terms the same way you write a fraction in lowest terms.

    11 to 1   11:1 $\frac{33}{3} = \frac{11}{1}$

These are equal ratios.

THINK ALOUD What other numbers can you use to write the same ratio as 33 to 3, or 11 to 1?

Another example: 6 jets to 8 airplanes   6 to 8   6:8 or 3:4 $\frac{6}{8}$ or $\frac{3}{4}$

GUIDED PRACTICE

Read the statement and answer the question.

> There are 2 pilots, 6 flight attendants, and 140 passengers aboard the Concorde.

1. What is the ratio of pilots to flight attendants?

2. What is the ratio of flight attendants to passengers?

320   LESSON 10–1

3. What is the ratio of flight attendants to pilots?

4. What is the ratio of crew members to passengers?

PRACTICE

Write the ratio in lowest terms.

5. $2:4$
6. 6 to 8
7. $8:9$
8. 10 to 2
9. $35:7$
10. $13:15$
11. 5 to 1
12. $4:40$
13. $\frac{25}{75}$
14. $\frac{100}{10}$

Write the ratio in three ways.

15. 135 passengers to 42 passengers
16. 17 jet planes to 4 helicopters
17. 2 h to 12 h
18. 1 pilot to 3 copilots
19. 1 engine to 2 engines
20. $325 to $2,350
21. 3 trips in 9 days
22. 15¢ per gallon

Write two other equal ratios.

23. $\frac{1}{4}, \frac{11}{44}$
24. $2:4, 3:6$
25. $\frac{100}{20}, \frac{50}{10}$
26. 16 to 12, 8 to 6

If the statement is a ratio, write it in lowest terms.
If the statement is not a ratio, write *not a ratio*.

27. A flight to Boston from Los Angeles is 2,600 mi.
28. An airplane has 15,000 lb of luggage for 150 passengers.
29. There are 7 flights per day from Dodge City to Kansas City.
30. It took 6 h to fly from New York to Paris in 1970.

PROBLEM SOLVING

Use the table to answer the question.

31. What is the ratio of the number of people on a Concorde to the number of people on a Boeing 747?

32. An interstate bus carries 43 passengers. How many buses are needed to carry the maximum number of people on a Boeing 707?

33. If an average person weighs 150 lb, what is the combined weight of the passengers on a full Boeing 747?

34. For which two airplanes is the ratio of the maximum number of people about 2 to 1?

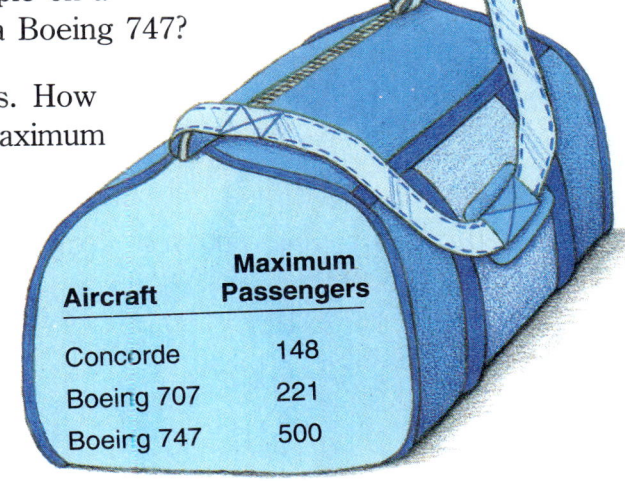

Aircraft	Maximum Passengers
Concorde	148
Boeing 707	221
Boeing 747	500

CHAPTER 10 321

Rates and Unit Rates

The world's fastest runner can cover 100 m in about 10 s. He runs at a rate of 100 m in 10 s. What is his rate per second?

A **rate** is a special type of ratio that compares different kinds of quantities, such as meters and seconds.

To find the rate per second, or the **unit rate**, write an equal ratio with a denominator of 1.

seconds → $\frac{100}{10} = \frac{10}{1}$ ← meters

The fastest runner runs at the rate of about 10 m per second, or 10 m/s.

THINK ALOUD What does 88 km/h mean?

Other examples of rates:
$4.25/h
70 heartbeats/min
98¢/L

GUIDED PRACTICE

THINK ALOUD Is the ratio a rate? Write *yes* or *no*. Explain how you know.

1. 78 m/min
2. 800 turns per 60 s
3. 57 heartbeats : 63 heartbeats
4. 100 g to 200 g
5. 15 breaths per minute
6. the speed limit on a highway

Write equal ratios to find the unit rate.

7. $20 in 5 h
8. 800 m in 2 min
9. 75¢ per 3 g
10. 4 for 60¢

PRACTICE

Find the unit rate.

11. 300 m in 30 s
12. $25 for 5 shirts
13. $.99 per 3 g
14. 12 km per 2 h
15. $21 for 3 h
16. 54 bats in 9 boxes

17. swimming 24 km in 3 h
18. hiking 80 km in 2 days
19. running 1,500 m in 4 min
20. cycling 244 km in 4 h
21. walking 78 m per minute
22. roller skating 1,500 m in 60 s
23. jogging 27 km in 3 h
24. driving 330 km in 4 h

NUMBER SENSE Choose *a*, *b*, *c*, or *d* as the best rate.

a. 33 km/h   b. 6 km/h   c. 7,297 km/h   d. 120 km/h

25. the speed of walking
26. the speed of wind in a hurricane
27. the speed of running
28. the speed of a rocket plane

PROBLEM SOLVING

Olympic Gold Medals Men's 200-Meter Run

Year	Time (in s)
1920	22.00
1948	21.10
1972	20.00
1988	19.75

Use the table to answer the question.

29. In 1920, what was the winning rate for the men's 200-m race?

30. What was the rate per second of the winner in 1972?

31. What is the difference in times between the 1988 record and the 1920 record?

32. In which year was the runner's rate per second the fastest? Was it more than or less than 10 m/s?

Critical Thinking

Tell which is the better buy. Find the unit price, or cost per unit, for each item. Then discuss with a partner why the cheaper item might not be the one you would choose to buy.

1. 4 L of Tropic juice for $3.20
   2 L of O. J. juice for $3.20

2. 2 cans of tomatoes for $.98
   24 cans of tomatoes for $11.04

3. 2 cups of fruit yogurt for $.89
   2 cups of plain yogurt for $.79

4. 6 McIntosh apples for $2.40
   5 Granny Smith apples for $2.50

CHAPTER 10   323

Exploring Proportional Thinking

Japan's flag was adopted in 1854.
Look at the two pictures of the flag below.

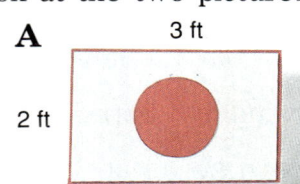

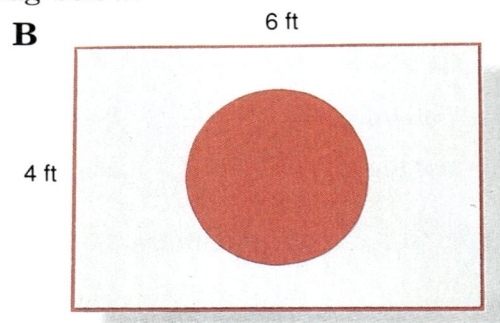

1. How do the ratios of the width to the length of flag A and flag B compare?

We use proportional thinking when we say that 2 ft is to 3 ft as 4 ft is to 6 ft. We can write 2:3 as 4:6 or $\frac{2}{3} = \frac{4}{6}$

$\frac{2}{3} = \frac{2 \times 2}{3 \times 2} = \frac{4}{6}$ $\frac{4}{6} = \frac{4 \div 2}{6 \div 2} = \frac{2}{3}$

Multiplication and division help us think proportionally.

Since the ratios of the dimensions of the flags are equal, we can say that flags A and B are proportional.

Look at these two flags.

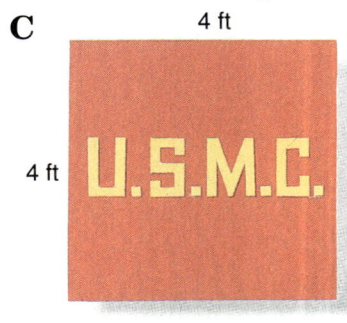

2. Find the ratio of the length to the width of flag C.

3. Find the ratio of the length to the width of flag D.

4. Does the ratio of the length to the width of flag C equal the ratio of the length to the width of flag D? Are flags C and D proportional?

LESSON 10–3

Do these examples show proportional thinking? Explain.

5. 7 flags is to 5 flags as 21 flags is to 15 flags.

6. 3 flags are to 1 pole as 8 flags are to 2 poles.

7. 4 is to 8 as 1 is to 2

8. 3:4 as 8:12

9. $\frac{16}{14} = \frac{8}{7}$

10. 10 is to 5 as 30 is to 15

11. 75:100 = 15:20

12. $\frac{7}{2} = \frac{28}{8}$

13. **IN YOUR WORDS** How can you tell whether two ratios make a proportion?

These flags are from different states, or *prefectures*, in Japan.

14. Work with a partner. Use proportional thinking to find two pairs of flags that are proportional.

Meaning of Proportions

Mount Rushmore is located in South Dakota.

Each head carved on Mount Rushmore is about 60 ft high.

For an average person, the ratio of the height of the head to the total height is about 1 to 8.

Suppose a statue of George Washington had a head 60 ft high and a total height of 480 ft. Is 60 to 480 the same as 1 to 8?

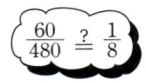

Two equal ratios are a **proportion**. There are several ways to tell whether two ratios are a proportion.

You can use division or multiplication.

height of head → $\frac{60}{480} = \frac{60 \div 60}{480 \div 60} = \frac{1}{8}$ $\frac{1}{8} = \frac{1 \times 60}{8 \times 60} = \frac{60}{480}$
total height →

You can also use cross multiplication. The arrows in the proportion at right show the **cross products** 60×8 and 1×480. For two ratios to be equal, the cross products must be equal.

$60 \times 8 \stackrel{?}{=} 1 \times 480$
$480 = 480$

Both methods show that the ratios are equal and form a proportion.

GUIDED PRACTICE

Use multiplication or division to write an equal ratio.

1. $\frac{6}{20}$
2. $\frac{3}{8}$
3. $\frac{100}{4}$

Compare the two ratios. Write *equal* or *not equal*.

4. $\frac{1}{6}$ ▤ $\frac{2}{12}$
5. $\frac{4}{8}$ ▤ $\frac{12}{16}$
6. $\frac{6}{5}$ ▤ $\frac{24}{20}$

PRACTICE

Are the ratios a proportion? Write = or ≠.

7. $\frac{12}{18}$ ▤ $\frac{4}{6}$
8. $\frac{10}{40}$ ▤ $\frac{4}{16}$
9. $\frac{4}{5}$ ▤ $\frac{6}{10}$
10. $\frac{3}{12}$ ▤ $\frac{5}{15}$
11. $\frac{9}{3}$ ▤ $\frac{81}{27}$
12. $\frac{36}{6}$ ▤ $\frac{12}{2}$
13. $\frac{3}{50}$ ▤ $\frac{6}{75}$
14. $\frac{21}{24}$ ▤ $\frac{7}{8}$
15. $\frac{2}{5}$ ▤ $\frac{10}{20}$
16. $\frac{7}{2}$ ▤ $\frac{72}{22}$
17. $\frac{150}{300}$ ▤ $\frac{1}{2}$
18. $\frac{18}{16}$ ▤ $\frac{9}{8}$
19. $\frac{75}{25}$ ▤ $\frac{3}{1}$
20. $\frac{5}{9}$ ▤ $\frac{5}{3}$
21. $\frac{22}{11}$ ▤ $\frac{2}{1}$

Choose **8**, **9**, or **10** to complete the proportion.

22. $\frac{4}{3} = \frac{▤}{6}$
23. $\frac{9}{81} = \frac{1}{▤}$
24. $\frac{5}{▤} = \frac{15}{24}$
25. $\frac{6}{60} = \frac{1}{▤}$

26. **CRITICAL THINKING** Decide whether this statement is true or false. Explain how you know.

> If $\frac{a}{b} = \frac{c}{d}$ and $\frac{c}{d} = \frac{e}{f}$, then $\frac{a}{b} = \frac{e}{f}$.

PROBLEM SOLVING

Read the statement. Then answer the question.

> The Statue of Liberty in New York harbor is about 110 ft tall. Its right arm is 42 ft long.

27. If a 55-in.-tall statue were built to the same proportion, would its right arm be 21 in. long?

28. The Statue of Liberty is about 20 times taller than the average woman. How tall is the average woman?

Mental Math

4 cans are to $1 as 8 cans are to ▤.

Think: 8 is twice as much as 4, so the price is twice as much as $1.

2 × $1 = $2, so 8 cans are $2.

Use proportional thinking to solve mentally.

1. 1 book is to $2 as 3 books are to ▤.
2. 12 c are to 3 qt as ▤ c are to 1 qt.
3. 3 apples are to $1.50 as 6 apples are to ▤.

CHAPTER 10

Solving Proportions

Goodnight's painting is about 96 in. high by 60 in. wide. An art student painted a smaller version, which was 48 in. high. How wide must the smaller painting be for the two paintings to be proportional?

96 in.

60 in.

A Culture Preserved, by Paul T. Goodnight

To find the width, you can solve the proportion:

$\frac{\text{height}}{\text{width}}$ $\quad \frac{96}{60} = \frac{48}{n}$

There are two ways to find n. You can use equal ratios.

$\frac{96 \div 2}{60 \div 2} = \frac{48}{n}$, so $n = 30$.

You can also use cross multiplication to solve a proportion.

Cross-multiply.	Find the value of n.
$\frac{96}{60} \diagup\!\!\!\!\diagdown \frac{48}{n}$ $96 \times n = 48 \times 60$	$96 \times n = 48 \times 60$ $96 \times n = 2{,}880$ $n = 2{,}880 \div 96$ $n = 30$

Use inverse operations.

The width of the smaller painting must be 30 in. for the paintings to be proportional.

CRITICAL THINKING Could these proportions have been used to find the width? Explain.

$\frac{60}{96} = \frac{n}{48}$ $\qquad \frac{96}{48} = \frac{60}{n}$

═══════ **GUIDED PRACTICE** ═══════

Cross-multiply to solve the proportion.

1. $\frac{3}{4} = \frac{n}{16}$ **2.** $\frac{5}{7} = \frac{n}{14}$ **3.** $\frac{5}{8} = \frac{25}{n}$ **4.** $\frac{n}{8} = \frac{9}{12}$ **5.** $\frac{2}{n} = \frac{3}{9}$

LESSON 10–5

MATH AND ART

PRACTICE

Solve the proportion. Use a calculator when needed.

6. $\frac{3}{3} = \frac{n}{6}$
7. $\frac{1}{2} = \frac{5}{n}$
8. $\frac{n}{4} = \frac{1}{2}$
9. $\frac{n}{2} = \frac{10}{1}$
10. $\frac{8}{n} = \frac{4}{5}$

11. $\frac{2}{3} = \frac{n}{9}$
12. $\frac{5}{4} = \frac{n}{8}$
13. $\frac{n}{15} = \frac{6}{10}$
14. $\frac{30}{n} = \frac{3}{2}$
15. $\frac{9}{13} = \frac{27}{n}$

16. $\frac{5}{7} = \frac{n}{35}$
17. $\frac{n}{8} = \frac{12}{3}$
18. $\frac{7}{2} = \frac{n}{6}$
19. $\frac{3}{4} = \frac{9}{n}$
20. $\frac{8}{12} = \frac{12}{n}$

21. $\frac{n}{5} = \frac{27}{45}$
22. $\frac{15}{15} = \frac{25}{n}$
23. $\frac{1}{3} = \frac{n}{33}$
24. $\frac{12}{n} = \frac{8}{20}$
25. $\frac{60}{1} = \frac{n}{3}$

26. $\frac{180}{60} = \frac{n}{5}$
27. $\frac{n}{4} = \frac{13}{26}$
28. $\frac{1.5}{15} = \frac{n}{10}$
29. $\frac{n}{1} = \frac{7}{3}$
30. $\frac{3}{8} = \frac{n}{20}$

PROBLEM SOLVING

Use the information about the paintings to solve.

— 39 in. —
50 in.

36 in.
— 25 in. —

Left: *Ngozi's Serendipity*
Above: *Shero*
both by Paul T. Goodnight

31. Are the paintings proportional? Explain how you know.

32. Which of the two paintings would fit into a box that is $2\frac{1}{2}$ ft by 3 ft?

33. **IN YOUR WORDS** How do the areas of the two paintings compare? Explain your answer.

34. A copy of *Shero* has the same ratio of height to width but is 10 in. wider. What is the height and the width of the copy?

Problem Solving: Scale Drawings

The planets Saturn, Jupiter, and Neptune all have moons that are larger than Earth's Moon.

These drawings show how the diameters of these four moons compare.

The ratio of the size in a drawing to the actual size is the **scale**. A drawing made using a scale is called a **scale drawing**.

The scale of the measurements in each drawing to the actual measurement is 1 mm to 100 km. What is the actual diameter of Neptune's moon Triton?

- What is the scale?
- What are you supposed to find?
- How can you use a proportion to solve the problem?

Solve. How can you check your answer?

CRITICAL THINKING Which planet has the largest moon? How large is it in kilometers?

Triton, a moon of Neptune
48 mm

Titan, a moon of Saturn
58 mm

Ganymede, a moon of Jupiter
53 mm

Earth's Moon
35 mm

GUIDED PRACTICE

1. Read the statement and answer the questions.

   The diameter of Mercury is about 4,900 km. This planet is smaller than Jupiter's moon Ganymede.

   a. What is the actual diameter of Mercury?

   b. What would you need to know to make a scale drawing of Mercury?

   c. Using the scale 1 mm to 100 km, what size would you draw Mercury?

330 LESSON 10–6

PRACTICE

This model of a spacecraft was built on a scale of 1 cm to 4 m. Use the scale drawing of the model to solve.

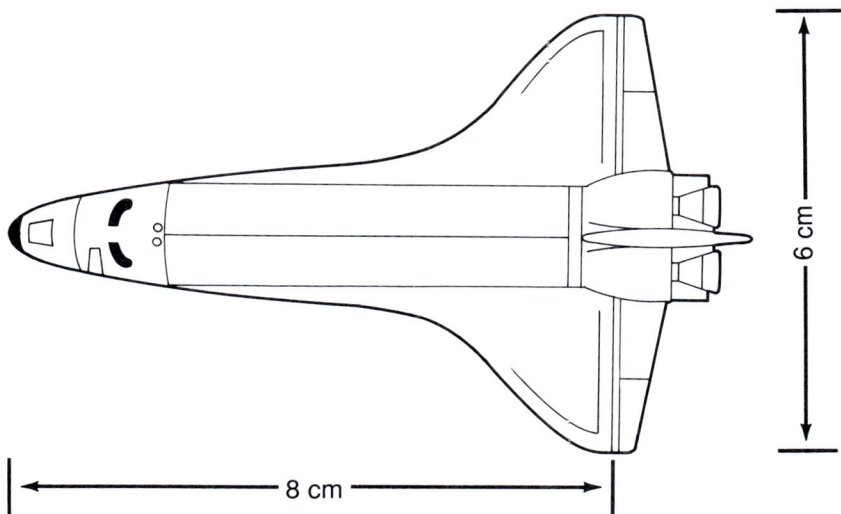

2. What is the actual length of the spacecraft from nose to wing?

3. What is the wingspan's actual width?

4. The actual payload bay is 18 m long. What is its length when drawn to scale?

5. A second model is 3 times as large as this model. What is the scale of the second model?

CHOOSE Choose any strategy to solve.

6. Mars has two moons. Phobos travels once around Mars in about 7 h. Deimos makes 1 revolution in about 31 h. In the time it takes Deimos to make 7 revolutions, how many will Phobos make?

7. Earth's Moon is about 380,000 km from Earth. The ratio of the distance between Earth's Moon and Earth to the distance between Phobos and Mars is 40:1. About how far is Phobos from Mars?

8. From 1957 through 1969, 870 spacecraft were sent into space. The United States successfully launched 102 more spacecraft than the U.S.S.R. How many spacecraft did the United States launch?

9. Neptune is about 10 times larger than its moon Triton. Earth's Moon is about $\frac{1}{4}$ the size of Earth. Earth's size is about $\frac{1}{3}$ the size of Neptune. Arrange these planets and moons from largest to smallest.

CHAPTER 10   331

MIDCHAPTER CHECKUP

LANGUAGE & VOCABULARY

Explain how rates and proportions are related to ratios. Then give an example of a rate and a proportion.

QUICK QUIZ

Write the ratio in three ways. *(pages 320–321)*
1. 37 fish to 4 aquariums

Write the ratio in lowest terms. *(pages 320–321)*
2. 15 to 3
3. 8:12

Find the unit rate. *(pages 322–323)*
4. $20 for 5 tickets
5. 240 km in 3 h

Are the ratios a proportion? Write = or ≠. *(pages 326–327)*
6. $\frac{6}{8}$ ▨ $\frac{18}{32}$
7. $\frac{28}{4}$ ▨ $\frac{7}{1}$

Solve the proportion. *(pages 328–329)*
8. $\frac{5}{9} = \frac{n}{36}$
9. $\frac{35}{n} = \frac{10}{8}$

Solve. *(pages 330–331)*
10. This model of a car was built on a scale of 2 in. to 3 ft.
    a. What is the length of the model of the car?
    b. How can you use a proportion to find the actual length of the car?
    c. What is the actual length of the car?

8 in.

4 in.

Write the answers in your learning log.
1. Your friend thinks that the numerator of a ratio will always be smaller than the denominator. Do you agree with this thinking? Explain why or why not.

2. Describe two ways you can test whether two ratios make a true proportion.

In 1647, the Massachusetts Bay Colony passed a law requiring each town of more than 50 households to open a school. What is the ratio of schools to households in your city or town?

1. Use the numbers 3 and 12 and another whole number to write two ratios equal to $\frac{1}{4}$.

2. Use the numbers 3 and 12 and another whole number to write two ratios equal to $\frac{1}{2}$.

Meaning of Percent

What is your favorite food? When the Harrison School newspaper asked 100 students to name their favorite dinner, 25 students chose pasta.

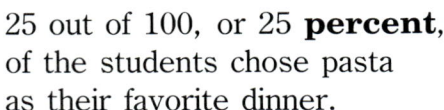

25 out of 100, or 25 **percent**, of the students chose pasta as their favorite dinner.

Percent means **per hundred**. The percent symbol is %.

THINK ALOUD How can you write 25% as a fraction?

Other examples.

5 out of 100

Fraction: $\frac{5}{100}$

Percent: 5%

Read as *five percent*.

78 out of 100

Fraction: $\frac{78}{100}$

Percent: 78%

Read as *seventy-eight percent*.

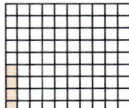

=== GUIDED PRACTICE ===

THINK ALOUD Tell what percent is shaded and what percent is not shaded. Explain how you got your answers.

1.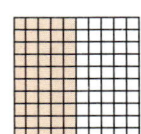
2.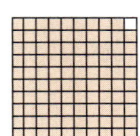
3.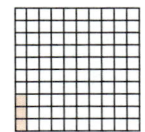

Write the percent.

4. 56 out of 100
5. 29 out of 100
6. 100 out of 100
7. $\frac{2}{100}$
8. $\frac{33}{100}$

334   LESSON 10–7

PRACTICE

What percent is shaded? not shaded?

9.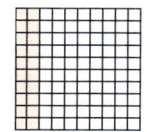
10.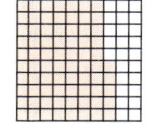
11.

Write the percent.

12. 45 out of 100
13. 17 out of 100
14. 29 out of 100
15. 6 out of 100
16. 50 out of 100
17. 99 out of 100
18. 87 out of 100
19. $10\frac{1}{2}$ out of 100

Write the percent as a fraction.

20. 25%
21. 30%
22. 40%
23. 62%
24. 95%
25. 75%
26. 50%
27. 18%
28. 35%
29. 5%
30. 80%
31. 20%

32. **CRITICAL THINKING** How would you explain what 100% means? Give an example of 100%.

NUMBER SENSE Compare. Write >, <, or =.

33. 50% ▇ 75%
34. 80% ▇ 8%
35. 9% ▇ $\frac{99}{100}$
36. 42.1% ▇ $\frac{42}{100}$

PROBLEM SOLVING

Use the graph to answer the question.

37. How many students chose chicken as their favorite dinner? Write your answer in a sentence using percent.

38. What percent of the students did not choose pizza or spaghetti as their favorite dinner?

39. **IN YOUR WORDS** Is it true that, based on this survey, the favorite American dinner is pizza? Explain.

40. Based on this survey, how many students would choose hamburger if 1,000 students were surveyed?

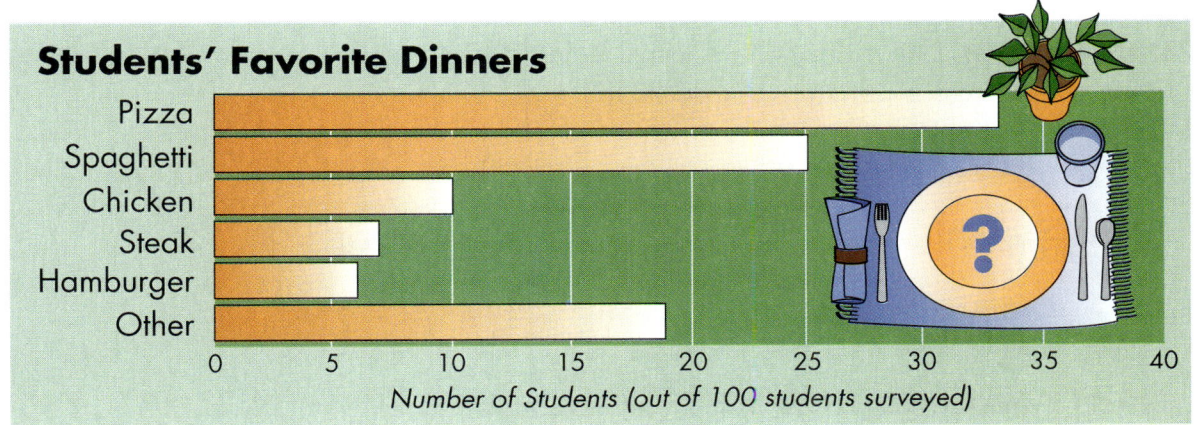

Students' Favorite Dinners — Number of Students (out of 100 students surveyed)

Percents and Decimals

Did you know that the human body is made up of chemical elements, such as oxygen, carbon, and hydrogen? It even has traces of copper, zinc, and iodine.

The periodic table gives information about all the known chemical elements.

In fact, 0.65 of the elements in your body is oxygen. It might be easier to understand what this means if 0.65 is written as a percent.

Start by writing a fraction with a denominator of 100. Why?

$0.65 = \frac{65}{100} = 65\%$

THINK ALOUD How does the decimal point move when you change a decimal to a percent?

Percents are easily changed to decimals. $18\% = \frac{18}{100} = 0.18$

THINK ALOUD How does the decimal point move when you change a percent to a decimal?

Other examples:

$10\% = \frac{10}{100} = 0.10$ $3.3\% = \frac{3.3}{100} = \frac{3.3 \times 10}{100 \times 10} = \frac{33}{1,000} = 0.033$

$0.01 = \frac{1}{100} = 1\%$ $0.075 = \frac{75}{1,000} = \frac{75 \div 10}{1,000 \div 10} = \frac{7.5}{100} = 7.5\%$

CRITICAL THINKING How would you write 0.5% as a decimal?

336   LESSON 10-8

GUIDED PRACTICE

Write the percent as a decimal.
1. 83%   2. 17%   3. 5%   4. 77%   5. 87.5%   6. 1.5%

Write the decimal as a percent.
7. 0.33   8. 0.62   9. 0.09   10. 0.2   11. 0.05   12. 0.025

PRACTICE

Write the percent as a decimal or the decimal as a percent.
13. 0.25   14. 0.07   15. 0.87   16. 0.4   17. 0.7   18. 0.5
19. 0.66   20. 0.15   21. 0.08   22. 0.155   23. 0.333   24. 0.007

Write the percent as a decimal.
25. 42%   26. 9%   27. 60%   28. 1%   29. 85%   30. 9%
31. 3%   32. 4.5%   33. 6.6%   34. 1.5%   35. 66.6%   36. 150%

NUMBER SENSE Compare. Write >, <, or =.
37. 59% ▆ 0.51   38. 0.35 ▆ 35%   39. 0.3 ▆ 3%   40. 80% ▆ 0.8
41. 16% ▆ 0.16   42. 0.1 ▆ 100%   43. 25% ▆ 2.5%   44. $\frac{67}{100}$ ▆ 6.7%

45. **CALCULATOR** How would you change 35% to a decimal using a calculator? Explain.

PROBLEM SOLVING

CHOOSE Choose mental math or pencil and paper to solve. Use the table.

46. What percent of all the chemical elements in your body is not oxygen?

47. Which element is 0.1 of the elements in your body?

48. You have 12 times as much carbon in your body as which other element?

49. Which element makes up less than 0.012 of the elements in your body?

50. **IN YOUR WORDS** Explain why the table lists 1.2% of the body's elements as *Other*.

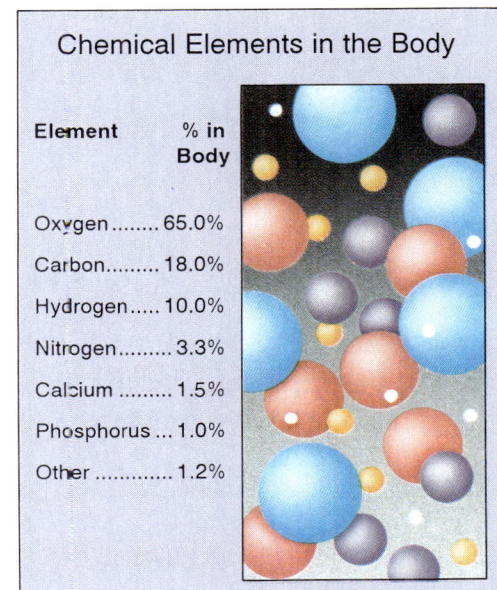

Chemical Elements in the Body

Element	% in Body
Oxygen	65.0%
Carbon	18.0%
Hydrogen	10.0%
Nitrogen	3.3%
Calcium	1.5%
Phosphorus	1.0%
Other	1.2%

Fractions and Percents

By the year 1900, the people of the United States had elected 25 Presidents. Of those Presidents, 7 were born in Virginia.

You can write *7 out of 25* as a fraction.

7 out of 25 = $\frac{7}{25}$

You know how to write a fraction as a decimal and a decimal as a percent. So you can make a MATH CONNECTION to write a fraction as a percent.

		Write the fraction as a decimal.	Write the decimal as a percent.
$\frac{7}{25}$	Think: $\frac{7}{25}$ means 7 ÷ 25.	$25\overline{)7.00}$ 0.28	0.28 = 28%

Another example:

$\frac{3}{8}$ $8\overline{)3.000}$ 0.375   0.375 = 37.5%

If you divide 8 by 11 on a calculator to write $\frac{8}{11}$ as a percent, you get the repeating decimal 0.727272 . . . . Round a decimal to the nearest hundredth to get the nearest whole percent.

$\frac{8}{11}$ = 8 ÷ 11 = 0.727272.   →   0.73 = 73%

You also can write a percent as a fraction in lowest terms.

60% = $\frac{60}{100}$ = $\frac{3}{5}$ 33% = $\frac{33}{100}$

338   LESSON 10-9

GUIDED PRACTICE

Write the fraction as a decimal and as a percent.

1. $\frac{4}{100}$
2. $\frac{18}{30}$
3. $\frac{21}{28}$
4. $\frac{1}{10}$
5. $\frac{8}{20}$
6. $\frac{5}{8}$

PRACTICE

Write as a percent. Use a calculator when needed.

7. $\frac{3}{10}$
8. $\frac{6}{25}$
9. $\frac{46}{100}$
10. $\frac{17}{50}$
11. $\frac{7}{8}$
12. $\frac{6}{16}$
13. $\frac{3}{15}$
14. $\frac{9}{12}$
15. $\frac{66}{75}$
16. $\frac{8}{80}$
17. $\frac{5}{40}$
18. $\frac{3}{2}$
19. 3 out of 12
20. 9 out of 30
21. 45 out of 60
22. 1 out of 8

Write the percent as a fraction in lowest terms.

23. 20%
24. 15%
25. 1%
26. 88%
27. 200%
28. 2.5%

MENTAL MATH The denominator of each fraction is a factor of 100. Find the percent mentally.

29. $\frac{1}{10}$
30. $\frac{1}{2}$
31. $\frac{1}{4}$
32. $\frac{1}{5}$
33. $\frac{2}{5}$
34. $\frac{3}{4}$

35. **CALCULATOR** If your calculator has no percent key, what steps do you follow to change a fraction to a decimal? to a percent?

36. Write as a percent. Round to the nearest whole percent.
    a. $\frac{5}{6}$
    b. $\frac{14}{21}$
    c. $\frac{1}{12}$

ESTIMATE Is the fraction closest to *10%, 50%,* or *100%*?

37. $\frac{9}{100}$
38. $\frac{2}{4}$
39. $\frac{89}{100}$
40. $\frac{1}{5}$
41. $\frac{7}{8}$
42. $\frac{2}{3}$

PROBLEM SOLVING

Solve the problem about the United States' Presidents.

43. Five of the first 25 Presidents were born in March. What percent of those Presidents were born in March?

44. Of the first 10 Presidents, 20% were born in Massachusetts. How many of those Presidents were not born in Massachusetts?

45. George Bush was born on June 12, 1924. He became the 41st President in January of 1989. How old was George Bush when he took office?

46. Eight Presidents out of 41 were born in Virginia. What percent of Presidents is that? Round the percent to the nearest tenth.

Exploring Percent

The sixth grade students at Webster Intermediate School started the Kids-for-Recycling Club. All 60 students in the sixth grade joined the club. We can show this with a percent bar diagram. Work from the top down.

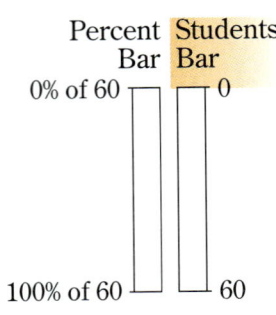

Each bar represents all of the students in the sixth grade. One bar represents the number of students and the other the percent of the students who joined the club.

1. How many students does the whole bar represent?

2. What percent does the whole bar represent?

3. Why are 0 and 0% across from one another on the bars?

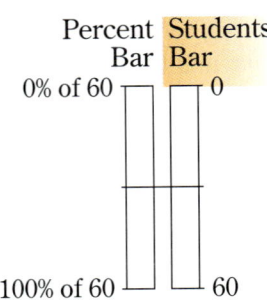

Fifty percent of the students are making posters about recycling. A line has been drawn across the diagram to show how many students are making posters.

4. How far down the bar was the line drawn?

5. By looking at the diagram, about how many students would you expect to be making posters?

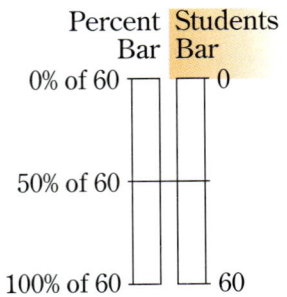

The percent bar has been labeled.

6. Why was 50% of 60 marked at the middle of the bar?

7. What would you mark across from *50% of 60* on the student bar?

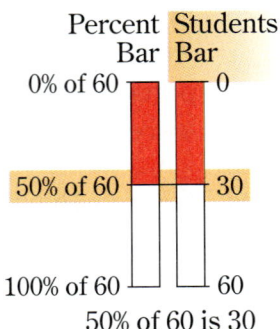

The complete diagram is shown at right. The math sentence that describes the situation is shaded.

8. What part of the bars would be shaded if all of the students were making posters? if no one signed up to make posters?

9. How would the numbers on the diagram change if there were 80 students in the sixth grade and 50% were making posters?

340   LESSON 10–10

10. The diagram at right shows how many sixth graders volunteered to put the posters around the community.

   a. How many students volunteered?
   b. What percent of the students volunteered?

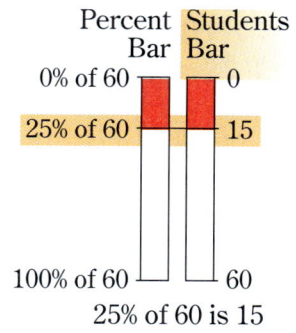

25% of 60 is 15

When making a percent bar diagram to fit a situation, it is helpful to be able to draw the line across the bars in *about* the right place. Knowing these fraction-percent relationships can help:

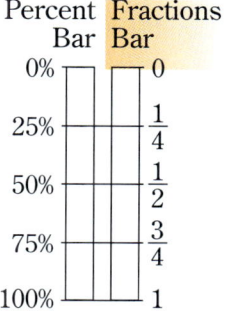

$\frac{1}{4} = 25\%$, $\frac{1}{2} = 50\%$, $\frac{3}{4} = 75\%$.

11. Describe where the line would cross the bars for the given percent.
    a. 20%   b. 60%   c. 80%

The Kids-for-Recycling Club plans to collect glass, aluminum, and paper for recycling from 200 homes.

12. By 10:00 A.M., they had picked up material from 40 people.
    a. What does the letter *A* represent?
    b. Which letter will 200 replace?
    c. Which letter will 40 replace?
    d. Which letter will the percent that 40 represents replace?

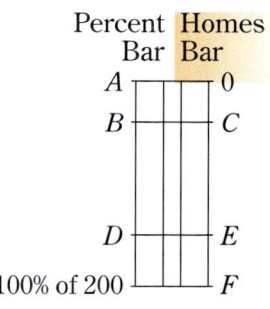

13. At 2:00 P.M. they had picked up recyclable material from 75% of the homes.
    a. Which letter will 75% replace?
    b. Which letter will the number of people that 75% represents replace?

14. At the end of the day recyclable material had been picked up from 200 homes. What percent represents the 200?

Draw and label a percent bar diagram for the sentence. Write the sentence under the diagram.

15. 80% of 300 is 240.   16. 30 is 25% of 120.   17. 80 out of 400 is 20%.

Finding a Percent of a Number

Many of the fruits you eat contain water. About 80% of an apple's weight is water. If an apple weighs 9 oz, how much of that is water?

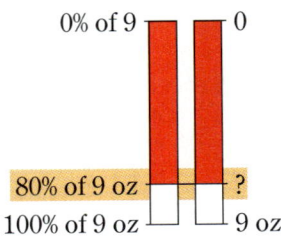

Use a percent bar diagram to represent the problem. You can use decimals or fractions to find 80% of 9 oz.

Write the percent as a decimal. Then multiply.	or	Write the percent as a fraction. Then multiply.
80% of 9 = ? 80% = 0.80 0.80 × 9 = 7.20		80% of 9 = ? 80% = $\frac{80}{100}$ $\frac{80}{100} \times 9 = \frac{720}{100} = 7.2$

A 9-oz apple has 7.2 oz of water. Use the percent bar to check.

CRITICAL THINKING If you know $\frac{1}{5}$ is the same as 20%, how can you tell what fraction is the same as 40%? as 60%? If 20% of 10 is 2, what is 40% of 10? 60% of 10?

GUIDED PRACTICE

Find the percent of the numbers.

1. 0% of 30 / 0 ; 30% of 30 / ? ; 100% of 30 / 30
2. 0% of 280 / 0 ; 15% of 280 / ? ; 100% of 280 / 280
3. 0% of 8 / 0 ; 60% of 8 / ? ; 100% of 8 / 8

4. What is 50% of 26?
5. What is 77% of 300?
6. What is 5.5% of 50?

342   LESSON 10-11

PRACTICE

Find the percent of the number. Use a calculator when needed.

7. 30% of 80
8. 60% of 70
9. 20% of $40
10. 45% of 50
11. 12% of 400
12. 35% of $160
13. 15% of 46
14. 80% of 80
15. 85% of 320
16. 26% of $13
17. 66% of 120
18. 25% of 50
19. 33% of 60
20. 29% of 82
21. 2% of 27
22. 18% of 20
23. 4% of 57
24. 9% of 900
25. 2.5% of 27
26. 6.5% of 200
27. 0.5% of 60

NUMBER SENSE Compare. Write >, <, or =.

28. 50% of 400 ▨ 75% of 200
29. 3% of 1,000 ▨ 3% of 100
30. 20% of 80 ▨ 100% of 60
31. 25% of 8 ▨ 50% of 6

CALCULATOR Round the amount to the nearest cent.

32. 25% of $32.57
33. 3% of $51.98
34. 83% of $130.50
35. 6.7% of $112

PROBLEM SOLVING

36. 97% of a watermelon is water. If a watermelon weighs 15 lb, how much of its weight is water?

37. A peach is 83% water, and an apple is 80% water. Which has more water, a 6-oz peach or a 7-oz apple?

Estimate

You can use compatible numbers to estimate a percent of a number.

23% of 80

Think: 23% is about 25%.

$25\% = \frac{1}{4}$

$\frac{1}{4} \times 80 = \frac{80}{4} = 20$

31% of $89

Think: 31% is about 30%. 89 is about 90.

$30\% = 0.3$

$0.3 \times \$90 = \$27$

Use compatible numbers to estimate.

1. 11% of 50
2. 52% of $200
3. 60% of $498
4. 75% of 404
5. 19% of $20
6. 62% of $81
7. 9% of 21
8. 23% of $160

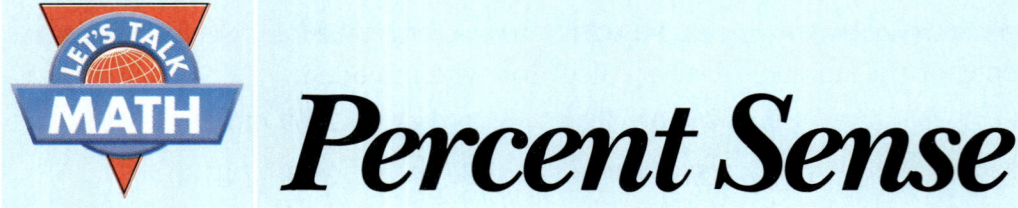

Percent Sense

When Do We Use Percent?

READ ABOUT IT

Percents are used in everyday life. In a newspaper you might see percents used in an ad for a sale or the interest rate of a bank. On television you might hear that a certain percent of people chose a particular product.

TALK ABOUT IT

Work with a partner to discuss what the percent means. Talk about how percent is making a comparison in terms of 100.

1. 70% of Taft School sixth graders chose pizza as their favorite food.

2. 50% of the students are girls.

3. 65% of pets are dogs.

4. Sales in computers increased 8% since last year.

5. The new student council president got 51% of the vote.

6. Bryce scored 98% on a math test of 100 questions.

7. There is a 20% chance of rain.

8. 80% of the students polled said they would use an ice-skating rink.

9. Kim gets a hit 20% of her times at bat.

10. The First Bank pays 6% interest on savings.

11. The Earth's atmosphere is 21% oxygen.

12. 75% of Jason's allowance is spent on books.

WRITE ABOUT IT

13. Cut out a percent example from a newspaper or magazine. Write a paragraph explaining how the percent is used in your example.

Problem Solving Strategy: Using Percent

Sales tax is often paid on food at a restaurant. The sales tax is a percent of the cost of the meal.

Teresa is writing out this bill for a customer. Sales tax is 8%. Use a percent bar diagram to find the sales tax.

The sales tax is 8% of the cost of the meal. Label the percent bar.

$12.00 is 100% of the cost. Label the money bar. Let t represent the amount of sales tax she will charge.

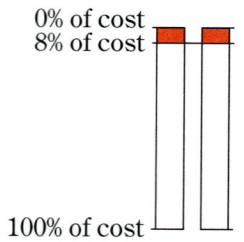

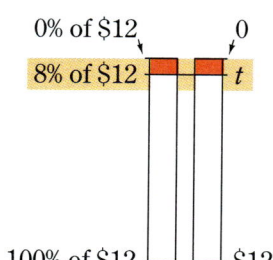

- What equation represents this problem?
- What decimal can you multiply the cost by to find the sales tax?
- Solve the problem.
- How can you check your answer?

GUIDED PRACTICE

Read the passage and answer the questions.

1. The cost of the food Zachary had at Doug's Diner was $8. The sales tax is 6%.
   a. What percent is the sales tax?
   b. How will you find 6% of the cost of the food?
   c. How much sales tax will Zachary pay?

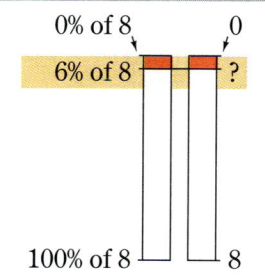

LESSON 10-13

PRACTICE

You may be expected to pay the waiter or waitress a tip. The tip is a percent of the cost of the meal. Use the menu to solve.

Star Diner
- Pizza — 8.50
- Oriental Chicken — 5.00
- Fish of the Day — 7.50
- Turkey Platter — 6.50
- Vegetable Platter — 2.95
- Beef Dinner — 9.00
- Salad — 3.00

Sales Tax 8%

2. Toni had the Beef Dinner. He decided to leave a 20% tip. What was the amount of the tip?

3. Tom had the Oriental Chicken. What was the sales tax?

4. Chris had the Fish of the Day and Bob had the Turkey Platter. They decided to leave a 15% tip. What was the amount of the tip?

5. Miles and Felicity had a pizza and 2 salads at the Star Diner. They left a 15% tip. What was the tip? What was the sales tax? What was the total cost of the meal?

6. James had a coupon for 10% off any one item on the menu at the Star Diner. What did he order if the amount he paid was $5.85?

CHOOSE Choose any strategy to solve. Use the menu above as needed. Use a calculator when needed.

7. At the end of the day Steve and Karen counted their tips. Karen had $2 more than Steve. Together they had $49. How much in tips did Steve make?

8. The owner of the Star Diner plans to raise the prices of the meals by 2% to cover rising costs. How much will she raise the price of a salad?

9. Luis needs a pizza ready for 5:00 P.M. He needs 30 min to prepare the pizza toppings. It takes 5 min to prepare the dough, 3 min to put the ingredients on the dough, and 20 min to bake the pizza. What time does he need to start?

10. A family's bill was $40.35. They decided to leave a 15% tip. About how much will the tip be?

CHAPTER 10   347

Mixed Review

Write the value of the underlined digit.

1. 7,842,130
2. 64,287
3. 1.74
4. 7.013
5. 14.29

Write in word form.

6. 7,804
7. 2,300,004
8. 67.5
9. 1.07
10. 3.084

Put the numbers in order from least to greatest.

11. $\frac{3}{4}, \frac{3}{5}, \frac{7}{8}$
12. $1\frac{7}{9}, 1\frac{1}{3}, \frac{11}{6}$
13. $\frac{8}{3}, \frac{30}{12}, \frac{11}{4}$

14. 58,047; 5,804; 58,074; 59,074
15. 0.58; 0.058; 0.85; 580.8

Find the answer.

16. 89.521 + 687
17. 375 × 88
18. 9,052 ÷ 73
19. 607.26 − 32.74
20. 72.48 × 8.5
21. 225 ÷ 7.5
22. 50.093 − 0.395
23. 7.4 + 15.023
24. 398.37 ÷ 7
25. 4.886 + 1.42
26. 8.5 × 50
27. 3 − 1.236
28. $\frac{1}{8} \div 4$
29. $\frac{3}{8} - \frac{1}{3}$
30. $12\frac{1}{5} + 8\frac{1}{4}$
31. $8 \times 1\frac{1}{2}$
32. $\frac{1}{3} \div \frac{2}{3}$
33. $1\frac{6}{7} \times 2\frac{1}{3}$
34. $9 - \frac{1}{6}$
35. $\frac{1}{2} + \frac{9}{10}$
36. $7\frac{1}{2} + 1\frac{5}{9}$
37. $4\frac{1}{2} \div 1\frac{3}{4}$
38. $9\frac{1}{2} - 8\frac{1}{3}$
39. $\frac{2}{7} \times 4\frac{1}{5}$

PROBLEM SOLVING

CHOOSE Choose estimation, calculator, mental math, or pencil and paper to solve.

40. Two thirds of the 9 planets are farther away from the Sun than Earth is. How many planets are closer to the Sun than Earth is?

41. Mars's diameter is about $\frac{1}{2}$ Earth's diameter. Earth's diameter is $\frac{1}{4}$ the diameter of Uranus. What fraction of the diameter of Uranus is the diameter of Mars?

42. Earth's diameter is about 8,000 mi. The Moon's diameter is about 0.25 times as large. About how large is the Moon's diameter?

43. The temperature on the Moon can reach 260°F. This is 3.25 times as hot as the temperature gets on Mars. How hot does it get on Mars?

44. Earth rotates every 24 h. One half of that time it is night. If you spend $\frac{2}{3}$ of the night asleep, how long do you sleep?

45. The Dogons of West Africa have used astronomy for 700 yr. They chose a leader when Jupiter (about 12 yr orbit) and Saturn (about 30 yr orbit) completed an orbit together. How often was that?

A person's weight is determined by a planet's gravity. Use the chart below to answer questions 46–50.

Gravity of Some Planets As Compared to Earth			
Mercury	$\frac{1}{3}$	Venus	$\frac{9}{10}$
Earth	1	Mars	$\frac{3}{8}$
Jupiter	$2\frac{3}{5}$	Saturn	$1\frac{1}{7}$

46. Marilyn weighs 90 lb on Earth. How much would she weigh on Venus?

47. Jerry weighs 91 lb on Earth. How much would he weigh on Saturn?

48. Rick would weigh 40 lb on Mercury. How much does he weigh on Earth?

49. Stephan would weigh 260 lb on Jupiter. How much would he weigh on Mars?

50. Jennifer weighs 77 lb on Earth. How much more would she weigh on Saturn than on Mercury?

CHAPTER CHECKUP

LANGUAGE & VOCABULARY

Use each of the following terms in a sentence to show that you understand the meaning.

1. ratio
2. percent
3. equal ratios

TEST

CONCEPTS

Write the ratio three ways in lowest terms. *(pages 320–321)*

1. 2 tables to 16 students

Are the ratios a proportion? Write = or ≠. *(pages 326–327)*

2. $\frac{3}{5}$ ▬ $\frac{9}{25}$

3. $\frac{400}{250}$ ▬ $\frac{8}{5}$

Write the percent. *(pages 334–335, 338–339)*

4. 27 out of 100
5. $\frac{4}{100}$

SKILLS

Find the unit rate. *(pages 322–323)*

6. 320 km in 4 h
7. $24 for 3 pizzas

Solve the proportion. *(pages 328–329)*

8. $\frac{4}{7} = \frac{n}{56}$
9. $\frac{n}{200} = \frac{6}{50}$
10. $\frac{16}{1} = \frac{n}{5}$

Write the percent as a decimal or the decimal as a percent. *(pages 336–337)*

11. 54%
12. 3%
13. 0.92
14. 0.6
15. 27.5%
16. 0.024
17. 0.07
18. 9.5%

Write the fraction as a percent. (pages 338–339)

19. $\frac{23}{100}$
20. $\frac{7}{25}$
21. $\frac{33}{60}$
22. $\frac{5}{8}$

Find the percent of the number. (pages 340–343)

23. 70% of 350
24. 42% of 86
25. 9% of 54

PROBLEM SOLVING

Solve. (pages 346–347)

26. The tip Arjay left for the waitress was 15% of the bill for his meal. The bill was $6.00.
    a. What percent tip did Arjay leave?
    b. How would you find the amount of the tip Arjay left?
    c. What is the amount of the tip Arjay left?

Solve. (pages 330–331, 346–347)

27. The Great Salt Lake in Utah is about 120 km long. Using the scale 1 cm to 10 km, what length would you draw the Great Salt Lake on a map?

28. Of the people who visit Ben's Hobby Shop, 70% make a purchase. If 350 people visit this shop on Friday, how many will probably make a purchase?

Write the answers in your learning log.
1. Explain how to find 50% of a number mentally.

2. Explain how a percent bar diagram can help you to estimate 77% of 365.

EXTRA PRACTICE

Write the ratio in lowest terms. *(pages 320–321)*

1. 3:30
2. 20 to 15

Write the ratio in three ways. *(pages 320–321)*

3. 40 players to 7 teams

Find the unit rate. *(pages 322–323)*

4. 36 songs on 3 tapes
5. $.84 for 4 bananas
6. 540 km in 6 h

Are the ratios a proportion? Write = or ≠. *(pages 326–327)*

7. $\frac{6}{1}$ ▬ $\frac{12}{6}$
8. $\frac{5}{9}$ ▬ $\frac{25}{45}$
9. $\frac{150}{200}$ ▬ $\frac{3}{4}$
10. $\frac{7}{8}$ ▬ $\frac{49}{64}$

Solve the proportion. *(pages 328–329)*

11. $\frac{1}{7} = \frac{n}{42}$
12. $\frac{90}{n} = \frac{5}{6}$
13. $\frac{n}{3} = \frac{9}{27}$
14. $\frac{12}{12} = \frac{16}{n}$
15. $\frac{n}{8} = \frac{25}{10}$

Write the percent. *(pages 334–335)*

16. 72 out of 100
17. 4 out of 100
18. 100 out of 100
19. $\frac{5}{100}$
20. $\frac{46}{100}$

Write the percent as a decimal. *(pages 336–337)*

21. 32%
22. 9%
23. 55.5%

Write the decimal as a percent. *(pages 336–337)*

24. 0.7
25. 0.96
26. 0.045

Write the fraction as a percent. *(pages 338–339)*

27. $\frac{13}{100}$
28. $\frac{6}{10}$
29. $\frac{23}{25}$
30. $\frac{6}{15}$
31. $\frac{3}{30}$
32. $\frac{44}{80}$

Find the percent of the number. *(pages 340–343)*

33. 20% of 75
34. 36% of 120
35. 6% of 36
36. 4.5% of 600

Solve. *(pages 330–331, 346–347)*

37. A rocket model was built using the scale of 2 cm to 5 m. If the height of the scale model is 14 cm, what is the actual height of the rocket?

38. The food the James's ordered for dinner totaled $24.50. Sales tax in their state is 5%. How much tax will be added to their bill?

ENRICHMENT

HOUSE PLANS

Before a house can be built, an architect or a builder must make a scale drawing of the layout of the house. Among other things, this drawing tells the builder where each room is located and what size each room is.

Follow the steps at right to make a scale drawing of your house or apartment.

- Measure the length and width of each room to the nearest $\frac{1}{2}$ ft.
- Choose a scale so that your drawing fits on your paper. You could try $\frac{1}{4}$ in. to 1 ft.
- Compute the dimensions of each room for your scale drawing.
- Using the dimensions you computed, draw each room in its proper location. If your home has more than one story, make a separate drawing for each story.

How *Tall* is the Flagpole?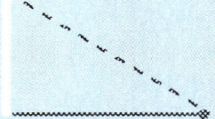

When two triangles are similar, the ratios of corresponding sides are equal. So you can find the missing length of a side by writing and solving a proportion.

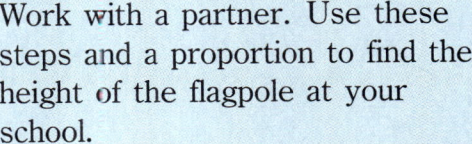

A flagpole and its shadow make two sides of a triangle.

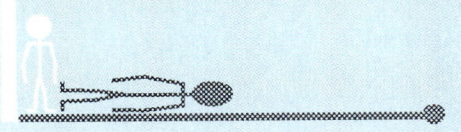

Work with a partner. Use these steps and a proportion to find the height of the flagpole at your school.

- Have your partner stand beside the flagpole that is casting a shadow. Find these lengths: your partner's height, your partner's shadow, and the flagpole's shadow.

- Use this proportion and your measurements to solve.

$$\frac{\text{partner's height}}{\text{partner' shadow}} = \frac{\text{flagpole's height}}{\text{flagpole's shadow}}$$

ENRICHMENT 353

CUMULATIVE REVIEW

Solve.

1. $5\frac{3}{4} + 4\frac{5}{6}$
   a. $9\frac{7}{12}$
   b. $9\frac{4}{5}$
   c. $10\frac{3}{4}$
   d. none of these

2. $8\frac{4}{5} - 5\frac{2}{3}$
   a. $2\frac{2}{15}$
   b. $3\frac{2}{15}$
   c. $3\frac{2}{2}$
   d. none of these

3. $9\frac{1}{8} - 4\frac{1}{2}$
   a. $4\frac{7}{8}$
   b. $5\frac{3}{8}$
   c. $4\frac{5}{8}$
   d. none of these

4. $4 \div \frac{3}{8}$
   a. $10\frac{2}{3}$
   b. $1\frac{1}{2}$
   c. $\frac{3}{32}$
   d. none of these

5. $\frac{5}{6} \div \frac{4}{9}$
   a. $\frac{10}{27}$
   b. $\frac{8}{15}$
   c. $\frac{5}{8}$
   d. none of these

6. $4\frac{1}{5} \div 1\frac{3}{4}$
   a. $4\frac{4}{15}$
   b. $2\frac{2}{5}$
   c. $7\frac{7}{20}$
   d. none of these

Choose the answer that best describes the figure.

7.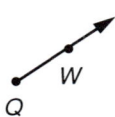
   a. $\overrightarrow{QW}$
   b. $\overline{WQ}$
   c. $\overrightarrow{WQ}$
   d. none of these

8.
   a. right angle
   b. obtuse angle
   c. acute angle
   d. none of these

9.
   a. congruent
   b. symmetrical
   c. similar
   d. none of these

Find the area. Use $\pi \approx 3.14$.

10. Circle:
    $r = 5$ yd
    a. 78.5 yd²
    b. 314 yd²
    c. 31.4 yd²
    d. none of these

11. Rectangle:
    $l = 4$ in., $w = 3.2$ in.
    a. 14.4 in.²
    b. 51.84 in.²
    c. 7.2 in.²
    d. none of these

12. Triangle:
    $b = 8$ ft, $h = 3$ ft
    a. 24 ft²
    b. 12 ft²
    c. 6 ft²
    d. none of these

PROBLEM SOLVING REVIEW

Remember the strategies and types of problems you have done so far. Solve.

Problem Solving Check List
- Choosing the operation
- Using estimation
- Making a table
- Using guess and check
- Using a graph
- Multistep problems

1. An all-terrain mountain bike can be bought for one payment of $175.99, or for 6 payments of $32.50 each.
   a. What is the bike's price with a single payment?
   b. How would you find the bike's total price using the installment plan?
   c. How much is saved by making a single payment?

2. In basketball, 1, 2, or 3 points can be scored with each basket. Rashida earned 8 points for her team. What are the different combinations of points she could have scored?

3. Each row of a theater seats 24 people. If 498 people attended a play and filled up each row in order, how many rows were completely filled?

4. It takes $2\frac{7}{8}$ yd of material to make each cheerleading uniform. About how much material is needed to make 18 uniforms?

5. It is 2,348 mi from Houston, Texas, to Seattle, Washington. About how many hours will it take to drive this distance at 55 mi/h?

6. Paperback books are on sale for $3 each and hardcover books for $7 each. Mr. Sung spent $32 at the book sale. How many of each type of book did he buy?

7. A round swimming pool has a diameter of 28 ft. To the nearest foot, how much fencing is needed to enclose the pool?

Use the graph to solve.

8. What fraction of the teens surveyed prefer playing sports or listening to music in their free time?

9. If 400 teens were surveyed, how many would probably name reading as their favorite activity?

Favorite Free-Time Activity of Teens

TECHNOLOGY

EAGER ESTIMATION

In the computer game "Making Sense of Percents," you win points for estimating percents. Do this pencil and paper activity to sharpen your estimation skills.

Work with a partner. Each of you chooses three of the percent problems below to estimate. You must make your estimate within the time limit. Points are given for each estimate that is no more than 5 greater than or 5 less than the exact answer. The player with the most points wins.

50 s 3 points	40 s 5 points	30 s 8 points
26% of 121	81% of 202	39% of 405
82% of 182	17% of 603	21% of 585

IS IT ENOUGH?

You have the choice of earning a weekly salary of $300, or $150 per week plus a percentage of sales. How much would you have to sell for the salary plus percentage to equal a straight salary?

1. 25% of all sales after the first $200
2. 2% of all sales
3. 8% of all sales up to $150; 12% of the next $150; 15% of all sales over $300

CHOICE PERCENT

Estimate the answer. Then choose the exact answer from the choices given in the box.

Use your calculator to check your work.

1. 18% of $326
2. 28% of $168
3. 53% of $258
4. 62% of $785
5. 22% of $984
6. 78% of $577
7. 32% of $648
8. 96% of $1,021

$216.48
$207.36
$58.68
$980.16
$47.04
$136.74
$450.06
$486.70

356   TECHNOLOGY

Statistics and Probability 11

DID YOU KNOW...?
Each year, over 22 million pounds of lobster are caught off the coast of Maine!

MAINE'S ANNUAL INCOME FROM SOME PRODUCTS

Products	Income (In Millions of Dollars)
Forest Products	1,320
Sand and Gravel	15
Potatoes	90
Poultry	22
Fishing	108
Eggs	95
Milk	95

MAINE AND ITS PRODUCTS

Make a graph using the data in the chart. In which representation, the chart or your graph, is it easier to tell which products bring in the most money to the state of Maine? Why?

The term "forest products" means paper and pulp products as well as lumber and wood products. Use an encyclopedia to find out how some forest products are made. Design a series of diagrams that shows how one of these products is made.

USING DATA
Collect
Organize
Describe
Predict

CHAPTER 11

Exploring Statistics

Mrs. Jiminez asked the students in her class how much time they spent watching television the day before. She recorded each student's time.

Time Spent Watching Television Yesterday

Girls						Boys			
$\frac{1}{2}$	$1\frac{1}{2}$	2	1	2	2	3	2	$1\frac{1}{2}$	$1\frac{1}{2}$
$2\frac{1}{2}$	2	$1\frac{1}{2}$	2	1	$\frac{1}{2}$	$2\frac{1}{2}$	3	2	2
2	$1\frac{1}{2}$	$1\frac{1}{2}$	$2\frac{1}{2}$	0		2	1	3	$2\frac{1}{2}$

Work in a group. Think about the following statements about the above data. Each one is false or gives an incorrect reason. Rewrite the statements so they are correct.

1. Since the total number of hours for each group is 26, on the average boys and girls watch about the same amount of television.

    > The **range** can be found by subtracting the least value in a data set from the greatest.

2. Girls watched from 0 to $2\frac{1}{2}$ h of television, a range of $2\frac{1}{2}$ hours. Boys watched from 1 to 3 hours, a range of 2 hours. So, since the range for girls is bigger, the girls must watch more television.

3. The **mode** for both boys and girls is 2 h. So, boys and girls watch about the same amount of television.

    > The **mode** is the number that occurs most often in a set of data.

4. The typical boy watches more television since the greatest number of hours for the boys is 3 and the greatest number of hours for the girls is $2\frac{1}{2}$.

Mrs. Jiminez suggested that the class make a tally chart to organize the data.

5. What do each of the tally marks mean?

6. How can you tell the range from a tally chart?

7. How can you tell the mode from a tally chart?

Look at the "shape" made by the numbers of tally marks in the chart.

8. Decide whether boys or girls watch more television.

9. Describe the reasons for your decision. Share with the class.

Now collect the same data about your own class. Have one person write each student's response about how much television he or she watched yesterday. Write the boys' and girls' responses in different charts.

10. Find the range of hours spent watching television for boys and for girls.

11. Find the mode of hours spent watching for boys and for girls.

12. Decide whether boys or girls in your class watch more television. Explain your answer.

13. Share your results with other groups. How do the results compare?

Mean and Median

Last month, the different members of the Wilson School Reading Club read these numbers of books: 9, 3, 7, 4, 9, 10. When interviewed for the school newspaper, Zahara reported that members of the club read about 7 books per month. She was reporting the average, or **mean**, number.

> **The mean is the sum of the items in a set of data divided by the number of items in the set.**

$$\frac{9 + 3 + 7 + 4 + 9 + 10}{6} = \frac{42}{6} = 7$$

Peter reported that the members of the club read about 8 books per month. He was reporting the **median**, or middle number, in the data set:

> **The median is the middle number in a set of data arranged in order.**

Order the data.   3, 4, 7, 9, 9, 10.

Find the middle   3, 4, 7, 9, 9, 10
of the data set.

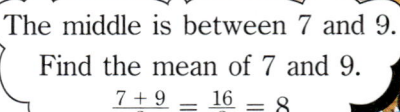

The middle is between 7 and 9.
Find the mean of 7 and 9.
$\frac{7 + 9}{2} = \frac{16}{2} = 8$

Joey read no books last month. Mrs. Cordara, the club leader, read 21 books.

CRITICAL THINKING Suppose this data, 0 and 21, was included in the data set above. Would the mean change? the median? Why or why not?

GUIDED PRACTICE

The 5 members of the Public Library Reading Club read these numbers of books last month: 4, 7, 2, 8, 9.

1. Explain how to find the mean.
2. Explain how to find the median.

3. **THINK ALOUD** How is finding the median different when there are an even number of items in the data set than when there are an odd number?

Suppose a sixth member had read 18 books last month.

4. Find the new mean.
5. Find the new median.

6. **THINK ALOUD** Which changed more, the mean or the median? Why?

PRACTICE

Mr. Hinkle asked members of the Current Events Club how many newspapers they read last week. The responses were:
4, 5, 5, 2, 4, 4

7. Find the mean, median, and the mode.
8. How do the mean, median, and mode compare?

Shani and Tim didn't read any newspapers last week.

9. Include their data in the set above. List the new data set.
10. What is the mean, median, and mode of the new data set?

11. a. Did the mean, the median, or the mode change?
    b. Explain why.

12. **CRITICAL THINKING** If you were reporting how many newspapers the members of the club read, which number would you use? Why?

PROBLEM SOLVING

The 5 members of the Public Library Reading Club reported the number of days they read the newspaper last week. Create a data set for the 5 members of the club in each of the following situations.

13. The median was 4.
14. The mode was 5.
15. The mean was 6.
16. The median was 4 and the mean was 5.

Problem Solving: Interpreting Graphs

In a recent year the ethnic background of teen-agers in the United States was tallied. The results are shown in the table.

Ethnic Background	Number
White	18 million
African American	3.5 million
Hispanic	2 million
Other	1.8 million

THINK ALOUD Are these numbers exact or estimates? Explain.

Data from a table can also be shown in a graph.

Pictograph

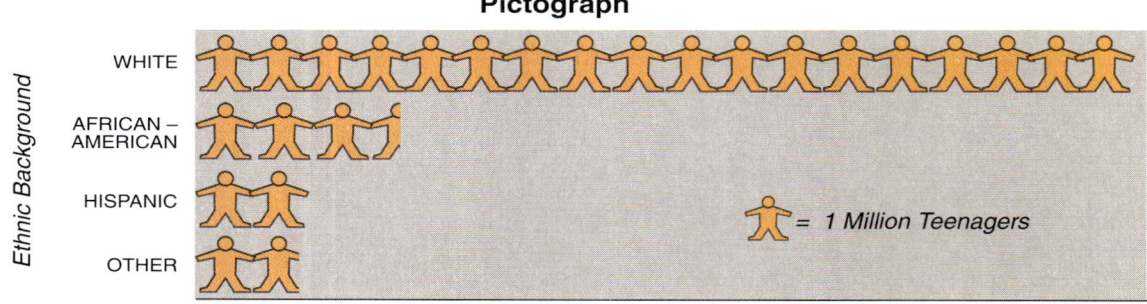

Bar graph

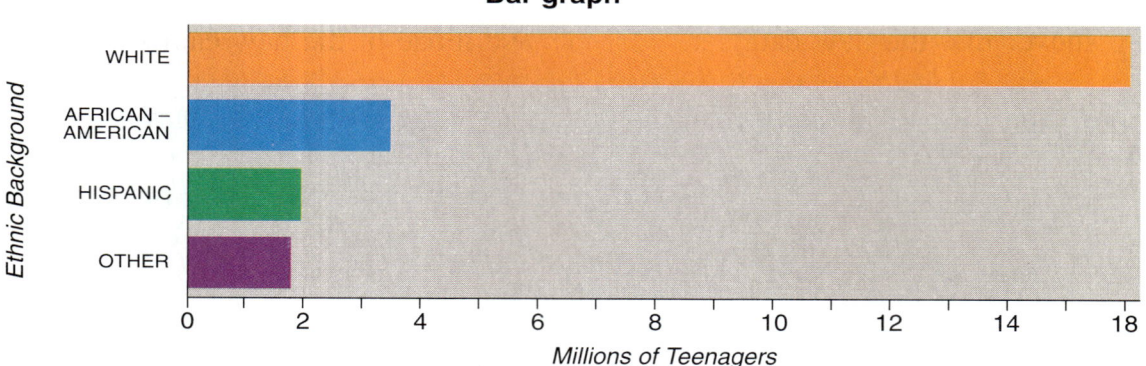

Tables and graphs represent information in different forms. Tables represent information through written numbers while graphs represent visually and with numbers.

THINK ALOUD What does it mean to say, "Graphs communicate visually"?

GUIDED PRACTICE

Use the table and graphs on page 362 to answer the questions.

1. **a.** In which type of representation is information found by just reading numbers?
   b. Which of the representations best communicates visually?
   c. About how many times more African American teen-agers were there than Hispanic teen-agers? Describe how you can estimate the answer from the pictograph, bar graph, and table.

PRACTICE

Use the table and two graphs below to answer Exercises 2–4.

Table

Families in the United States in a Recent Year			
Ethnic Background	White	African American	Hispanic
Number of Families	53 million	6 million	3 million

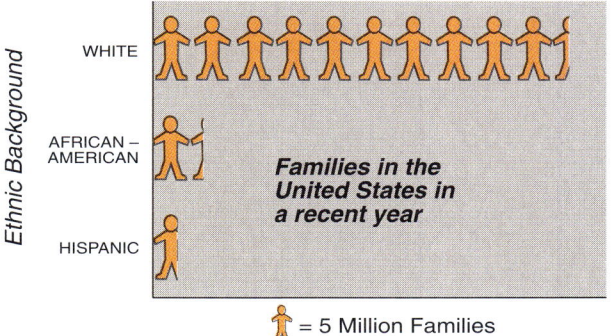

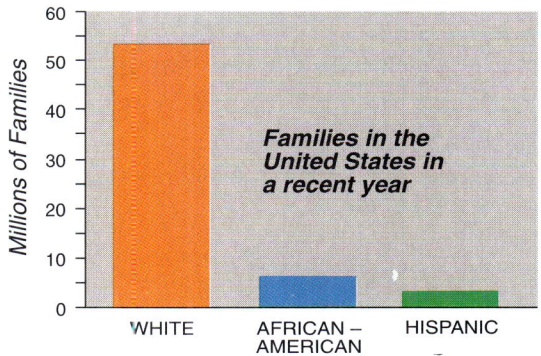

2. Describe how you would use each representation to estimate the number of Hispanic families.

3. About how many times more white families were there than African American families? Describe how you estimated this.

4. In which representations is the large difference between the number of white and the number of African American families communicated visually?

5. **CREATE YOUR OWN** Find a graph in a newspaper or magazine. Write four questions that can be solved by reading the graph. Trade with a partner and solve.

Making Double Bar Graphs

Which do you choose to read daily—a book, a magazine, or a newspaper? The bar graph below shows information about what these two groups of students read for pleasure almost daily.

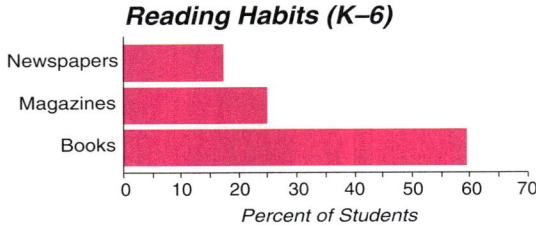

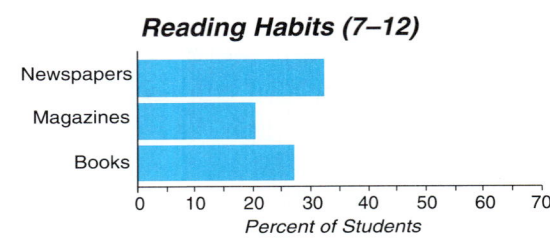

You can combine the two bar graphs above into one **double bar graph**. Use separate kinds of bars for K-6 and 7-12.

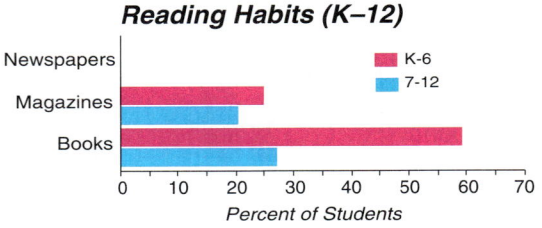

THINK ALOUD Is it easier to compare the data about the two groups of students with the single bar graphs or the double bar graph? Why?

CRITICAL THINKING Explain why an individual student could be represented more than once on the graph.

GUIDED PRACTICE

1. Copy and complete the double bar graph shown above.

2. What is the most popular type of pleasure reading for the K-6 students? for the 7-12 students?

3. What is the least favored type of pleasure reading for K-6? for 7-12?

4. **THINK ALOUD** By looking at the graph, can you tell the number of students surveyed? Explain your answer.

PRACTICE

Write *true* or *false* using the double bar graph. Explain.

5. More students like to read books than like to read magazines.

6. As students get older they like to read different things than when they were younger.

7. K-6 students like magazines more than they like newspapers.

364   LESSON 11-4

Use the two bar graphs below.

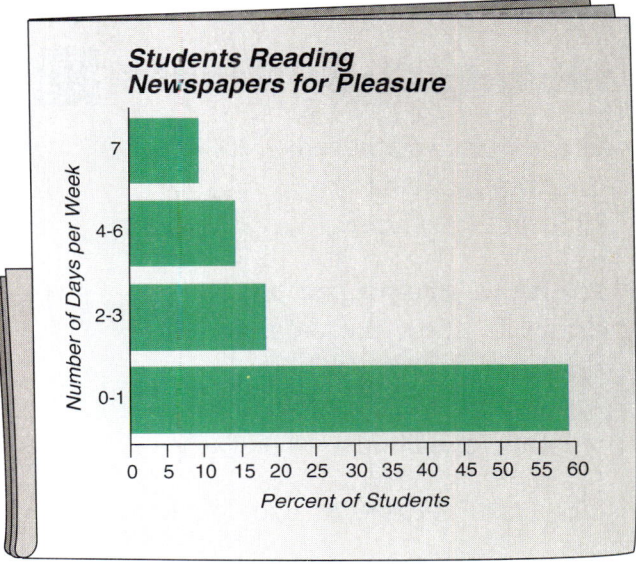

8. Make a double bar graph comparing how often students read books and newspapers for pleasure.

9. **IN YOUR WORDS** What comparisons can you make from your graph?

The graphs above represent results from about 20,000 surveys. Estimate the number of students who

10. read a book for pleasure once a week or less.

11. read a newspaper for pleasure every day.

Write *true*, *false*, or *can't tell*. Explain your answer.

12. Over half the students read a newspaper 2 or more times a week.

13. Over half the students read a book 2 or more times a week.

14. More students read books for pleasure than read newspapers for pleasure.

15. Every student who reads the newspaper every day also reads a book every day.

PROBLEM SOLVING

16. Use the information from the table at the right and the single bar graph about newspapers above to make a double bar graph.

17. **CREATE YOUR OWN** Write three questions about the data in the bar graph you made in exercise 16. Trade with a partner and solve.

Magazine Reading	
Once a week or less	45%
2 to 3 days a week	32%
4 to 6 days a week	15%
Every day	8%

CHAPTER 11 365

Making Circle Graphs

Erick's class gathered the weather data at right in November.

A **circle graph** can be drawn to show the data. Shaded sections will represent each type of weather conditions.

The class decided to use this key.

(*gray*) cloudy
(*blue*) partly cloudy
(*yellow*) sunny

Start with a circle divided into ten equal sections.

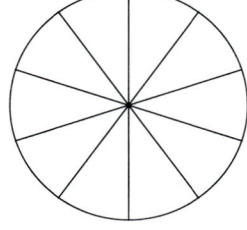

THINK ALOUD What does the whole circle represent? Each section of the circle represents how many days?

Here are the first two steps to make a circle graph for the data:

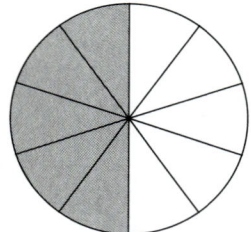

Step 1 To show 15 of the cloudy days, shade five sections gray. Why?

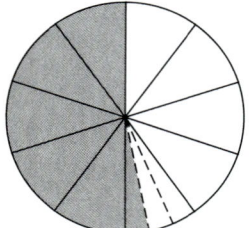

Step 2 Julie needs to show one more cloudy day. What part of a section is one day? Shade $\frac{1}{3}$ of the next section gray.

=== GUIDED PRACTICE ===

Answer the questions to complete the circle graph started above.

1. How many sections should be shaded to show the partly cloudy days?

2. How many sections should be shaded to show the sunny days?

3. Finish shading the graph with blue and yellow.

PRACTICE

Julie's class also kept track of the amount of rain for each day in November.

Use a circle divided into ten equal sections to make a circle graph for the data at right.

Type of Day	Tally	Frequency
No Rain	|||| |||| ||||		16
Drizzle (less than 1 cm)	|||| ||||	9
Rain (between 1 and 5 cm)	||||	4
Heavy Rain (over 5 cm)			1

4. How many days does each section represent?

5. How many colors will be needed to represent the different rain conditions?

6. How many sections should be shaded to show the days with drizzle? with heavy rain? with rain? with no rain?

7. Shade the circle graph to show the four rain conditions.

8. IN YOUR WORDS How could you make the graph using a circle divided into six sections?

Use your completed circle graph to answer these questions.

9. Which of the rain conditions occurred about half the time?

10. Which condition occurred about half as often as drizzle conditions?

PROBLEM SOLVING

Complete the table. You may wish to use a calculator. Then use the information from the table to answer exercises 11 and 12.

11. Use a circle divided into ten equal sections. What percent is represented by each section?

12. Use estimation to make a circle graph showing the percent of days with each type of weather.

August Weather			
type of day	number of days	fraction of days	percent of days
partly cloudy	14	$\frac{14}{31}$	45%
cloudy	7	$\frac{7}{31}$	23%
sunny	10	$\frac{10}{31}$	

CHAPTER 11   367

Making Line Graphs

The chart shows the Fahrenheit temperature recorded every 2 h for 1 day.

Time	Temp.	Time	Temp.
12:00 midnight	31°	12:00 noon	44°
2:00 A.M.	28°	2:00 P.M.	48°
4:00 A.M.	26°	4:00 P.M.	42°
6:00 A.M.	27°	6:00 P.M.	34°
8:00 A.M.	36°	8:00 P.M.	26°
10:00 A.M.	41°	10:00 P.M.	29°

You can show changes in temperature over time with a line graph.

Step 1 Draw the horizontal and vertical axis on grid paper.

Step 2 Label the horizontal axis for the times 12:00 midnight to 10 P.M.

Step 3 Label the vertical axis for the temperatures from 0° to 50°. Let each division represent 2°.

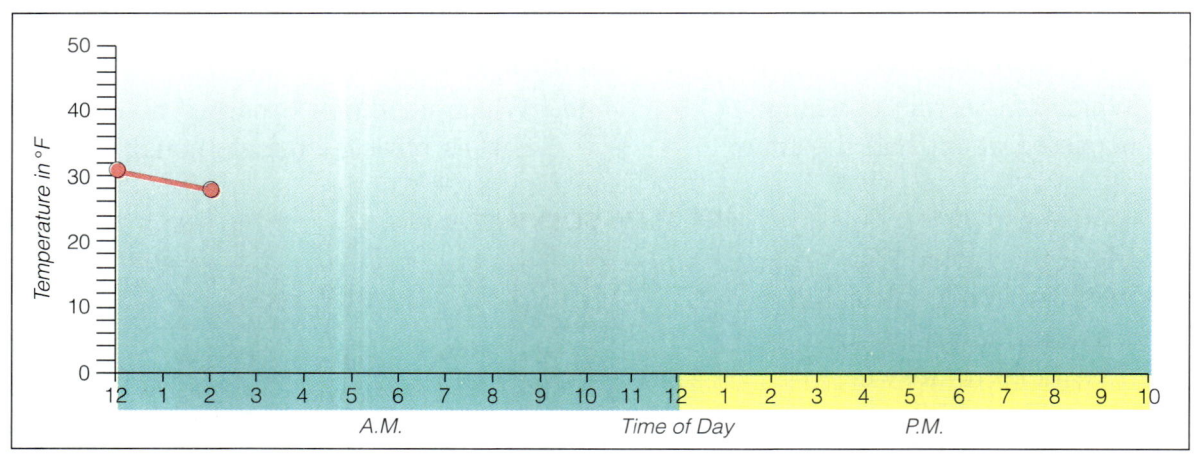

THINK ALOUD Why do you think the scale for the vertical axis is divided into 2° increments and not 10°?

Step 4 Start at 12:00 midnight and locate 31° on the vertical axis. Make a dot up from 12:00 midnight and across from 31.

Step 5 Continue plotting the temperatures and join the dots to make the line graph.

368 LESSON 11-6

GUIDED PRACTICE

Use the graph you made from the data on page 368.

1. a. What was the temperature at 12:00 noon?
   b. How would you locate 1:00 P.M. on the graph?
   c. Estimate the temperature at 1:00 P.M.

PRACTICE

Time	Temp. in °F	Time	Temp. in °F
12:00 midnight	14	12:00 noon	26
1:00 A.M.	12	1:00 P.M.	28
2:00 A.M.	11	2:00 P.M.	27
3:00 A.M.	11	3:00 P.M.	25
4:00 A.M.	13	4:00 P.M.	17
5:00 A.M.	12	5:00 P.M.	14
6:00 A.M.	14	6:00 P.M.	14
7:00 A.M.	13	7:00 P.M.	11
8:00 A.M.	19	8:00 P.M.	8
9:00 A.M.	23	9:00 P.M.	7
10:00 A.M.	25	10:00 P.M.	5
11:00 A.M.	24	11:00 P.M.	5

2. Make a line graph of the above time and temperature data. Put the times on the horizontal axis. Put the temperatures on the vertical axis. Let each division represent 2°.

Estimate the temperature at the given times.

3. 1:30 A.M.
4. 8:30 A.M.
5. 3:30 P.M.
6. 9:45 P.M.

7. In which 1-hour periods did the temperature stay the same?

8. In which 1-hour period did the temperature increase the most?

9. **IN YOUR WORDS** Look at the *shape* of the graph. Describe it.

10. Would you expect a similar shape for a summer day? Explain.

PROBLEM SOLVING

11. Explain how to use the line graph that you made for the time and temperature data above to estimate the time the sun rose and the sun set. How could you predict the temperature at 1:00 A.M. the next day?

12. **CREATE YOUR OWN** Use newspaper, television, or radio weather reports to collect the daily high temperature of your city or any city. Collect data for two weeks. Draw a line graph to represent your data.

CHAPTER 11   369

More Than Meets the Eye

Problem Solving: Misleading Graphs

READ ABOUT IT

A magazine published a series of articles on how America has changed from the 1920s to the 1980s. One article included this graph showing what percent of Americans lived in urban areas (cities or suburbs) during each decade.

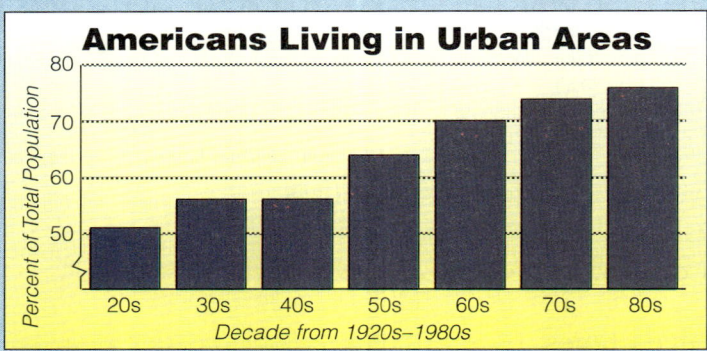

TALK ABOUT IT

Use the graph to answer Exercises 1–5.

1. What trend does the graph show?

2. The bar for the 1980s is about how many times as long as the bar for the 1920s?

3. What percent of the population lived in urban areas in the 1920s? in the 1980s?

4. The percent for the 1980s is about how many times as large as the percent for the 1920s?

5. Why might someone say that this graph is misleading?

Compare this graph of the same data to the graph on page 370.

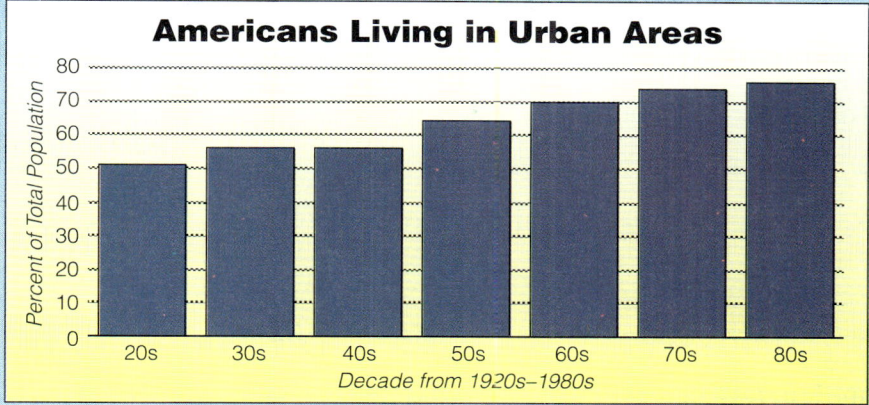

7. How is the vertical scale different on the two graphs? What effect does that have on the graph?

6. Which graph gives a more accurate impression of how fast the urban population is growing? Explain.

8. Why might the author have chosen to use a graph which exaggerates how fast the urban population is growing?

9. It is important that a reader realize when the vertical scale does not start at 0, since just comparing the lengths of the bars is misleading. How was the vertical scale of the first graph marked to show that there are skipped numbers?

WRITE ABOUT IT

Suppose this list shows the estimated population of your town over the last five years: 35,000; 36,000; 38,000; 39,500; 42,000

Work with a partner. Each of you should draw one of the following graphs of this data:

- a graph that might be used to attract new business to the town.
- a graph that might be used in an ad for a "countrylike" housing development in the town.

10. Exchange graphs. Write a paragraph describing the impression given by your partner's graph. Does the graph accomplish the desired purpose? Is it misleading?

MIDCHAPTER CHECKUP

LANGUAGE & VOCABULARY

What type of graph would you use to
  a. to show a change over time?
  b. to compare 2 sets of data?
  c. to represent fractional parts of a whole?

QUICK QUIZ

Find each of the following for this set of numbers.
(pages 358–361)   **7, 12, 9, 8, 6, 9, 7, 4, 8, 7**

1. the range
2. the mode
3. the median
4. the mean

Use the line graph. *(pages 368–369)*

5. Explain why the vertical scale was set up with each division representing 2 in.

6. Between which two months was there the greatest increase in the amount of snow on the ground?

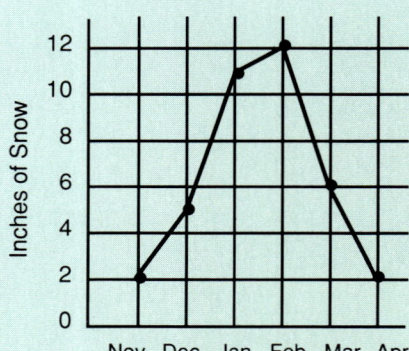

Use the tally chart. *(pages 364–367)*

7. How many students were surveyed?
8. Make a bar graph of the data.
9. On a circle graph with 6 equal sections, how many sections would you shade to show students with 4 siblings?

Number of Siblings	Tally
0	IIII
1	TTTT TTTT
2	TTTT
3	III
4	II

10. Use the tally chart on siblings and the bar graph you drew for Exercise 8. *(pages 362–363)*
    a. What number of siblings was most common?
    b. Which representation best communicates visually?
    c. Use each representation to estimate how many times as many students have 1 sibling as have 3 siblings.

Write the answer in your learning log.

1. The mean of a group of numbers is 95. What do you know about the numbers?

2. The median of a group of numbers is 16. What do you know about the numbers?

DID YOU KNOW . . . ? The family of John Adams was important in Massachusetts history. A family tree is a kind of tree diagram. Choose a famous person in your state. Draw this person's family tree showing at least three generations.

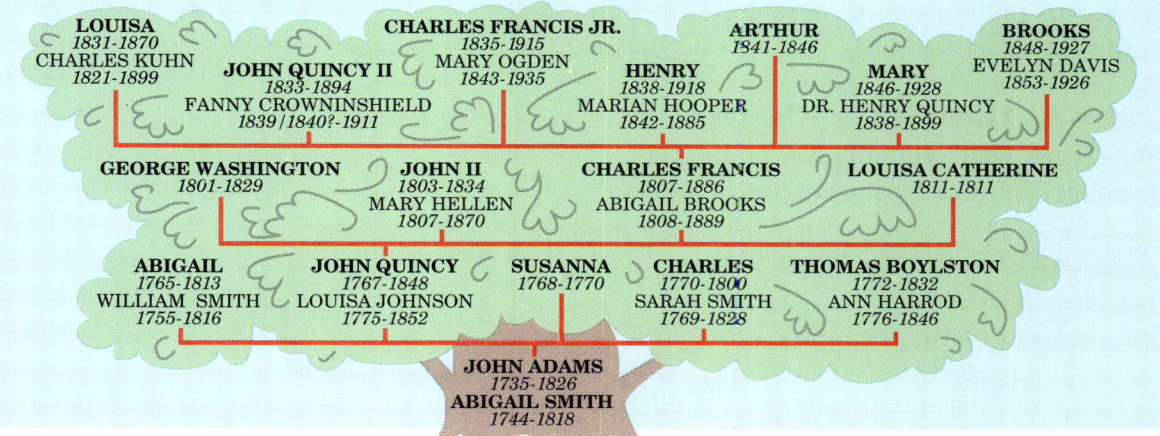

Do you think the students in the classes surveyed for the graphs of siblings are typical? Collect data on the number of siblings from about 25 students in your school.

- What are the mean, median, and mode for your data?
- Use your data to make a bar graph. Does your graph have about the same shape as the one you drew for Exercise 8.

Sampling and Predicting

A national weekend magazine published a readers' questionnaire on fitness.

One of the items is shown here:

How Many Sit-ups Can You Do?

Number of Sit-ups	Percent of Response
None	17%
1-10	39%
11-25	21%
26-50	15%
More than 50	8%

The **population** is the entire group of people about which information is wanted.

The population for this survey is all the readers of the magazine.

CRITICAL THINKING Why might the magazine want information about the numbers of sit-ups its readers can do?

The **sample** is the part of the population actually used to gain information about the population.

The sample for this survey is made up of readers who sent in questionnaires.

Sometimes the sample is very similar to the population. Then it is called **representative**. A representative sample can be used to make predictions about the population.

GUIDED PRACTICE

1. Suppose the sample above is representative. Decide whether each prediction can reasonably be made based on the data above. Write *yes* or *no*. Explain your answer.

    a. Over half the readers of the weekend magazine can do ten or fewer sit-ups.

    b. Over half of all office workers can do ten or fewer sit-ups.

2. **THINK ALOUD** Why might this sample not be representative? (*Hint*: Do you think everyone responded?)

3. A school newspaper wants information about the sports interests of the students in the school. The editors pull names out of a hat to question 100 students.
   a. What is the population?
   b. What is the sample?
   c. Do you think the sample is likely to be representative? Explain.

PRACTICE

Mrs. Johung used this chart in the parent newsletter to tell about all the sixth graders in the school. The information was actually from only one of the sixth grade classes.

4. What is the population?
5. What is the sample?

Suppose the sample is representative. Decide whether each prediction can be reasonably made based on the data above. Write *yes* or *no*. Explain your answer.

6. Over half of all the sixth graders at the school can do more than 25 sit-ups in 1 min.

7. Over half of all the teachers at the school can do more than 25 sit-ups in 1 min.

8. Over half of all the students at the school can do more than 25 sit-ups in 1 min.

9. Over half of all the sixth graders in the United States can do more than 25 sit-ups in 1 min.

PROBLEM SOLVING

Which situation is more likely to give a representative sample? Choose *a* or *b*. Explain.

10. A middle school principal wants to know what playground equipment the students want.
    a. Ask all students in one sixth grade class.
    b. Ask the fifth student on each alphabetical attendance list.

11. The gymnastics coach wants to know what color the team prefers for new uniforms.
    a. Interview the first 5 students to enter the lunchroom.
    b. Interview the first 5 team members arriving for practice.

CHAPTER 11 375

Measuring Chance

"I don't have my homework because my pet rock ate it."

"I don't have my homework because my dog ate it."

There is no chance that this story is true.

There is some chance that this may be true.

The possibility that something may or may not happen is called **chance**. **Probability (P)** is a measure of chance. We use 0, 1, or any number between 0 and 1 to show the chance of some event happening.

A probability of 0 means there is *no chance* for an event to happen.

A probability of 1 means that an event is *certain* to happen.

A probability between 0 and 1 means an event may or may not happen.

THINK ALOUD Which event above has a probability of 0?

The diagram below helps you picture probabilities.

| Impossible | Maybe/Maybe not | Certain |

0 ⊢―――――――――――――――――――⊣ 1

THINK ALOUD What is an event that you would place at 0? at 1? between 0 and 1? closer to 1 than 0?

GUIDED PRACTICE

1. Think of tossing one number cube. The 6 sides are labeled 1–6. Draw a probability line showing 0 to 1. Place each event on the line according to its chances of occurring.

   a. landing on 10
   b. landing on 4
   c. landing on an even number
   d. landing on a number less than 7

THINK ALOUD Can an event have a probability of $\frac{4}{3}$? Explain.

PRACTICE

Choose *0*, *1*, or *between 0 and 1* to describe the probability.

2. Someone you know may be in the Olympics someday.

3. If you jump up, you will come back down again.

4. The temperature will be 200°F tomorrow.

5. The next new student at school will have brown eyes.

Choose the event that has the greater chance of happening.

6. a. You will be asleep at 2 A.M. tomorrow.
   b. You will be asleep at 2 P.M. tomorrow.

7. a. You will find a penny on the street tomorrow.
   b. You will find a penny on the street some day.

NUMBER SENSE Could the number represent a probability? Write *yes* or *no*. Explain.

8. 35%   9. $\frac{3}{2}$ 10. 2.6   11. 0.013   12. $\frac{7}{8}$ 13. 5

CRITICAL THINKING Give the probability of the event happening. Then change the sentence so that the probability is between 0 and 1.

14. You will find books in the library.

15. You will have Math this week.

16. There will be two Sundays in this week.

17. The coins in your pocket will be worth more than $100.

PROBLEM SOLVING

Use the pets pictured to answer. Write **0**, **1**, or **between 0 and 1**.

18. What is the chance that all the pets can fly?

19. What is the chance that some of the pets have 4 legs?

20. **CREATE YOUR OWN** Use the information to describe an event that is sure to happen and one that may happen.

Equally Likely

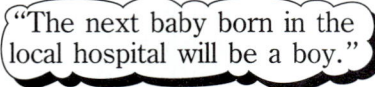 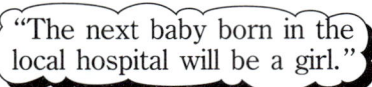 

"The next baby born in the local hospital will be a boy."

"The next baby born in the local hospital will be a girl."

Which statement has a greater chance of happening?

Since the number of boys and girls born is about the same, the birth of a boy and the birth of a girl are said to be **equally likely**.

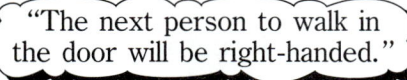

"The next person to walk in the door will be right-handed."

"The next person to walk in the door will be left-handed."

Since there are more right-handed people than left-handed people, these events are **not equally likely**.

The possible results of an experiment are called **outcomes**. The outcomes for both spinners below are red, blue, and green.

Are the chances of landing on blue the same as the chances of landing on red?

The parts shown are the same size so the chances of landing on each color are equally likely.

The chance of landing on blue is greater since the blue section is larger. The outcomes are not equally likely.

GUIDED PRACTICE

1. Are the chances **equally likely**? Explain your answer.
   a. That the next person you meet will have brown hair or red hair?
   b. That the next coin you toss will land on heads or tails?

2. **THINK ALOUD** Give a situation with equally likely outcomes.

PRACTICE

Use a number cube with sides labeled 2, 3, 4, 5, 6, and 7.
Are these events **equally likely**? Explain.

3. rolling a 2 or a 6.   4. rolling an odd or even number?

5. rolling a prime or composite number?

Look at these spinners:

A.   B.

Are the chances **equally likely**? Explain.

6. Spinning a red or blue
   a. on Spinner A   b. on Spinner B

7. Spinning a red or yellow
   a. on Spinner A   b. on Spinner B

Are the chances equally likely? Explain your answer.

8. Without looking, you will pick a muffin or an apple when you reach into your lunch bag with your eyes closed.

9. Without looking, you will pick Gary, Roberta, or Manuel if each name is on a 3 by 5 file card inside a bag.

10. Someone's last name will begin with S or U.

A dog has six puppies. Four are black and two are brown. Three are male and three are female.

11. Describe two outcomes that are equally likely.

12. Describe two outcomes that are not equally likely.

PROBLEM SOLVING

Is the statement reasonable based on the survey?
Write *yes* or *no*. Explain.

13. The chances are equally likely that a student does or does not read a magazine almost every day.

Survey Results

- $\frac{3}{4}$ of all students have a radio in their bedroom.
- $\frac{1}{4}$ of all students read a magazine almost daily.
- $\frac{1}{2}$ of all students read for pleasure almost daily.

14. About $\frac{1}{4}$ of all students do not have a radio in their bedroom.

15. If 200 students were surveyed, about 50 of them said no to the question "Do you read a magazine almost every day?"

16. The survey shows that out of the 40,000 voters in a town, about 20,000 read for pleasure almost every day.

CHAPTER 11   379

Exploring Probability

Mrs. Jelenik ran an experiment with her class. She put 6 marbles in a bag. Some were red, some were blue, and some were yellow.

Each student took 4 turns drawing a marble out of the bag without looking. After each draw the marble was returned to the bag and the bag was shaken.

1. Did each marble have an equally likely chance of being drawn? Explain why or why not.

The table shows the results of the experiment.

Color	How often
red	33
blue	36
yellow	31
	Total: 100

The **experimental probability** of drawing a yellow marble is:

$$\frac{\text{the number of times a yellow marble was drawn}}{\text{the total number of marbles drawn}} = \frac{31}{100} = 0.31 \text{ or } 31\%$$

Let's find the experimental probability of drawing a red marble.

2. a. How many times was a red marble drawn?
   b. What was the total number of trials?
   c. What is the experimental probability of drawing a red marble?
   d. Write the experimental probability of drawing a red as a decimal; as a percent.

3. Give the experimental probability of drawing a blue marble.
   a. as a fraction
   b. as a decimal
   c. as a percent

4. Use the results of the experiment to try to guess how many marbles of each color were in the bag.

The picture shows what was actually in the bag.

The **actual probability** of an event is found by looking at the possible outcomes.

Probability (P) of an event = $\frac{\text{number of ways the event can occur}}{\text{total possible outcomes}}$

The actual probability of drawing a red marble is: $P(\text{red}) = \frac{\text{the number of red marbles}}{\text{the total number of marbles}} = \frac{2}{6}$ or $\frac{1}{3}$

5. Give the actual probability of drawing a blue marble
   a. as a fraction in lowest terms. **b.** as a decimal rounded to hundredths.

6. Were the experimental probabilities and actual probabilities close in value?

Students spun the spinner at the right 1,000 times. Red resulted 267 times, blue 249 times, and yellow 484 times.

7. Write the experimental probability as a fraction and a decimal:
   a. of spinning a red.
   b. of spinning a blue.
   c. of spinning a yellow.

8. Write the actual probability as a fraction and a decimal:
   a. of spinning a red.
   b. of spinning a blue.
   c. of spinning a yellow.

9. Is the experimental probability close to the actual probability?

10. CRITICAL THINKING In exercises 7 and 8, what could you do to make the experimental probability closer to the actual probability?

The more data you gather, the more likely your experimental probability will be close to the actual probability.

For some events, such as a paper cup landing upside down, you cannot find the actual probability.

11. IN YOUR WORDS Find the experimental probability of a paper cup landing upside down. Explain.

Independent Events

Ken is one of ten students who have signed up to raise the flag at school for a one week period. One student and an alternate will be selected by drawing names from a hat. The probability of being selected on the first draw is $\frac{1}{10}$.

Alicia's name was drawn from the hat. She is going to be the flag raiser for the week.

Now Ken is one of the 9 students whose names are still in the hat. The probability that he will be selected increases to $\frac{1}{9}$ because Alicia's name was not returned to the hat.

> **When one event causes the probability of another to change, they are called dependent events.**

The selection of the alternate is dependent on the selection of the representative. Why?

There is a new drawing each week. All ten names are put back in the hat.

THINK ALOUD   What is the probability that Ken's name will be chosen the first week? the second week? the third week? Did the probability change? Explain.

> **For independent events, one event does not affect the chances of the other event.**

The selection of the second week is independent from the selection for the first week. Why?

GUIDED PRACTICE

Read the passage and Events A and B. Write *independent* or *dependent*. Explain your choice.

Each week 1 student from a class of 32 is Recycling Monitor. The student is selected by drawing a name from a hat.

1. A: In the first week the drawing is made with all names in a hat.
   B: The first Recycling Monitor's name is not included in week 2.

2. A: In the 3rd week the drawing is made with all names in a hat.
   B: In the 4th week the drawing is made with all names in a hat.

3. **THINK ALOUD** Explain how independent and dependent events are different.

PRACTICE

Name Events A and B as *dependent* or *independent*.

4. A: You turn on the television.
   B: A picture will appear on the screen.

5. A: The Bears win the Super Bowl this year.
   B: The Cubs will win the World Series next year.

6. A: You roll a number cube and the top number is 4.
   B: You roll the number cube again and the top number is 4.

7. A: Carol gets an A on her spelling quiz.
   B: Carol will get an A on her spelling unit test.

CRITICAL THINKING Read Event A. Then use the cards to write Event B.

8. Event A: All cards are in the bag. A 4 is drawn. Write an independent event for Event B.

9. Event A: All cards are in the bag. A 5 is drawn. Write a dependent event for Event B.

PROBLEM SOLVING

The Volunteer Club fills helper requests by selecting member's names from a hat.

10. Mrs. Abrams needs 2 helpers to shovel snow today at 10 A.M. Will the selection of the second helper be dependent or independent on the selection of the first helper? Explain.

11. **IN YOUR WORDS** Mrs. Jones needs one helper on the 15th and one helper on the 30th of this month. How could the selections of two helpers be made independently?

Sample Space

How many different slacks and shirt outfits is it possible to make? You can find out by combining each color of slacks with each color of shirt.

Slacks: brown, navy
Shirts: white, pink, yellow

A list of all possible outcomes is called the **sample space**.

Here are some ways to find the sample space.

You can make a table.

	white	pink	yellow
brown	brown, white	brown, pink	brown, yellow
navy	navy, white	navy, pink	navy, yellow

You can also make a tree diagram:

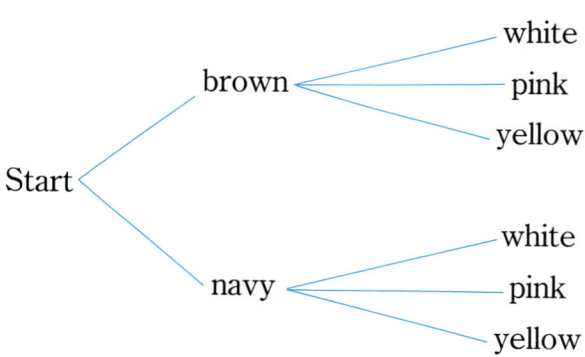

Sample Space
brown, white
brown, pink
brown, yellow

navy, white
navy, pink
navy, yellow

THINK ALOUD Describe what part of the table gives you the sample space. Is the sample space the same list as in the tree diagram?

CRITICAL THINKING How can you use multiplication to find the number of all possible outcomes? Does multiplication help you list the sample space? Why or why not?

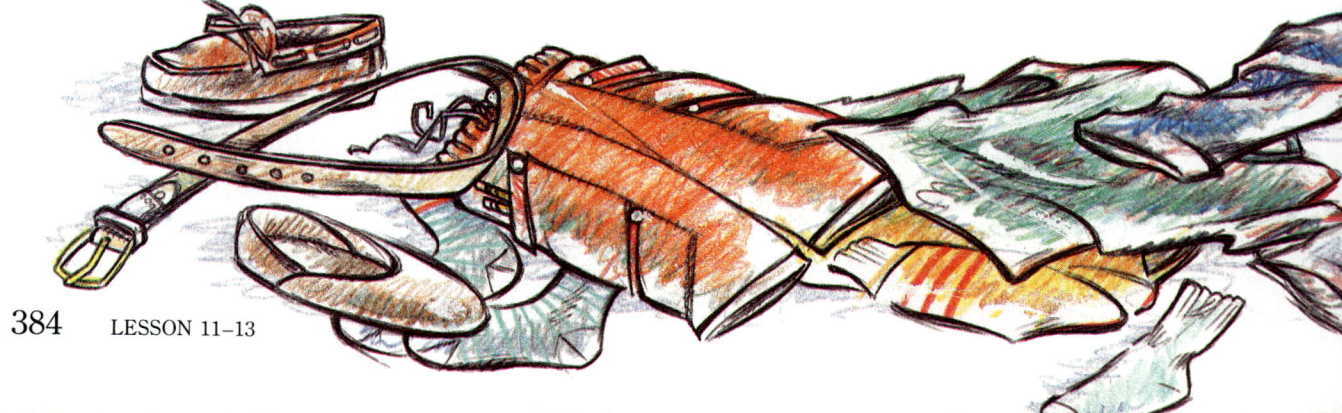

LESSON 11-13

GUIDED PRACTICE

1. Use the tree diagram to list the sample space.

2. How many possible combinations are there?

3. Suppose you added tan shorts to the list of items. Draw a table to find the sample space.

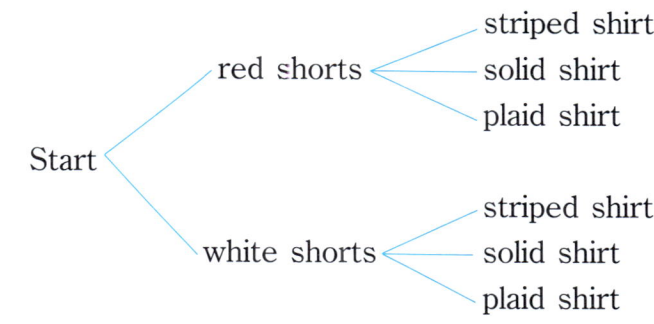

PRACTICE

Ayo has blue jeans and black jeans. His sweatshirts are yellow, purple, green, and red.

4. Use a tree diagram or table to list the sample space of possible outfits.

5. How many possible outfits are there?

Add a white sweatshirt to Ayo's list above.

6. What is the sample space now?

7. How many possible outfits does Ayo have now?

8. How many more new outfits did he get by adding the white sweatshirt?

PROBLEM SOLVING

Jane's mother will buy her six new items of clothing to make coordinated outfits. How many different outfits can she make from

9. 3 pairs of slacks, 3 blouses?

10. 2 shirts, 4 pairs of slacks?

11. 1 skirt, 5 sweaters?

12. **CREATE YOUR OWN** Show how you can make 12 different outfits from a combined total of 7 shirts and slacks.

CHAPTER 11   385

Exploring Combinations, Permutations

Mrs. McFarland is advisor to a pen pal club. When the pen pals finally met, each of the 4 members shook hands with all of the other members.

You can make an organized list to find the number of handshakes that took place.

You can also draw a diagram to find the number of handshakes.

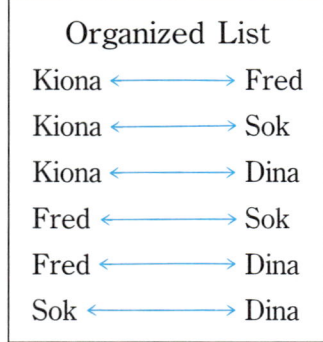

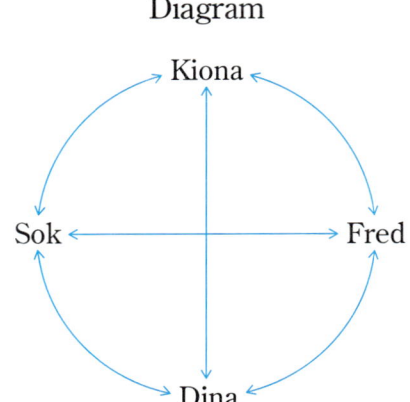

1. How many handshakes for 4 pen pals?

Suppose there were 5 pen pals in the club.

2. How many handshakes would take place? Use an organized list or draw a diagram to find out.

Each of the 4 members of the club is required to send a letter to all the other members.

3. How many letters were sent? Use an organized list or a diagram.

4. **THINK ALOUD** Did you expect the number of handshakes and the number of letters to be the same for 4 students? Were they? Explain why or why not.

> Kiona shaking hands with Fred is the same as Fred shaking hands with Kiona. The order does not matter. Count 1 handshake.

> Kiona sending a letter to Fred is not the same as Fred sending a letter to Kiona. The order of their names is important. Count 2 letters sent.

Combinations refer to the ways of arranging objects when the order of the objects does not matter.

Permutations refer to the arrangements of objects when the order of the objects does matter.

386   LESSON 11–14

CRITICAL THINKING Think of the list of handshakes and the list of letters sent. Which list shows combinations? Which list shows permutations? Explain.

Work with a group to decide if the list of events is a combination or a permutation.

5. At a party with 6 people, enough games were played so that each student was the partner of every other student once.
   a. How many games were played? Make an organized list or draw a diagram.
   b. Does it matter in which order the names are listed? Explain.
   c. Is this situation a permutation or combination?

6. Another game required each of 6 students to give every other student a different secret password.
   a. How can you find the total number of passwords used for 6 students?
   b. Is this situation a permutation or combination? Explain.

7. **CREATE YOUR OWN** Work with a group. Write a word problem that uses combinations to solve. Then write one that uses permutations. Trade with other groups. Solve.

Investigating Pendulums

Have you ever seen a clock like this? It works because of the swinging of a hanging weight called a pendulum.

Will the big swing of a pendulum take a longer time to go back and forth than a smaller one? Does the length of the pendulum rod affect the time to make one swing?

Galileo, a famous scientist in the 1600s, asked these questions.

Work in a small group.

1. Get a string 40 in. long. Tie a small object to one end of the string. Also, find a clock with a second hand.

   Tape the end of the string to the edge of a table so the pendulum can swing freely.

2. Let the pendulum make 6 full small swings, as shown in the picture. Record the total time for the 6 swings in seconds. Then figure out the average time for 1 full swing. Do this again for two other sizes of swings.

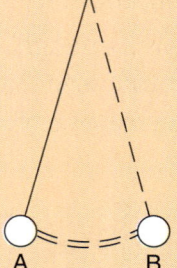
small swing

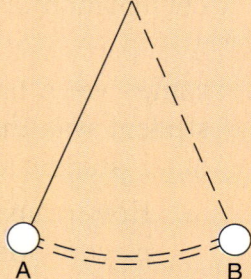
larger swing

One full swing goes from A to B, then back to A.

Size of Swing	Time (seconds) 6 Swings	Time (seconds) 1 Swing
Small		
Medium		
Large		

3. Explain your findings in Exercise 2. Does the size of the swing affect the time it takes to make 1 swing?

4. Does shortening a pendulum string affect the time it takes to make 1 swing? Find out. Gather information.

Length of String (inches)	Time (seconds) 6 Swings	Time (seconds) 1 Swing
40 in.		
30 in.		
20 in.		
10 in.		

What is your conclusion?

5. How does the weight of the pendulum affect the time of a swing? Discuss your strategy to solve this problem, then solve it.

Because a pendulum swings with a regular beat, it has been used by clock makers for over 300 years. The pendulum swings and the workings inside the clock tick out the seconds, one by one.

Mixed Review

Round to the place of the underlined digit.

1. 8̲7,024
2. 6̲,984
3. 8.71̲4
4. 0.02̲8
5. 7̲.076
6. 13.9̲7
7. $8\frac{2}{\underline{5}}$
8. $6\frac{5}{\underline{7}}$

Write in standard form.

9. two billion, six thousand
10. seven and two thousandths
11. 6 + 0.7 + 0.05 + 0.004
12. 5 + 0.8 + 0.009

Compare. Write >, <, or =.

13. 7.5 ▨ 7.46
14. 0.85 ▨ 0.850
15. 3.174 ▨ 3.147
16. $\frac{7}{8}$ ▨ $\frac{4}{5}$
17. $1\frac{5}{9}$ ▨ $1\frac{2}{3}$
18. $2\frac{8}{10}$ ▨ $2\frac{3}{4}$

Find the answer.

19. $5\frac{1}{8} - 3\frac{2}{3}$
20. $\frac{11}{12} + \frac{5}{6}$
21. $2\frac{1}{4} \times 1\frac{3}{5}$
22. $\frac{5}{6} \div 10$
23. $\frac{7}{12} - \frac{1}{3}$
24. $2\frac{2}{5} \div \frac{1}{3}$
25. $6\frac{1}{2} + 2\frac{4}{5}$
26. $1\frac{7}{8} \times 4$
27. 14 − 1.079
28. 1.4 × 50
29. 25.92 ÷ 3.6
30. 0.85 ÷ 2.5
31. $10.01 − $9.98
32. 0.35 × 0.7
33. 4.2 ÷ 0.6
34. 24.2 × 505
35. 8.71 + 20
36. $2\frac{2}{3} = \frac{n}{9}$
37. 25% of 40
38. 75% of 60
39. 1.96 + 0.174 + 15 + 1.09
40. 8.2 + 0.174 + 1.74 + 9

PROBLEM SOLVING

CHOOSE Choose estimation, calculator, mental math, or pencil and paper to solve.

41. The U.S. formally obtained Florida from Spain in 1821. Twenty-four years later Florida became a state. When did Florida become a state?

42. Florida's 1,350-mile coastline is the longest of any state. Along the Atlantic Ocean, it covers 580 mi. How long is Florida's coastline along the Gulf of Mexico?

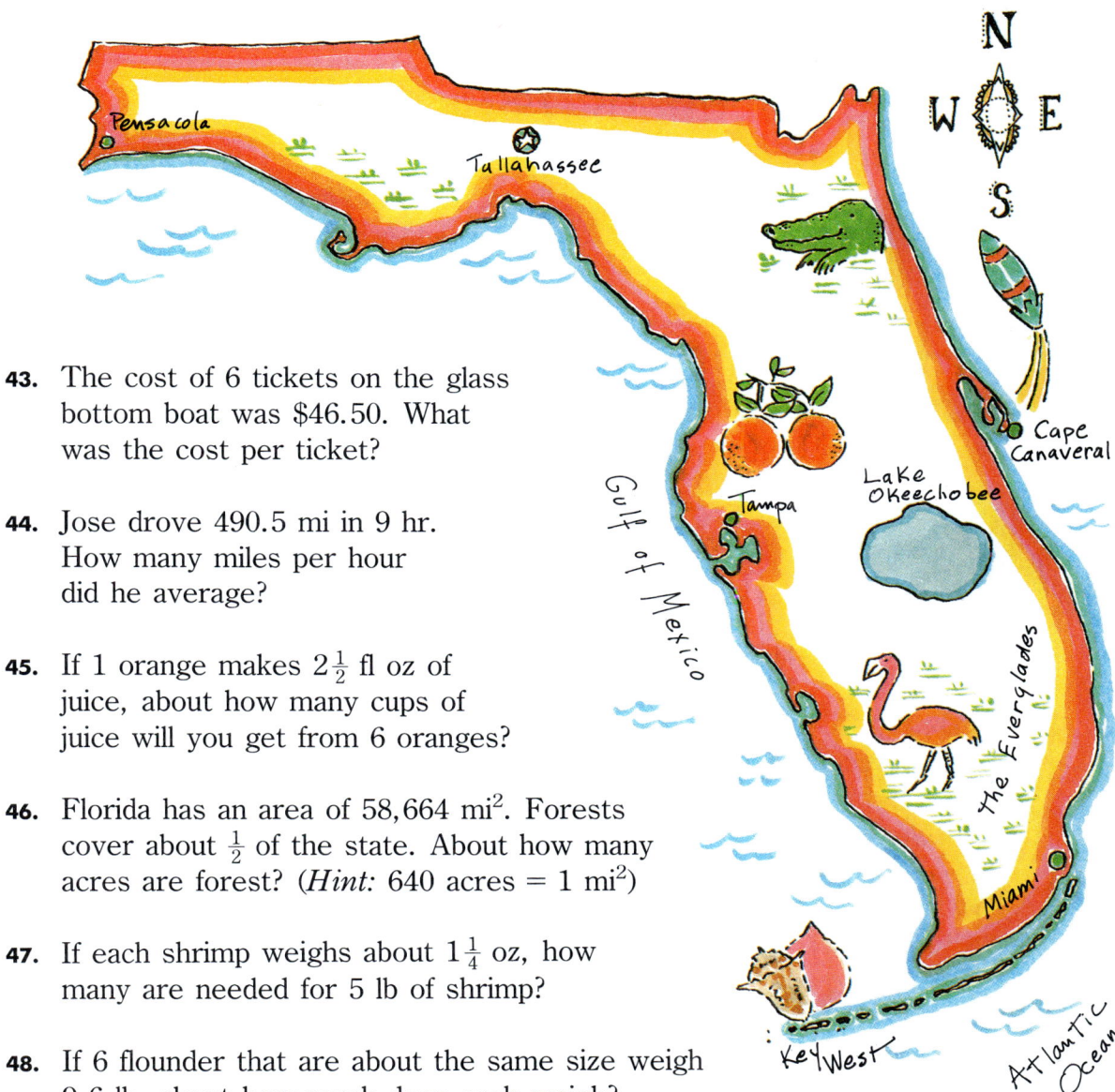

43. The cost of 6 tickets on the glass bottom boat was $46.50. What was the cost per ticket?

44. Jose drove 490.5 mi in 9 hr. How many miles per hour did he average?

45. If 1 orange makes $2\frac{1}{2}$ fl oz of juice, about how many cups of juice will you get from 6 oranges?

46. Florida has an area of 58,664 mi². Forests cover about $\frac{1}{2}$ of the state. About how many acres are forest? (*Hint:* 640 acres = 1 mi²)

47. If each shrimp weighs about $1\frac{1}{4}$ oz, how many are needed for 5 lb of shrimp?

48. If 6 flounder that are about the same size weigh 9.6 lb, about how much does each weigh?

49. Kala lives 650 mi from Orlando. If she drives 55 mi/h, about how long will it take her to get from her home to Orlando?

50. The dolphin show lasts $\frac{3}{4}$ h. The whale show lasts $\frac{1}{3}$ h. Which show is longer? How many minutes longer is it?

51. Alligators grow at a rate of about 1 ft per year. A $5\frac{1}{2}$-ft long alligator was $\frac{3}{4}$ ft long when newly hatched. How old is the alligator?

52. In 1930 the population of Florida was 1.5 million. By 1985 it reached 11.4 million. By how much did the population increase?

CHAPTER CHECKUP

LANGUAGE & VOCABULARY

Write *true* or *false*. Explain.

a. If an event is very unlikely to occur, it can have a probability less than zero.

b. If two events are dependent, then when one event occurs, the other event must occur.

TEST

CONCEPTS

Both bar graphs show the same class. *(pages 364–367)*

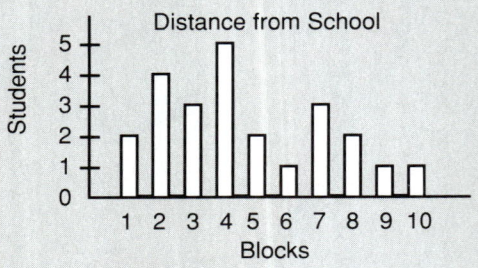
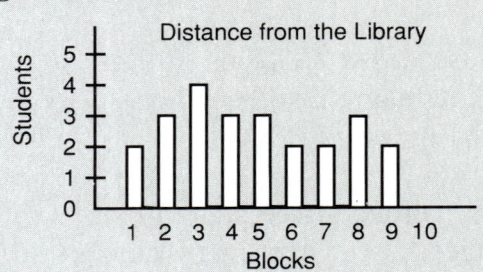

1. Do more students live at least 5 blocks from school or less than 5 blocks from school?

2. Why might you want to make a double bar graph of the data?

3. Could you make a circle graph of the data on distances from the library? Explain.

Name the population and the sample. *(pages 374–375)*

4. To find out if sixth grade students were using the library, the librarian asked questions of every tenth sixth grader who came into the library one week.

SKILLS

Use the chart at right. *(pages 368–369)*

5. Make a line graph of the data.

6. Between which 2 months was there the greatest decrease?

Harry's Bank Balance			
Jan.	$38	Feb.	$29
Mar.	$42	Apr.	$61
May	$53	June	$74
July	$74	Aug.	$85

Use **79, 85, 82, 85, 89** to solve. *(pages 358–361)*

7. Find the range.
8. Find the mode.
9. Find the median.
10. Find the mean.

Use a number cube labeled 1 through 6. *(pages 378–381)*

11. What is the actual probability of rolling
    a. a 4?
    b. a 7?

12. Are the chances equally likely of rolling a number greater than 4 or a number less than 4? Explain.

Find the sample space. *(pages 384–385)*

13. A restaurant serves 2 sandwiches and 3 drinks. How many combinations can you make?

PROBLEM SOLVING

Solve. *(pages 360–363, 370–371)*

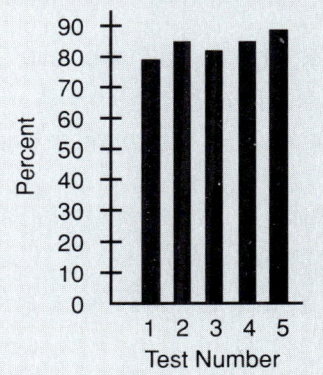

14. Use the graph of Yolanda's test scores.
    a. How many times did her scores drop?
    b. Describe how Yolanda's scores changed.
    c. Redraw the graph to give the impression that her scores improved drastically.

15. What is Yolanda's median test score?

16. School records show how many students are enrolled in each grade. How might school officials graph this data to plan how supplies will be shared among the grade levels?

Write the answer in your learning log.

1. Describe an event that has a probability of 1.
2. Explain what is meant by experimental probability.

EXTRA PRACTICE

Use the bar graphs below. *(pages 360–361, 364–367)*

Meals Served One Week

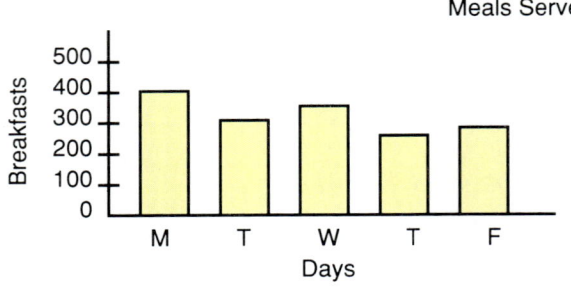 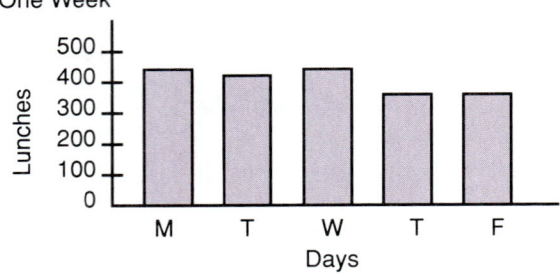

1. Estimate the number of lunches served each day. Find the mean.

2. On what day was the median number of breakfasts served?

3. Make a double bar graph of the data.

4. About 1,600 breakfasts were served during the week. In a circle graph with 8 sections, how many sections would be used to represent the breakfasts served on Tuesday?

Name the population and the sample. *(pages 374–375)*

5. To find out how the sixth graders like the school lunch, a lunchroom worker asks all students in one class.

Choose *0, 1,* or *between 0 and 1* to describe the probability. *(pages 376–377)*

6. You will be famous one day.

7. Everyone you meet will be blond.

Are the events dependent? Explain. *(pages 382–383)*

8. A. You spin a spinner and it stops on 6.
   B. You spin the same spinner and it stops on 5.

Find the number of combinations. *(pages 384–385)*

9. 3 coats and 5 hats

10. 3 main courses and 3 desserts

Answer the questions. *(pages 362–363, 370–371)*

11. Rajiv did research on the populations of major cities around the world. How should he represent his data
    a. to show as exact numbers as possible?
    b. to show how the sizes compare?
    c. to exaggerate the differences in size?

ENRICHMENT

Probability Around Us

For the next several days, go on a probability hunt. Listen and look for examples of the idea of chance. You might hear them in conversation, read them in a magazine, or see them on television. Whenever you do find one, write it down. Then answer these questions about each statement.

- Did the statement claim an event was certain, impossible, or between certainty and impossibility?
- Did you agree with the statement?
- Did the outcome of the event (if it has been completed) agree with the claim?

Share your findings with others in your class. Discuss whether the probabilities that were given were reasonable. Must the outcome of an event agree with the probability in order for the probability to have been reasonable?

Sample Space at Home

Have you ever thought that you do not have enough clothes or that there is not enough interesting food in the house? You might be surprised by the actual number of outfits or meals you can make from what you have at home.

Find each of the following sample spaces.

- the number of outfits you can make if an outfit consists of a pair of pants or a skirt together with a shirt or blouse.
- the number of meals you can make if a meal consists of a main dish together with a salad or vegetable.

Are the results what you expected?

Could you really use all the combinations you found? Explain.

CUMULATIVE REVIEW

Find the ratio or rate.

1. ☐☐☐ : ○○○○
   a. 4:3
   b. 3:4
   c. 3:7
   d. none of these

2. ★★★★ : ☐☐☐☐☐
   a. $\frac{4}{5}$
   b. 5:4
   c. 4 to 9
   d. none of these

3. 30 pages in 15 min
   a. 2 min per page
   b. 15 min per page
   c. 2 pages/min
   d. none of these

Solve for x.

4. $\frac{1}{8} = \frac{x}{32}$
   a. 16
   b. 8
   c. 4
   d. none of these

5. $\frac{2}{5} = \frac{x}{35}$
   a. 7
   b. 9
   c. 15
   d. none of these

6. $\frac{9}{4} = \frac{27}{x}$
   a. 3
   b. 12
   c. 14
   d. none of these

Write as a percent.

7. 0.82
   a. 82%
   b. 18%
   c. 8.2%
   d. none of these

8. 0.06
   a. 6%
   b. 60%
   c. 0.6%
   d. none of these

9. $\frac{7}{20}$
   a. 7%
   b. 20%
   c. 35%
   d. none of these

Find the percent of the number.

10. 30% of 60
    a. 30
    b. 18
    c. 50
    d. none of these

11. 25% of 200
    a. 100
    b. 75
    c. 50
    d. none of these

12. 90% of 70
    a. 63
    b. 160
    c. 630
    d. none of these

Find the circumference to the nearest whole number.

13. diameter = 6 cm
    a. 9 cm
    b. 15 cm
    c. 19 cm
    d. none of these

14. radius = 15 m
    a. 47 m
    b. 94 m
    c. 141 m
    d. none of these

15. radius = 32 m
    a. 201 m
    b. 101 m
    c. 64 m
    d. none of these

PROBLEM SOLVING REVIEW

Remember the strategies and types of problems you have had so far. Solve.

Problem Solving Check List
- Choosing the operation
- Using guess and check
- Using proportion
- Too much information
- Too little information

1. A new dome formed inside Mt. St. Helens following its eruption in 1980. The dome is about 270 m high. This is about $\frac{2}{3}$ the height of the World Trade Center in New York.
   a. About how high is the dome?
   b. What proportion could you use to find the height of the World Trade Center?
   c. About how high is the World Trade Center?

2. Sonia and Leslie together weighed 105 kg. When Sonia and Karen got on the scale, it showed 112 kg. How much does Sonia weigh?

3. Math is the favorite subject of 3 out of every 7 students in a class. If 12 students prefer math, how many students are in the class?

4. What is the greatest product that can be obtained by using the digits 3, 4, 5, 6, and 7 in the expression ▦▦▦ × ▦▦ if each digit is used only once?

5. Malkah runs a bagel delivery service. One weekend she delivered 18 dozen bagels. How much did she earn for the weekend if she charges $1.50 per delivery?

6. A balance scale has a rock on one side and the following weights on the other: 2 kg, 500 g, 200 g, and 50 g. Another 25 g is needed to balance the rock. What is the mass of the rock in grams?

7. Walter has promised that he will save $.25 out of every dollar he earns from his after-school job. If Walter earns $50 and keeps his promise, how much will he save?

8. Charley has saved $50 to buy presents. He buys 2 books, each of which cost $13.95, and a magazine for $1.95. How much change will he get if he pays with two twenty-dollar bills?

9. A triangular plot of land has an area of 100,000 ft². The plot was cut from a rectangular plot with the same base and height (length and width). What was the area of the rectangular plot?

TECHNOLOGY

IT'S PROBABLE

In the computer game "Probable Urnings," teams guess how many black and white balls are in an Urn. This activity will sharpen your probability skills.

You need 10 of the same objects of each of two different colors and a paper bag.

Work in teams of two. One partner puts 10 of the objects in the bag. The ratio of one color object to the other color object may change each time the game is played. Take turns preparing the bag and guessing.

The other partner draws from the bag 2 objects at random at a time. The objects are put back after each draw. Repeat. On the fifth draw, guess the number of each color object in the bag. Continue drawing and guessing two more times.

Number of Draws and Guesses	Points
5	3
6	2
7	1

In Albuquerque, New Mexico, the average monthly temperature for January is 34.8°F. The average monthly temperature for July is 78.8°F.

Find the daily temperature for your city for the last month. Was your city more like Albuquerque in July, or in January?

HOW HIGH IS UP

The chart shows the 6 highest elevations in the U.S. and Canada. Estimate which list has the highest mean. Then find the exact means. Find the difference between the means.

U.S.	Feet	Canada	Feet
Mt. McKinley	20,320	Mt. Logan	19,850
Mt. Whitney	14,494	Mt. Luciana	17,147
Mt. Elbert	14,433	Mt. King	16,971
Mt. Ranier	14,410	Mt. Steele	16,644
Gannett Peak	13,804	Mt. Wood	15,885
Mauna Kea	13,796	Mt. Walsh	14,780

Geometry and Measurement 12

DID YOU KNOW...?
One full-grown electric eel can produce enough electricity to light up ten 40-watt lightbulbs.

LENGTHS OF SOME TROPICAL FISH

Fish	Length (in.)
Stoplight Parrotfish	24
Imperial Angelfish	12
Spotted Goatfish	10
Saddleback Butterfly Fish	6
Clown Anemone	2

To find out how many tropical fish can safely live in a fish tank of a certain size, think:

The total length of the fish should be equal to or less than the number of gallons of water in the tank.

Research capacities of tropical fish tanks by calling or visiting a pet store. Plan several combinations of the fish listed above that could safely live in one of the tanks.

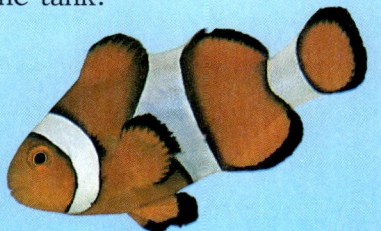

USING DATA
Collect
Organize
Describe
Predict

CHAPTER 12   399

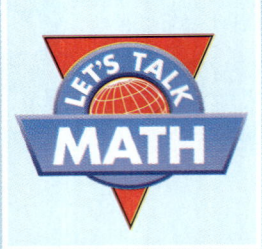

Shaping Our World

Geometry and Bridges

Did you ever cross a stream on a log? If you did, you traveled on the simplest type of span bridge.

Span Bridges

Arch bridges of brick and stone were built by many early cultures, including the Chinese, Greeks, Egyptians, and Romans.

Arch Bridges

Above left: Shinkyo Bridge, Japan
Above right: Ganter Bridge, Switzerland, 1980
Below left: Bridge over Colorado River in Utah
Below right: Richmond Bridge, Tasmania, 1823, the oldest existing bridge in Australia

Suspension bridges have roads or walkways suspended from cables.

Suspension Bridges

Left: Verrazano Narrows Bridge, New York, 1964, one of the world's longest suspension bridges
Right: Straw bridge over Apurimac River in Peru, rebuilt each year

TALK ABOUT IT

Bridges contain many geometric figures.
What types of figures do you see on these bridges?

1. span bridges
2. arch bridges
3. suspension bridges
4. Do you see differences between the older and newer bridges? Explain.
5. Describe a distance around your home or school that is about the length of the Verrazano Narrows Bridge—4,261 ft.

Arch bridges can have one of the three basic shapes shown at the right.

6. Describe how the three types would be different if you were driving on them.
7. Which types of arch bridges are shown on page 400?

WRITE ABOUT IT

8. Engineers use mathematics to design bridges. What are some of the mathematics skills that engineers need to use?

CHAPTER 12   401

Exploring Prisms and Pyramids

The first glassmaking factory in America was built in 1608 in the colony of Jamestown, Virginia. Today some glassmaking factories make see-through cases used to display delicate statues. The cases are in the shape of **space figures**. Here are some of the shapes.

Figure A

Figure B

rectangular prism

triangular prism

Figure C

Figure D

rectangular pyramid

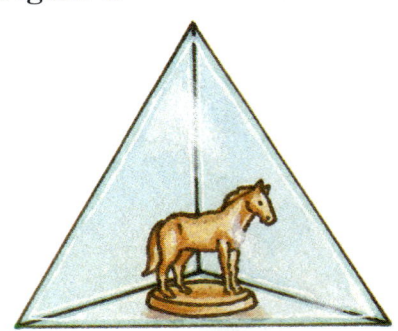

triangular pyramid

1. Each glass case sits on a wooden stand. What is the shape of the stand for each of the figures above?

2. The flat surfaces of a space figure are called **faces**. What types of polygons are used as faces in each figure above?

402   LESSON 12-2

Copy the table at right. Fill it in as you answer Exercises 3–6.

3. Use Figures A–D on page 402.
   a. Fill in the numbers of faces.
   b. Two faces meet at a line segment called an **edge**. Fill in the numbers of edges.
   c. Two edges meet a **vertex**. Fill in the numbers of vertexes.

	Number of Each		
	Face (F)	Edge (E)	Vertex (V)
A			
B			
C			
D			
E			

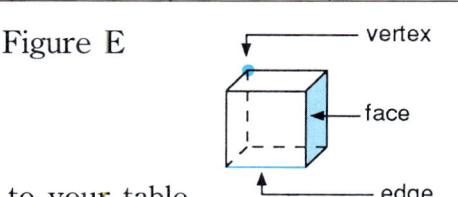

Figure E

4. A **cube** is a special rectangular prism in which all faces are squares. Fill in the row for Figure E.

5. Find other prisms or pyramids. Add them to your table.

6. **CRITICAL THINKING** Use the data in your table. Describe a relationship among the number of faces, edges, and vertexes. (*Hint*: Think about adding and subtracting.)

Use the patterns at the right for Exercises 7–13.

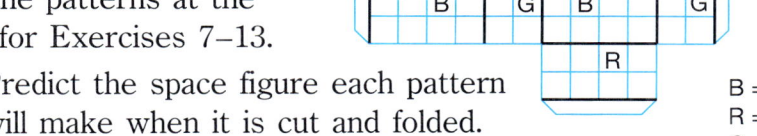

7. Predict the space figure each pattern will make when it is cut and folded.

B = Blue
R = Red
G = Grey

Use one-centimeter grid paper. Draw and cut out the patterns. Color them as shown.

8. Describe which faces of the rectangular prism are congruent (have the same size and shape).

9. Fold and tape the rectangular prism. Where are the congruent faces?

10. If all the faces of the rectangular prism were congruent, what kind of space figure would it be?

11. Which faces of the rectangular pyramid are congruent?

12. Fold and tape the pyramid. Where are the congruent faces?

13. Can a pyramid have all faces congruent? If so, what type of pyramid is it?

Exploring Surface Area

Alva is making a box to house her shell collection. She is drawing and cutting out a pattern for a rectangular prism.

To find the amount of cardboard to buy, Alva needs to find the sum of the areas of the faces, or the **surface area**, of the box.

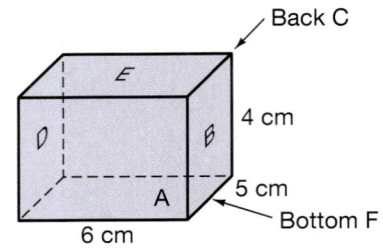

You can find the area of face A by counting squares to get 24 cm². You can also use this formula:

area (A) = length (l) × width (w)
$l = 6$ cm $w = 4$ cm
$A = 6 \times 4$, or 24 cm²

Both ways are correct.

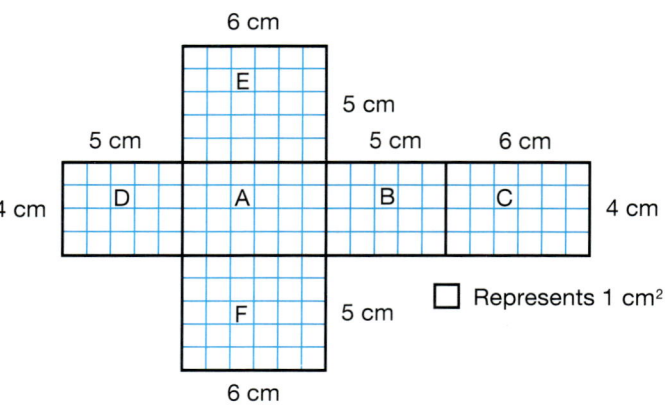

1. Find the areas of faces B–F on the pattern. Copy and complete the table at right.

2. Find the surface area by adding the areas of the faces.

3. **IN YOUR WORDS** Study the table you made. Is there an easier way to find the surface area? If so, describe it.

Face	Area	Face	Area
A	24 cm²	C	
B		D	
E		F	

Draw the pattern of the cube at right on grid paper. Cut it out. Tape the cube together.

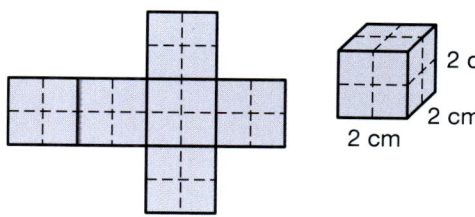

4. **IN YOUR WORDS** What is an easy way to find the surface area of a cube? Find the surface area of the cube you made.

Find the surface area. Use a calculator when needed.

5.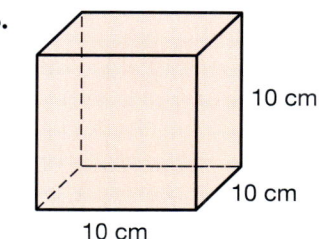

6.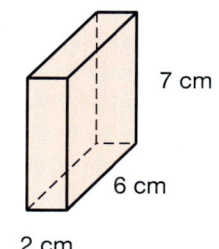

7.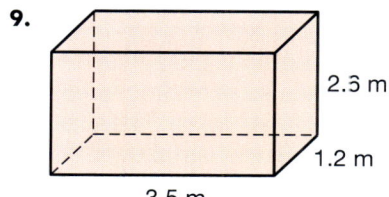

8.

9.

10.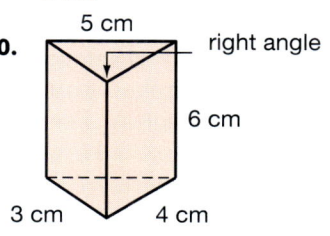

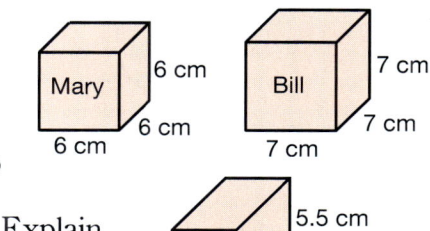

Use the figure at right.

11. Mary, Paul, and Bill are using boxes to display their shell collections. They want to cover the four sides and the bottoms of their boxes with contact paper. Whose box requires the most paper?

12. Can the three boxes be stored inside one another? Explain.

13. Use grid paper. Make a box with a surface area of 54 cm².

CHAPTER 12   405

EXPLORE

Exploring Constructions

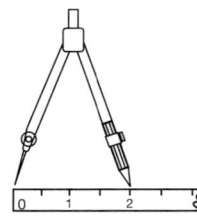

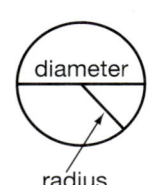

Geometric figures made using only a compass and a straightedge are called **constructions**. A straightedge is like a ruler without measurements on it.

1. Use a compass to draw a circle with a radius of 2 in. What is the diameter of the circle?

The steps for the construction of an arc of a circle are shown below.

Construction: Arcs of a Circle

Draw a circle with C as the center. The radius is $\overline{CA}$.	Without changing the radius, put the compass point at A. Make a mark as shown.	Put the compass point on the mark. Strike again. Repeat around the circle.
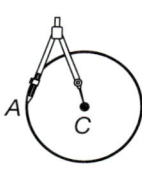	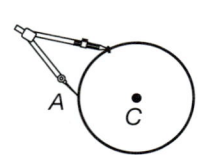	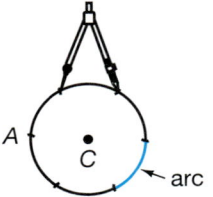

2. Practice the construction a few times.

Follow these steps to make the polygons and the designs shown below. Then color the design.
- Use a compass to make a circle.
- Use the arc construction to mark the points of the polygon.
- Use a straightedge to draw the polygon and the design.

3. regular hexagon

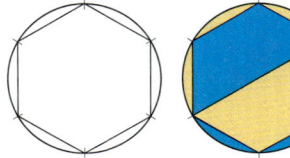

4. rectangle

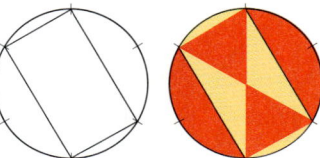

5. equilateral triangle

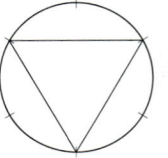 

406   LESSON 12-4

Use the arc construction.

6. Construct the design in Figure B. Use Figure A to help.

7. Use the arc construction to construct a design of your own.

Figure A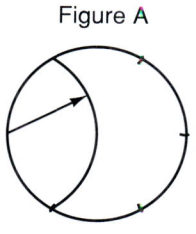

Figure B

The steps for the construction of perpendicular lines are shown below.

Construction: Perpendicular Lines

Draw a circle with P as the center. Draw $\overline{AB}$.	Make the opening of the compass a little longer than the radius. With the compass point at A, draw marks as shown.	Repeat for B so the marks intersect. Draw a line perpendicular to $\overline{AB}$.
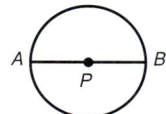	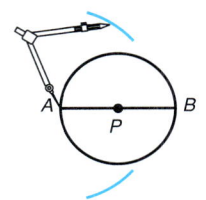	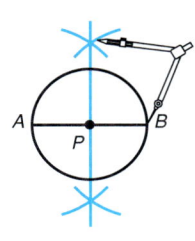

8. Practice the construction a few times.

Use the construction of perpendicular lines to make the polygons and designs shown below.

9. square

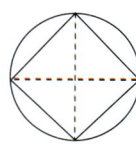 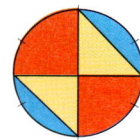

10. isosceles triangle

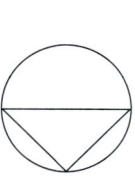 

Critical Thinking

Match the direction from which you look at the object with what you see. Write a, b, or c.

1. top
2. side
3. bottom

a. 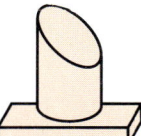   b. 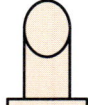   c.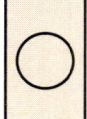

CHAPTER 12   407

Spheres, Cylinders, Cones

Spacecrafts remind us of space figures with curved surfaces.

Sputnik 1, the first satellite to orbit Earth, is in the shape of a **sphere**.

Helios 1, a 1974 solar probe, looks like a **cylinder** between parts of two **cones**.

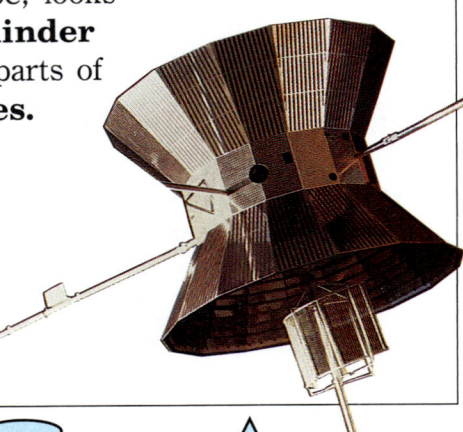

All points of a sphere are the same distance from its center.

 sphere  cylinder 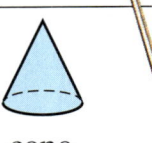 cone

Cylinders and cones have flat, circular faces as well as a curved surface. The circular faces are called **bases**. The bases of cylinders and prisms are on parallel planes.

GUIDED PRACTICE

Which space figure is most like the object? Explain.

1.
2.
3.

Describe the flat faces and the curved surfaces of the figure.

4.
5.
6.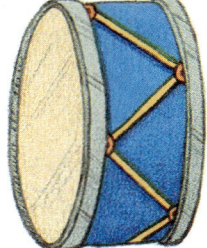

408 LESSON 12-5

PRACTICE

Use the list to the right. Name the figure.

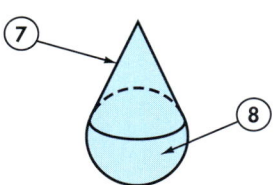 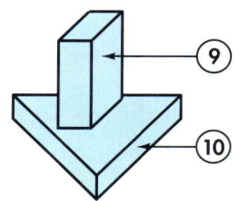 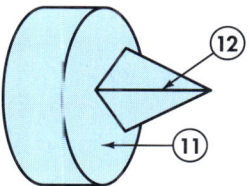

> pyramid
> cylinder
> cone
> sphere
> prism

13. It has two bases that are congruent circles.

14. It has one base only. The base is a circle.

15. It has two bases that are rectangles.

16. It has no bases.

Use the sphere to the right. Imagine points O, A, B, and C on the drawing. Is the point inside, outside, or on the sphere?

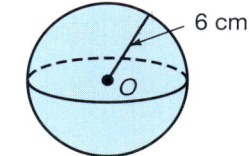

17. Point O

18. Point C when $\overline{OC} = 6$ cm

19. Point B when $\overline{OB} = 4$ cm

20. Point A when $\overline{OA} = 8$ cm

PROBLEM SOLVING

21. What space figure or parts of space figures does the command module remind you of?

22. IN YOUR WORDS Explain why you think the command module was designed with the shape or shapes that you named in Exercise 21.

Astronauts Armstrong, Collins, and Aldrin returned to Earth in the command module of *Apollo 11*.

Problem Solving: Visual Perception

The first book of puzzles was published about 1,500 yr ago by a Greek named Metrodorus.

When you try to solve a puzzle, it helps to picture the puzzle in your mind, or to **visualize** it. You may need to turn or change the puzzle in some way.

The two figures at right are the same size and shape.

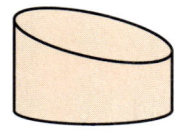
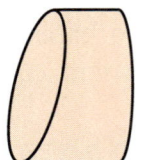

Which of the puzzles below can be made from the two figures above?

Figure A 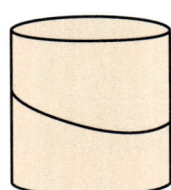   Figure B 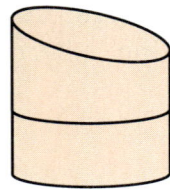   Figure C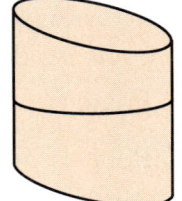

CRITICAL THINKING Describe another way you can put the two figures together.

Here is a more complicated figure to visualize. How many small cubes are used? These hints may help you.

- Some cubes are hidden.
- How many cubes are in one layer?
- Are all layers alike?

GUIDED PRACTICE

How many small cubes are used? Explain how you know.

1.

2.

410   LESSON 12-6

PRACTICE

How many small cubes are used?

3.    **4.**    **5.**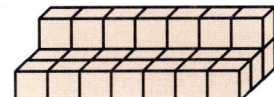

6.    **7.**    **8.**

How many of each piece will it take to make the cylinder?

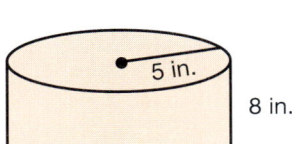
5 in.
8 in.

9.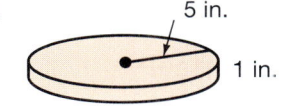
5 in.
1 in.

10.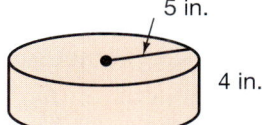
5 in.
4 in.

11.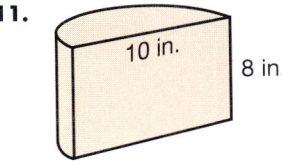
10 in.
8 in.

12.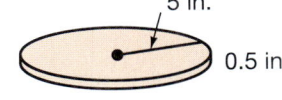
5 in.
0.5 in.

13. If the pieces in Exercises 10 and 11 are made of the same material, how do their weights compare?

An animal is shown on each of the puzzle's four sides. By twisting the pieces, other animals can be made.

14. How many animals can you make that have the feet and body of a lizard?

15. How many different animals can you make that have the head of a porcupine?

CHAPTER 12   411

MIDCHAPTER CHECKUP

LANGUAGE & VOCABULARY

Write a sentence that shows how the two terms are related.
  a. faces—edge   b. edges—vertex   c. base—cylinder

QUICK QUIZ

Use the construction shown. *(pages 406–407)*
1. Name a radius of the circle.
2. What line segments could be drawn to make an equilateral triangle?

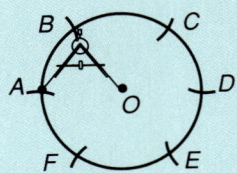

In the construction shown, $\overline{RS}$ and $\overline{XY}$ are perpendicular. *(pages 406–407)*
3. How many right angles are formed?
4. What figure is made by connecting R, X, S and Y in order?

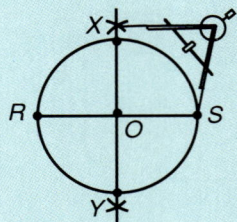

Name the figure. *(pages 402–403, 408–409)*
5. It has only one base. The base is a polygon.
6. It rolls. It has one base that is a circle.
7. It has six congruent square faces.

Find the surface area. *(pages 404–405)*
8. of the rectangular prism at right
9. of a cube with edges 5 in. long

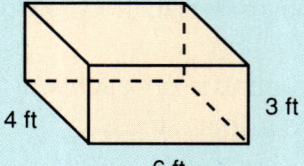

Solve. *(pages 410–411)*
10. a. How many small cubes are in the top layer?
    b. How can you find how many small cubes are used altogether?
    c. How many small cubes are used?

412   MIDCHAPTER CHECKUP

Write the answer in your learning log.

1. Describe the shapes you would need to make a triangular prism.

2. Explain how finding the surface area of a rectangular prism is related to finding the area of a rectangle.

DID YOU KNOW . . . ? Lee County, Virginia is nearer to six other state capitals than to Richmond, the capital of Virginia. How far is it from your city or town to your state capital? Are you closer to the capitals of any other states?

Answer each question for both figures.

1. How many small cubes are used?
2. How many small cubes are in the bottom layer?
3. If you double the length and width of both layers, and keep the height of the figure the same, how many small cubes will be used?
4. When you multiply the length and width by 2, how many times is the number of small cubes increased?

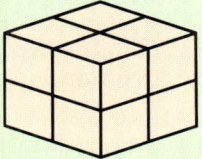

Volume: Rectangular Prisms

The sculpture by Donald Judd is made up of a series of rectangular prisms.

Untitled, 1987. The sculpture is made of galvanized iron and green plexiglass.

The **volume** (*V*) of a space figure is a measurement of the space inside it. Volume is measured in cubic units.

For a rectangular prism that is 5 ft long by 4 ft wide by 2 ft high, think of each layer of cubic units.

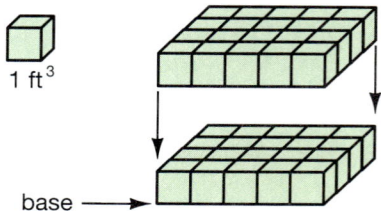

Since 1 layer contains 20 cubic ft (ft^3), 2 layers contain 40 ft^3.

You can also use this formula to find the volume:

Volume of a rectangular prism (*V*) = area of base (*B*) × height (*h*)

$V = B \times h$
$ = (5 \times 4) \times 2$
$ = 20 \times 2$
$V = 40$ cubic feet, or 40 ft^3

Since a cube is a rectangular prism with all sides congruent, you can use this formula to find the volume:

$V = s \times s \times s$

GUIDED PRACTICE

Read. Then write the measurement using words.

1. 45 ft^3
2. 50 in.^{3}
3. 5 yd^3
4. $7\frac{1}{2}$ ft^3

Find the volume of the prism.

5. area of base = 24 ft^2
   height = 9 ft
6. B = 12 in.^{2}
   h = 3 in.
7. B = 1.5 ft^2
   h = 2.4 ft

PRACTICE

Find the volume of the prism. Use a calculator when needed.

	8.	9.	10.	11.	12.	13.
area of base (B)	72 in.²	44 ft²	6.5 yd²	100 in.²	24 yd²	8 yd²
height (h)	15 in.	6 ft	4 yd	10 in.	1 yd	6.5 yd

14. If the length is 3 ft, the width 2 ft, and the height 2 yd, how do you find the volume? Find it.

ESTIMATE What is the volume of the shaded part?

15.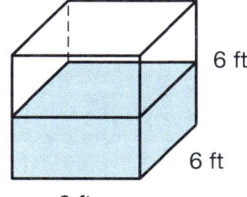
6 ft
6 ft
6 ft

16.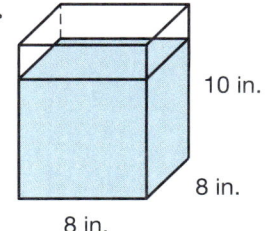
10 in.
8 in.
8 in.

17.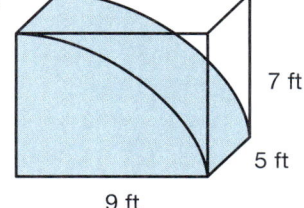
7 ft
5 ft
9 ft

18. Give a length, width, and height for a prism so that its volume is between 20 in.³ and 30 in.³.

PROBLEM SOLVING

The base of the sculpture is a 7-in. cube.

19. Estimate the height of the entire sculpture. Explain how you found your answer.

20. What is the volume of the cube section of the sculpture?

21. **CREATE YOUR OWN** Use geometric figures to draw a picture of a sculpture of your own.

Construction No. 557, a sculpture made of tin, brass, and iron, was made by Kasimir Meduniezky in 1919.

Volume: Cylinders

The tank is used to store water. About how many cubic meters (m³) of water can it hold?

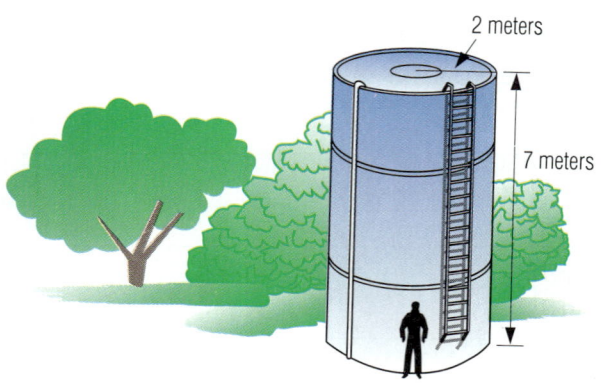

We are looking for the amount of space inside the cylinder, or the volume.

First find the volume of 1 layer. Find the area of the circular base and multiply by a height of 1 m.

Then multiply the volume of 1 layer by the height of 7 m to find the volume of the whole cylinder.

Remember: $A = \pi \times r^2$ and $\pi \approx 3.14$.

$V = (\pi \times r^2) \times 1$
$\approx (3.14 \times 2^2) \times 1$
$\approx 12.56 \text{ m}^3$

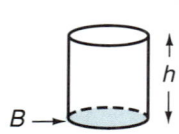

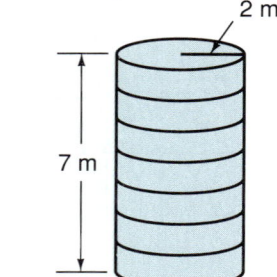

$V \approx 12.56 \text{ m}^3 \times 7$
$\approx 87.92 \text{ m}^3$

You can also use a formula to find volume.

volume of a cylinder = area of circular base (B) × height (h)

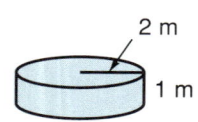

$V = B \times h$
$\approx (3.14 \times 2^2) \times 7$
$\approx 12.56 \times 7$, or 87.92 m^3

The tank can hold about 88 m³ of water.

CRITICAL THINKING Compare the formulas for the volume of a rectangular prism and the volume of a cylinder.

GUIDED PRACTICE

Find the volume. Let B stand for area of base, h for height.

1. $B = 10 \text{ m}^2$
   $h = 6 \text{ m}$
2. $B = 5 \text{ m}^2$
   $h = 3 \text{ m}$
3. $B = 4 \text{ cm}^2$
   $h = 7 \text{ cm}$

4. Explain how to find the volume of a cylinder when $r = 3$ m and $h = 4$ m.

PRACTICE

Find the volume of the cylinder. Use 3.14 for π.
Let r stand for radius, h for height, d for diameter.

5. $r = 2$ m
   $h = 3$ m
6. $r = 3$ m
   $h = 9$ m
7. $r = 18$ cm
   $h = 24$ cm

8. $r = 8$ cm
   $h = 10$ cm
9. $r = 7$ m
   $h = 7$ m
10. $d = 10$ m
    $h = 12$ m

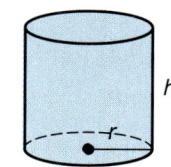

Find the volume of the figure.

11. the cylinder
12. the prism
13. the cylinder and prism combined

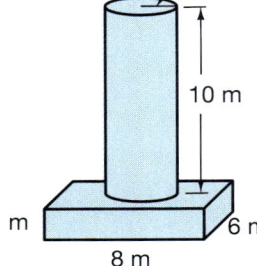

14. the cylinder
15. the prism
16. the part of the prism outside the cylinder

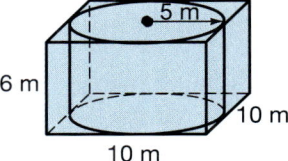

Find the volume of the shaded part.

17.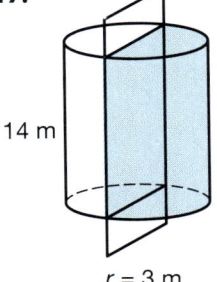

18.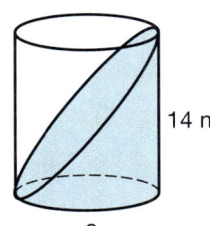

19.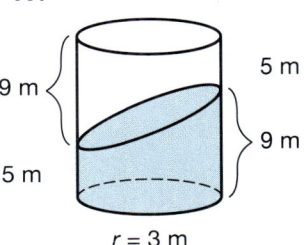

PROBLEM SOLVING

20. How much water will the trough hold?

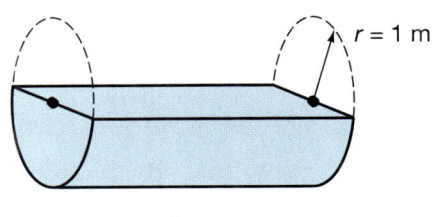

21. The farmer is concerned that only $\frac{3}{4}$ in. of rain fell in June. Last year in June $2\frac{1}{2}$ in. fell. How much more rain fell last year than this year?

22. The cost of a tractor part increased from $25.98 last year to $38.50 this year. How much was the increase this year?

CHAPTER 12   417

Problem Solving Strategy: Using Formulas

Mike is replacing a concrete slab 3 ft long, 3 ft wide, and 2 in. thick. How much concrete does he need?

Here is how Mike thinks:

> A slab is a rectangular prism with a large base and a short height. I need to change 2 in. to $\frac{1}{6}$ ft so that all measurements are in the same unit.

Which of these does Mike need to find?

Area → the amount of surface inside a plane figure

Perimeter → the distance around a plane figure

Volume → the amount of space inside a space figure

Mike's work is shown at right. Does Mike use the correct formula? Are his steps correct?

Special Formulas

Area of a Rectangle
$A = l \times w$

Area of a Triangle
$A = \frac{1}{2} \times (b \times h)$

Area of a Circle
$A = \pi \times r^2$

Volume of a Rectangular Prism and Cylinder
$V = B \times h$

$V = B \times h = (3 \times 3) \times \frac{1}{6}$
$= 9 \times \frac{1}{6} = 1\frac{1}{2} \text{ ft}^3$

I need $1\frac{1}{2}$ ft³ of concrete.

GUIDED PRACTICE

Read the passage. Answer the questions.

1. Mike bought an indoor-outdoor rug to put on the porch. The porch measures 10 yd by 3 yd. The cost of the rug was $11.95 per square yard.

    a. What are the dimensions of Mike's porch?
    b. How can you find the area of the porch?
    c. How much did the rug cost?

Match the measurement with the formula you would use.

2. distance around a concrete slab
3. surface of a rug
4. amount of liquid in a can
5. amount of concrete in a brick

a. volume of a cylinder
b. area of a rectangle
c. perimeter of a rectangle
d. volume of a prism

LESSON 12-9

PRACTICE

Solve. Use the formulas to help you. Use a calculator when needed.

6. Choose which of the methods, *a*, *b*, *c*, or *d*, show a correct way to find the area of the stepping stone.

   a. [2][2][.][5][×][1][5][=]
   b. [.][5][×][2][2][.][5][×][1][1][=]
   c. [2][5][.][2][×][1][5][÷][2][=]
   d. [1][5][×][2][2][.][5][÷][2][=]

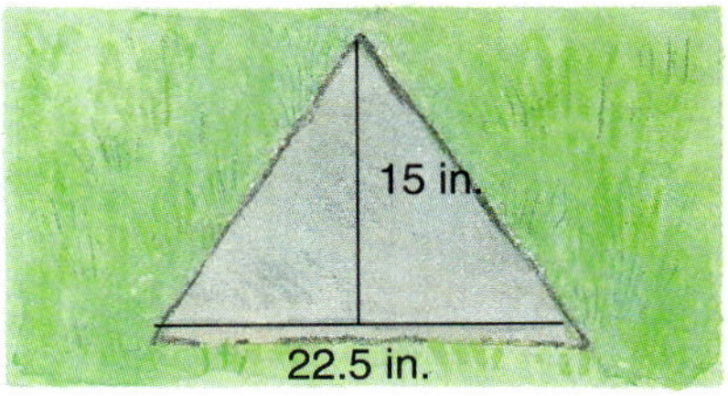

15 in.
22.5 in.

7. Unused concrete is kept in a container shaped like a cylinder. It has a radius of 20 in. and a height of 3 ft. How many cubic inches of concrete will the container hold?

8. Mike's dog is tied to a pole in the yard. The length of the rope is 14 ft. How much surface can Mike's dog cover?

CHOOSE Choose any strategy to solve.

9. Mike wants to paint his garage to match his house. The house is painted tan, white, and black. If Mike uses two of those colors, how many combinations can he choose from?

10. Mike stores his ladder in the garage. The area of the floor is 108 ft^2. The volume of the garage is 900 ft^3. Can Mike stand his 8 ft ladder in the corner?

Investigating Point of View

Do you recognize the common everyday object at the right? It's something you would be likely to find in any kitchen or restaurant.

The object is drawn from an unusual point of view—from the top. Thinking about objects from different viewpoints can help in planning building projects and rearranging furniture. It can also improve your visual awareness.

Work with a partner.

1. Draw a side view and a top view of a cylinder. Use a can or a drinking glass if you need help.

2. Draw as many different views as you can of a cube.

3. Draw a side view, a top view, and a bottom view of a pencil.

4. The figure at the right is called a **truncated cone**. Many lampshades have this shape. Draw a side view, a top view, and a bottom view of a truncated cone.

5. Sketch a side view of a tree. Then draw a bird's-eye view from the top and a bottom view, as if you were under the tree.

6. Look back at your drawings for Exercises 1-5. Which of the objects, if any, can be clearly identified by only one view?

7. Choose several common objects. Draw each from an unusual point of view. Try to make them difficult to recognize. Exchange drawings with your partner and guess what each other's objects are.

8. Make a poster showing one object from three different points of view. Label each view and write a short description of each diagram.

9. From which of your three views is it easiest to identify your object? From which is it hardest? Explain.

10. Share your poster with others in the class. What is the most common shape? What is the least common?

11. All the objects on this page are pictured from an unusual point of view. Can you identify the objects?

Mixed Review

Write the value of the underlined digit.
1. 8,754,012
2. 38,154
3. 0.182
4. 6.024
5. 14.79

Write in word form.
6. 27,014
7. 0.017
8. 3.75
9. 0.123
10. 7.7

Put the numbers in order from the greatest to the least.
11. $\frac{3}{5}, \frac{3}{7}, \frac{5}{9}$
12. $3\frac{5}{8}, 3\frac{3}{4}, \frac{7}{2}$
13. $\frac{7}{3}, \frac{29}{12}, \frac{5}{2}$
14. 29,704; 29,407; 29,740; 2,974
15. 0.028; 0.281; 0.82; 2.8

Find the answer.
16. 18,274 + 3,472
17. 1,081 − 984
18. 289 × 45
19. 1,034 ÷ 22
20. 3.2 × 0.04
21. 7.1 − 5.99
22. 2.872 + 3.01
23. 63 ÷ 1.5
24. 14.9 − 0.124
25. 0.4 × 0.08
26. 7.484 + 2.049
27. 56.28 ÷ 42
28. $2\frac{11}{12} + 1\frac{5}{6}$
29. $\frac{4}{5} \div \frac{7}{10}$
30. $\frac{9}{10} \times \frac{2}{3}$
31. $3\frac{3}{8} \div \frac{3}{4}$
32. $\frac{7}{9} + \frac{5}{6}$
33. $4 - 1\frac{5}{8}$
34. $2\frac{1}{2} \times 1\frac{3}{5}$
35. $3\frac{1}{4} \div 1\frac{1}{2}$
36. $5\frac{1}{7} - 3\frac{5}{7}$
37. $\frac{4}{8} = \frac{n}{12}$
38. $\frac{2}{n} = \frac{3}{9}$
39. 50% of 20

PROBLEM SOLVING

CHOOSE Choose estimation, calculator, mental math, or pencil and paper to solve.

40. A U.S. passport photo must measure 2 in. by 2 in. The length from chin to top of the head must be no more than $\frac{11}{16}$ of the photo's height. Can a photo showing a head length of $1\frac{1}{2}$ in. be used for a passport?

41. About 600,000 immigrants come to the U.S. each year. One sixth of them name New York as their final destination. How many immigrants each year give New York as their final destination?

42. It cost Mr. Ling $16.25 to buy lunch for his family. If each lunch cost $3.25, how many lunches did he buy?

43. It takes Ms. Jackson $\frac{1}{4}$ h to process a passport application. How many applications can she process in $6\frac{1}{2}$ h?

44. In recent years as many as 2,500 immigrants arrive per week at Kennedy terminal in New York. About how many arrive at Kennedy terminal in 1 yr?

45. Applicants for U.S. citizenship must actually be in the U.S. for at least $\frac{1}{2}$ of the required 5-yr residence. How many months is this?

Solve. Use the chart.

46. How many more immigrants came from Europe than from the Americas?

47. About how many times as many immigrants came from the Americas as from Asia?

48. About how many times as many immigrants came from Europe than from all other regions combined?

49. Angel Island, an immigration station near San Francisco, was open for 30 yr. If it closed in 1940, when did it open?

50. About 10 million immigrants came to the U.S. from 1910 to 1940. About $\frac{1}{20}$ of the immigrants, mostly Chinese and Japanese, came through Angel Island. About how many immigrants was this?

Immigrants to U.S. (1820 – 1988)	
Region	Millions
Europe	36.85
Asia	5.42
Americas	11.21
Africa	0.29
Australia and New Zealand	0.14
Other	0.33

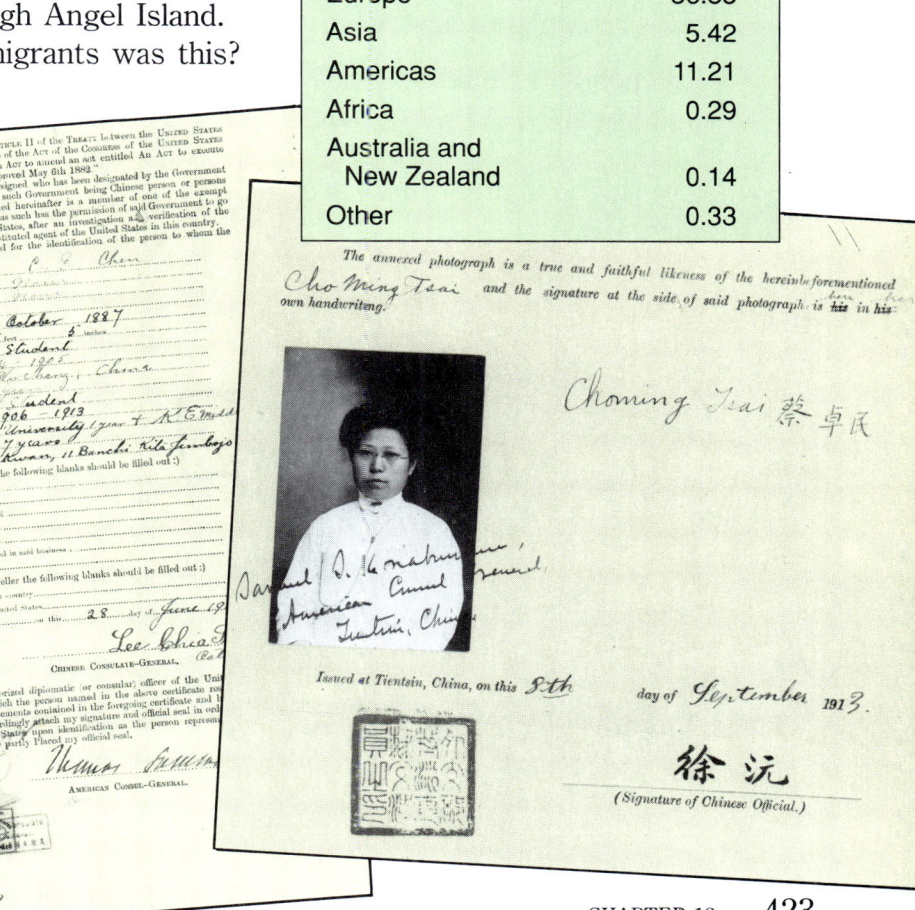

CHAPTER CHECKUP

LANGUAGE & VOCABULARY

Explain the difference between

a. a prism and a cylinder

b. surface area and volume

TEST

CONCEPTS

Use the construction shown. The compass opening is equal to the radius of the circle. *(pages 406–407)*

1. How many arcs of length MN will complete the circle?

2. What polygon would be made by drawing segments MN, NO, and MO?

3. What polygon would be made by drawing segments MN, NP, PQ, QR, RS, and SM?

Name the space figure. *(pages 402–403, 408–409)*

4.

5.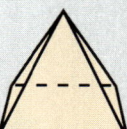

6.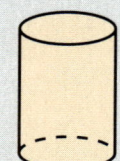

SKILLS

Find the volume of the cylinder. Let B stand for area of base and h for height. *(pages 416–417)*

7. $B = 75$ m^2
   $h = 8$ m

8. $B = 48$ cm^2
   $h = 12$ cm

9. $B = 19$ m^2
   $h = 6$ m

Find the volume of the cylinder. Use 3.14 for π. Let r stand for radius, d for diameter, and h height. *(pages 416–417)*

10. $r = 4$ m
    $h = 5$ m

11. $r = 2$ m
    $h = 4$ m

12. $d = 12$ cm
    $h = 6$ cm

Find the volume of the prism. *(pages 414–415)*

13. cube:
 length of each edge = 4 yd

14. Area of base = 36 ft²
 height = 9 ft

15. $B = 1.8$ ft²
 $h = 2$ ft

16. $B = 25$ in.²
 $h = 10$ in.

17. $B = 115$ in.²
 $h = 12.7$ in.

Find the surface area of the prism. *(pages 404–405)*

18.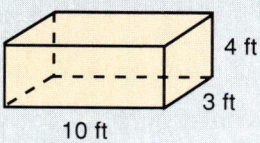

19.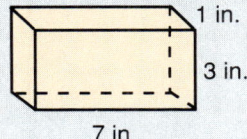

20.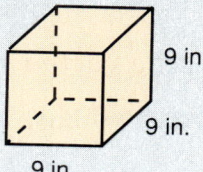

PROBLEM SOLVING

Use the figure at right. *(pages 410–411)*

21. a. What are the length and width of the bottom layer?
 b. Explain how visualizing the figure as two smaller space figures can help you find the number of small cubes used.
 c. How many small cubes are used?

Use a formula to help you solve. *(pages 418–419)*

22. A storage tank in the shape of a cylinder has a base with a radius of 2 m. If the tank is 4 m high, what is its volume?

LEARNING LOG

Write the answer in your learning log.

1. Does a cylinder with a diameter of 10 have the same volume as a cylinder with a radius of 5? Explain.

2. Does an 8 in. cube have a larger volume than a cylinder with a diameter of 8 in. and a height of 8 in? Explain.

EXTRA PRACTICE

$\overleftrightarrow{NP}$ has been constructed perpendicular to $\overline{LM}$.
(pages 406–407)

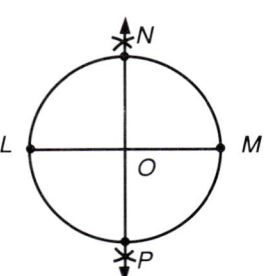

1. Is segment LO congruent to segment ON? How do you know?

2. Segment LN can be drawn creating triangle LNO. What type of triangle is LNO? How do you know?

Use the chart. Name all the figures that fit each description. (pages 402–403, 408–409)

| pyramid | prism | cone |
| cylinder | sphere | |

3. It has a curved surface.
4. It has at least one circular base.
5. It has two bases.
6. It has no circular faces.

Find the surface area. (pages 404–405)

7. [prism: 4 ft, 2 ft, 5 ft]
8. 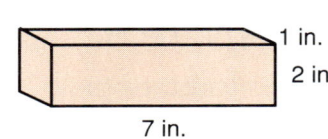 [1 in., 2 in., 7 in.]
9. 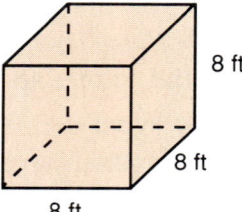 [8 ft, 8 ft, 8 ft]

Find the volume of the prism. (pages 414–415)

10. Area of base = 20 ft² height = 6 ft
11. $B = 144$ in.² $h = 12$ in.
12. $B = 28.6$ ft² $h = 6$ ft

13. Length = 20 in., width = 10 in., height = 5 in.

Find the volume of the cylinder. Use 3.14 for π. Let r stand for radius, d for diameter, and h for height. (pages 416–417)

14. $B = 120$ cm² $h = 12$ cm
15. $B = 40$ m² $h = 3$ m
16. $B = 7$ m² $h = 2$ m

17. $r = 5$ m $h = 1$ m
18. $r = 3$ cm $h = 6$ cm
19. $d = 10$ m $h = 2$ m

Solve. (pages 402–403, 410–411, 418–419)

20. What space figure will the pattern make when cut and folded?

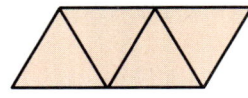

21. A liter cube is 10 cm on each edge. How many cubic centimeters are there in a liter cube?

ENRICHMENT

A Geometric Mobile

Work with a group. Decide on a rule for separating space figures into two categories. The rule may be related to the number of flat or curved surfaces, the number of bases, the types of polygons used as faces, or something else.

Collect some space figures that demonstrate your rule and some that do not. You will also need a wire coat hanger, string, and tape. Hang your figures so that those that meet the rule are together and separated from those that do not.

Your group might want to make several mobiles, each illustrating a different rule. When you have finished, challenge other groups to guess the rule you used.

The Ferris Wheel

One of the most popular sights at the 1893 exposition in Chicago, Illinois, was a wheel with a 250-ft diameter. This ride, by G. W. Ferris, was built to rival the Eiffel Tower. Each of its cars could carry 60 people at a time.

What is the largest Ferris wheel in your area? Find out the ride's diameter and compute its circumference.

PENTOMINOES

A pentomino consists of 5 connected squares. Each square joins its neighbor at a side, not just a corner.

The pentominoes at right are considered to be the same pentomino since the first one can be flipped to make the second one.

Draw as many different pentominoes as you can on grid paper. How many did you find?

Predict which ones will form an open cube. Check yourself by cutting and folding. Were you right?

CUMULATIVE REVIEW

Find the answer.

1. 12, 14, 19, 23
   mean = ▨
   a. 14
   b. 19
   c. 17
   d. none of these

2. 72, 94, 81, 89
   mean = ▨
   a. 81
   b. 84
   c. 85
   d. none of these

3. 3, 1, 0, 6, 4, 7, 5
   median = ▨
   a. 6
   b. 5
   c. 4
   d. none of these

Use the graph at right.

4. The highest temperature was on
   a. Monday
   b. Tuesday
   c. Wednesday
   d. none of these

5. The largest 1-day change was about
   a. 3°
   b. 5°
   c. 9°
   d. none of these

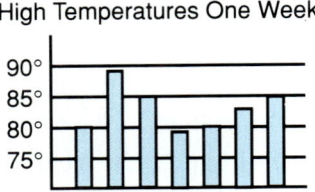
High Temperatures One Week

Which pair of outcomes is equally likely?

6. meeting someone who is
   a. brown-eyed or hazel-eyed
   b. left-handed or right-handed
   c. brown-haired or red-haired
   d. none of these

7. rolling a number cube labeled 1–6 and having the top number be
   a. a prime or a composite
   b. an even or an odd
   c. greater than 3 or less than 3
   d. none of these

Name the figure.

8.
   a. prism
   b. cone
   c. pyramid
   d. none of these

9.
   a. pyramid
   b. prism
   c. cylinder
   d. none of these

Find the surface area.

10. 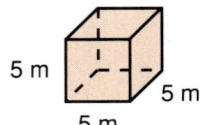 5 m, 5 m, 5 m
    a. 25 m²
    b. 150 m²
    c. 125 m²
    d. none of these

11. 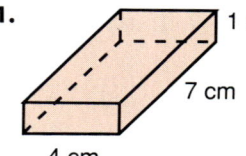 1 cm, 7 cm, 4 cm
    a. 25 cm²
    b. 28 cm²
    c. 78 cm²
    d. none of these

PROBLEM SOLVING REVIEW

Remember the strategies and types of problems you've had so far. Solve.

Problem Solving Check List
- Too little information
- Too much information
- Multistep problems
- Making a table
- Drawing a diagram
- Using guess and check

1. Karen began a job at 8:45 A.M. The job lasted until 4:30 P.M. She took 1 h for lunch and a 15-min break.
   a. What time did Karen begin?
   b. How much time did she take off from work altogether?
   c. How long did she actually work?

2. Roberto, Sandra, Jason, and Marlu are sitting along one side of a table. In how many different ways can the 4 children be arranged?

3. The area of the library is 2,700 ft². There are 15 tables, each with an area of 27 ft². How much floor space is available for walking?

4. Doris has an 8-oz cup and a 5-oz cup, both unmarked. How can she measure exactly 2 oz of water using only the two cups?

5. What is the largest number of Saturdays that there can possibly be in the month of February?

6. A juice machine accepts all coins except pennies. How many ways can Lars pay for a 60¢ drink with exact change?

7. Biographies on the library shelves are placed in alphabetical order. How many shelves are there between the *B*'s and *H*'s?

8. Each rectangular table in the study hall can seat 8 students, 3 on each long side and 1 on each short side. If 2 tables are placed with short sides together, what is the largest number of students the 2-table combination can seat?

9. Arlen, Bowse, Carle, and Dawson are towns along the same straight road in the order given. From Arlen to Bowse is 27 mi, from Carle to Dawson is 32 mi, and from Arlen to Dawson is 91 mi. How far is it from Bowse to Carle?

10. What is the mystery number? It is a 2-digit square number. When you add its digits, you get a sum of 10.

11. Jamala bought 2 books for $2.95 each and a puzzle for $2.89. What was her change?

TECHNOLOGY

JUST FOR KICKS

In the computer game "Soccer to Scale," you win points by estimating scale distances to appoximate the actual distance. Do this activity to sharpen your estimation skills.

Work with a partner. Take turns being the estimator and measurer. Use estimation and the scale to find the distance between the ball and the kicker.

Have your partner measure using a centimeter ruler, and then compute the actual distance. The difference between your estimate and the actual distance is your score. The person with the *lower* score wins.

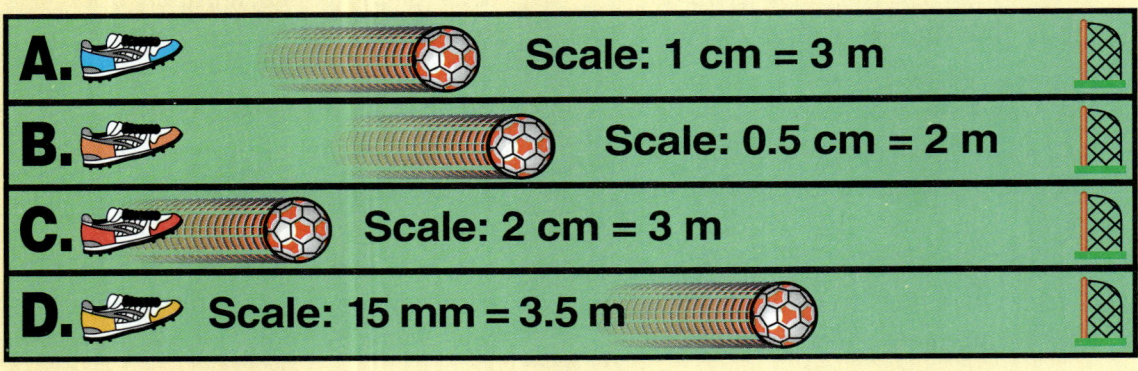

PACK IT UP

Your client wants to know what volume each box or cylinder will hold. Use the volume formulas and the dimensions given below.

Rectangular Prism: $V = l \times w \times h$
Cylinder: $V = \pi \times r^2 \times h$

1. length = twice the width
   height = half the width
   length = 34 cm

2. diameter = twice the height
   height = 9 cm

3. length = half the height
   width = twice the length
   height = 2.5 m

430 TECHNOLOGY

Integers, Coordinate Graphs 13

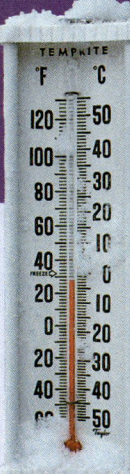

DID YOU KNOW...?

The highest recorded temperature in the United States, 134°F, occurred on July 10, 1913 in Death Valley, CA. The lowest recorded temperature in the United States, −80°F, occurred on January 23, 1971 in Prospect Creek, Alaska. That's a difference of 214°F!

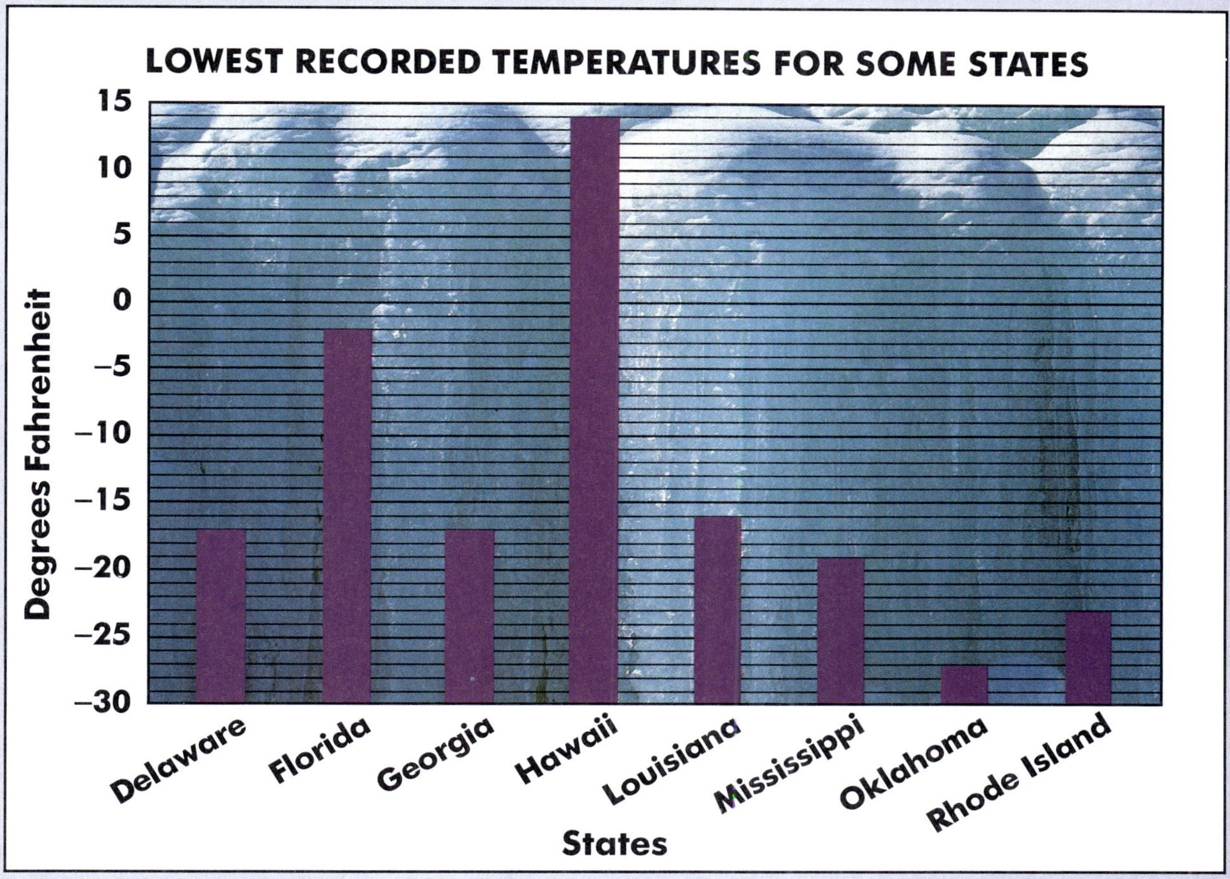

LOWEST RECORDED TEMPERATURES FOR SOME STATES

Degrees Fahrenheit vs. States: Delaware, Florida, Georgia, Hawaii, Louisiana, Mississippi, Oklahoma, Rhode Island

USING DATA
- Collect
- Organize
- Describe
- Predict

For which two states is the lowest recorded temperature the same? What is this temperature?

Work with a partner. Collect data to make a similar graph about highest temperatures for some states.

Write 3 questions about the data shown in your graph. Exchange with another group and solve.

CHAPTER 13   431

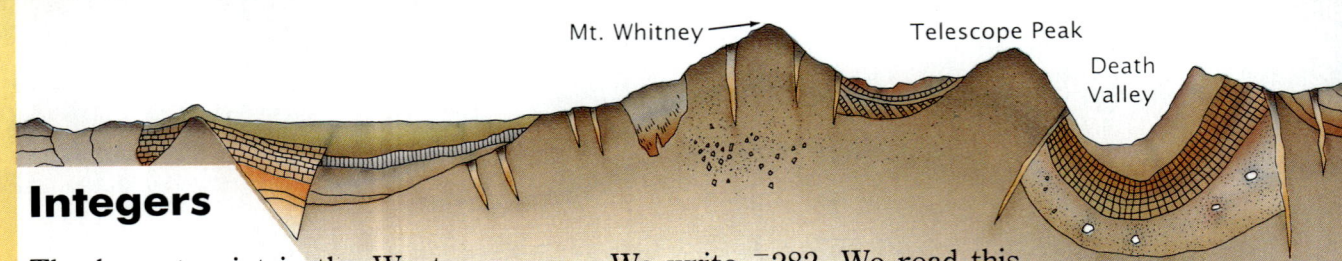

Integers

The lowest point in the Western Hemisphere is in Death Valley, in California. It is 282 ft below sea level.

We write ⁻282. We read this number as *negative 282*.

Telescope Peak in the nearby Panamint Mountain Range is 11,049 ft above sea level.

We write +11,049, or just 11,049. We read this number as *positive 11,049*.

THINK ALOUD What number represents sea level?

You can graph positive and negative numbers on the number line by pairing a point with a number. Numbers to the left of 0 are negative, and numbers to the right of 0 are positive.

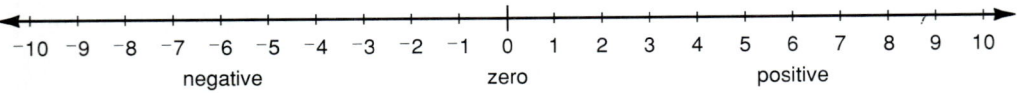

Two numbers that are the same distance from 0 but are on opposite sides of 0 are called **opposites**. For example, 6 and ⁻6 are opposites.

We can say that the opposite of owing someone four dollars (⁻$4) is having four dollars (⁺$4) in your pocket.

THINK ALOUD What is the opposite of 5 s after liftoff?

Whole numbers and their opposites are called *integers*.

GUIDED PRACTICE

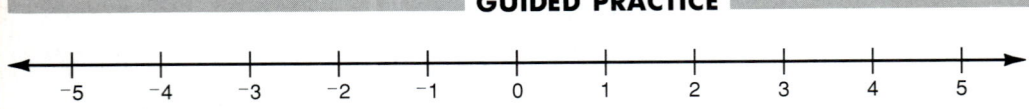

Use the number line. Write the integer.

1. positive one
2. negative five
3. the integer between ⁻3 and ⁻5
4. 5 units to the right of 0
5. 5 units to the left of 2

Write the opposite.

6. 15 floors up
7. negative eleven
8. 3
9. ⁻12

PRACTICE

Write the integer. Use the number line.

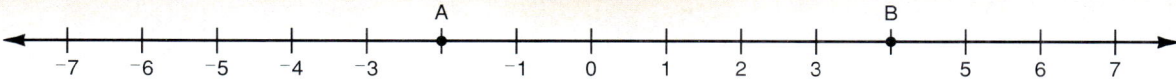

10. A **11.** B **12.** negative eight **13.** positive seven

14. 2 units to the right of 0   **15.** 1 unit to the left of 0

16. 4 units to the left of ⁻3   **17.** 4 units to the right of ⁻1

Write the opposite.

18. 7 mi west   **19.** 5 yd lost in football   **20.** 10° above zero

21. 9 units to the right of 0   **22.** negative twenty   **23.** positive eight

24. +6   **25.** ⁻24   **26.** 1   **27.** ⁻10   **28.** ⁻30   **29.** 0

NUMBER SENSE Write the next three integers in the pattern.

30. 15, 10, 5, ▨, ▨, ▨   **31.** ⁻10, ⁻7, ⁻4, ▨, ▨, ▨

PROBLEM SOLVING

32. The average rainfall in Death Valley is only about 2 in. per year. If the rainfall one year is 4 in., the change from the average is +2. What is the change from the average for a year with only 1 in. of rain?

33. Death Valley was named in 1849 by pioneers struggling across it. We can call that year ⁻84 since it was 84 years before Death Valley National Monument was established in 1933. What can we call the current year?

34. The record high temperature for the United States is 134°F, which occurred in Death Valley in 1913. The record low of ⁻80°F occurred in Prospect Creek, Alaska, in 1971. How many degrees apart are the two records?

Comparing and Ordering Integers

In the metric system, temperature is measured in degrees Celsius (°C).

The temperature at which water freezes is 0°C. Temperatures above freezing are positive, and those below freezing are negative.

Which is colder, ⁻10°C or ⁻20°C?

The thermometer shows that ⁻20°C is colder because it is more degrees below 0.

The number line looks like a thermometer turned sideways. The farther left we go, the smaller the number.

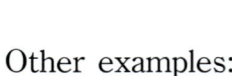

Other examples:

    5 is to the right of 3. → 5 > 3
    ⁻2 is to the left of 6. → ⁻2 < 6
    ⁻6 is to the left of ⁻3, which is to the left of 1. → ⁻6 < ⁻3 < 1

THINK ALOUD How can you compare ⁻6, ⁻3, and 1, using >?

GUIDED PRACTICE

Use the number line above to compare. Choose left or right for the first blank, and < or > for the second blank.

1. 3 is to the ▨ of ⁻2; 3 ▨ ⁻2.
2. ⁻5 is to the ▨ of 0; ⁻5 ▨ 0.
3. ⁻7 is to the ▨ of ⁻2; ⁻7 ▨ ⁻2.
4. 6 is to the ▨ of 4; 6 ▨ 4.

Order the integers from the least to the greatest.

5. 4, 2, 6
6. ⁻4, ⁻2, ⁻6
7. 3, ⁻2, 5
8. ⁻1, ⁻5, 2

434   LESSON 13-2

PRACTICE

Compare. Write >, <, or =.

9. 6 ▧ 2
10. 14 ▧ ⁻3
11. ⁻13 ▧ ⁻2
12. ⁻2 ▧ 3
13. 7 ▧ 0
14. 5 ▧ ⁻4
15. ⁻9 ▧ ⁻6
16. ⁻4 ▧ ⁻8
17. 4 ▧ ⁺4
18. ⁻7 ▧ 6
19. a positive integer ▧ a negative integer
20. a negative integer ▧ 0

Order the integers from the least to the greatest.

21. 16, 8, 12
22. 5, ⁻2, ⁻5
23. ⁻16, ⁻7, 4
24. ⁻11, ⁻2, ⁻10
25. 6, ⁻3, ⁻8, 5
26. ⁻6, ⁻8, 0, ⁻2

Write *true* or *false*.

27. ⁻6 is less than 4.
28. 4 is less than ⁻4.
29. ⁺11 is equal to 11.
30. ⁻12 is greater than ⁻21.
31. The opposite of the opposite of ⁻7 is equal to 7.

Write three integers that are between the two given integers.

32. ⁻2, ⁻7
33. 0, 4
34. ⁻3, 2
35. 0, ⁻4

36. Use the thermometer on page 434 to order these temperatures from the highest to the lowest.

    a. room temperature
    b. body temperature
    c. 30°C
    d. the temperature of cold milk
    e. 15°C
    f. the temperature on a snowy day

PROBLEM SOLVING

37. A cold winter day had a temperature of ⁻8°C. If the temperature changed to ⁻12°C, did it get warmer or colder? by how many degrees?

38. *Celsius* is also called "centigrade." Why is the prefix *centi* used? (*Hint:* Think about the key water temperatures.)

39. **IN YOUR WORDS** On the Fahrenheit scale, the boiling point of water is 180° higher than the freezing point of water. On the Celsius scale, the boiling point of water is only 100° higher than the freezing point of water. Explain how this is possible.

CHAPTER 13

Exploring Adding Integers

You can use charged particles to understand integers.

Suppose 🟥 means +1 or 1 and 🟦 means ⁻1.

If you have $3 (+3), you show it like this:

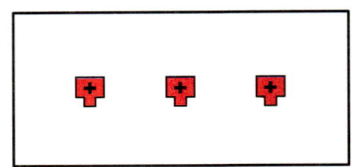

If you owe someone $5 (⁻5), you show it like this:

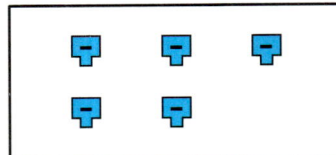

Use the charged particles on the integers workmaster (or two colors of chips). A sheet of paper can be used as a mat.

Place charges on your mat to show the integer.

1. ⁻1   2. +2   3. ⁻6   4. +5

5. If you have $3 and are given another $4, you can represent your money by placing 3 positive charges on your mat and then putting another 4 positive charges on it.

   a. How many charges are on the mat altogether?
   b. What type of charges are they?
   c. Write an equation to describe the result.

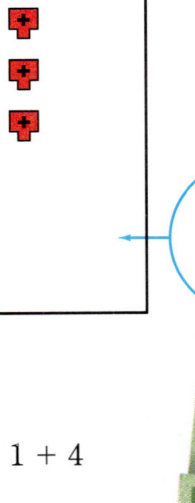

6. Now use your mat and 🟥 charges to add.

   a. 5 + 2   b. 3 + 7   c. 1 + 4

7. The sum of two positive integers is a ? integer.

436   LESSON 13-3

8. If you owe $5 and then get another $2 debt, you can show your total debt by placing 5 negative charges on your mat and then putting another 2 negative charges on it.

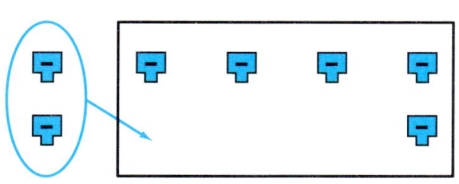

   a. How many charges are on the mat altogether?
   b. What type of charges are they?
   c. Write an equation to describe the result.

9. Now use your mat and 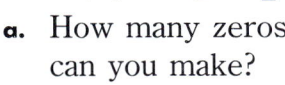 charges to add.

   a. ⁻4 + ⁻3   b. ⁻1 + ⁻8   c. ⁻2 + ⁻6

10. The sum of two negative integers is a __?__ integer.

11. Suppose you have $1, but you also owe $1. How much money will you have left after you pay off your debt?

When you put a 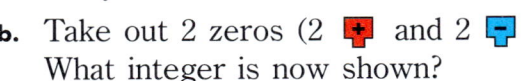 charge and a charge together, they cancel each other out and the result is the integer 0.

12. What integer do you get when you put 3 charges and 3 charges together?

13. If you have $8 and owe $2, you can show this by placing 8 and 2 on your mat.

   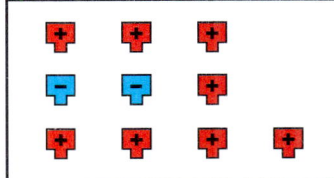

   a. How many zeros can you make?
   b. Take out 2 zeros (2 and 2). What integer is now shown?
   c. **IN YOUR WORDS** Does removing the zero charges change the integer represented? Explain.
   d. You have just found that 8 + ⁻2 = ▬.

   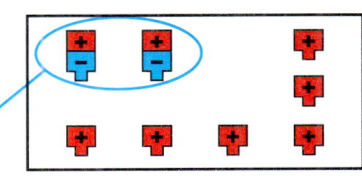

14. Place both the positive and the negative charges on your mat. Remove the zero charges. Then name the integer that is the sum.

   a. 4 + 7
   b. 9 + 3
   c. 5 + 6
   d. 4 + 2

15. **CRITICAL THINKING** What can you say about the sum of a positive integer and a negative integer?

CHAPTER 13   437

Adding Integers

Ardis, a research diver, goes down 8 ft to study pearl oysters, then goes down 4 ft more to study scallops. How far down does she go?

You can use integers to solve the problem.

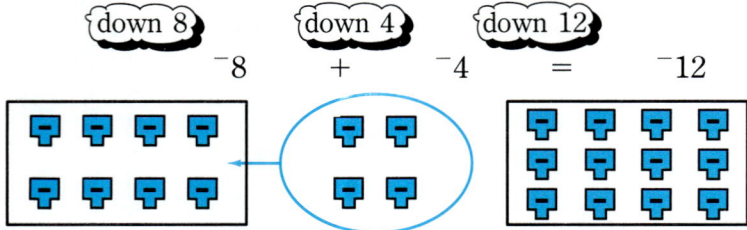

She goes down a total of 12 ft.

Another example:

> **If both numbers have the same type of charge, just add the numbers and write the sign to show the type of charge.**

Sometimes the numbers take you in opposite directions.

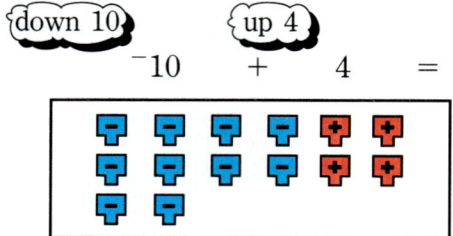

THINK ALOUD How many of the 10 negative charges are canceled by the 4 positive charges? How many remain?

Going down 10 ft and up 4 ft is equal to going down 6 ft.

CRITICAL THINKING Explain how you can mentally cancel charges to find the sums $^-10 + 12$ and $6 + {}^-14$.

> **The sum of a positive and a negative is positive if you have more positive charges and is negative if you have more negative charges.**

The [+/-] calculator key changes an integer to its opposite.

$50 + {}^-25 \rightarrow 50 + 25$ [+/-] [=] $^-10 + 20 \rightarrow 10$ [+/-] [+] 20 [=]

438   LESSON 13-4

GUIDED PRACTICE

Add. Think of charged particles or scuba diving.

1. ⁻4 + ⁻2
2. 11 + 15
3. ⁻11 + ⁻9
4. ⁻6 + 6
5. ⁻2 + 1
6. ⁻2 + 4
7. ⁻4 + 2
8. ⁻13 + 15

PRACTICE

Add. Think of charged particles or scuba diving if needed.

9. ⁻4 + ⁻5
10. 2 + 5
11. 4 + 3
12. ⁻4 + ⁻3
13. 7 + 7
14. ⁻3 + ⁻8
15. ⁻12 + 12
16. 5 + ⁻8
17. ⁻3 + 4
18. 6 + ⁻4
19. ⁻8 + ⁻3
20. 8 + ⁻17
21. ⁻1 + 6
22. 7 + ⁻12
23. 24 + ⁻16
24. ⁻2.8 + ⁻1.4

MIXED REVIEW Find the answer.

25. $2\frac{7}{8} + 3\frac{3}{5}$
26. 2.5 + 8.63
27. 15% of 60
28. $2\frac{1}{3} \times 1\frac{1}{2}$

CALCULATOR Find the answer.

29. ⁻365 + 189
30. ⁻134 + ⁻78
31. 329 + ⁻584
32. 967 + ⁻839

Find the rule. Then complete the table.

33.

In	5	⁻4	100	⁻6	0
Out	⁻5	4	⁻100		

34.

In	⁻10	⁻9	⁻8	⁻5	⁻2	1
Out	⁻5	⁻4	⁻3			

PROBLEM SOLVING

35. An open-sea diver dove down 42 ft to harvest sponges. After rising 28 ft in one minute, how far was he from the surface?

36. A commercial diver needs to repair the bridge supports 35 ft under water. He dives down 19 ft and then 12 ft. Is he at the right level?

Mental Math

Grouping for opposites can help you add a series of integers.

3 + 4 + ⁻8 + ⁻7 = ⁻8   3 + 4 = 7 and 7 + ⁻7 = 0

Use grouping for opposites to add mentally.

1. 5 + ⁻4 + 7 + 4 + ⁻5
2. ⁻13 + 28 + 6 + 7
3. ⁻126 + ⁻9 + 100 + 26

CHAPTER 13   439

Problem Solving Strategy: Using Patterns

Triangular numbers are numbers that can be drawn as triangles.

1	3	6	10
1	1 + 2 = 3	1 + 2 + 3 = 6	1 + 2 + 3 + 4 = 10

- Describe how triangular numbers increase.
- How would you draw the fifth triangular number?
- What numbers would you add to find the value of the fifth triangular number?
- What is the fifth triangular number?
- How can you use the picture of a triangular number to tell which triangular number it is?
- How could you find the 11th triangular number?

GUIDED PRACTICE

Square numbers are numbers that can be drawn as squares.

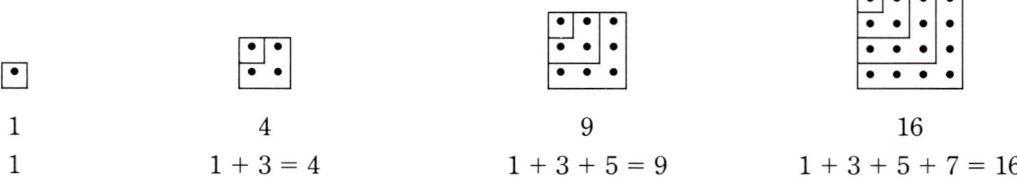

1	4	9	16
1	1 + 3 = 4	1 + 3 + 5 = 9	1 + 3 + 5 + 7 = 16

1. Answer these questions about square numbers.
   a. What is the third square number?
   b. How could you find the fifth square number using addition? using multiplication?
   c. What is the sixth square number?

PRACTICE

Use patterns to solve.

2. What is the tenth number in this sequence? 1, 1, 2, 3, 5, 8, 13, 21, . . .

3. What is the 7th number in this sequence? ⁻1, ⁻2, ⁻5, ⁻10, ⁻17, . . .

4. A company is putting prizes in these boxes of popcorn: the first, the sixth, the eleventh, and so on. Will there be a prize in the 26th box?

5. The white keys on a piano keyboard follow this pattern of notes: CDEFGAB. What note is the 18th white key above middle C on the piano?

6. Suppose you had a job that paid 1¢ the first day, 2¢ the second day, 4¢ the third day, 8¢ the fourth day and so on. What would you be paid on the 13th day?

7. Susan is making a quilt. The odd-numbered rows will repeat this pattern of colored squares: blue, white, blue, blue, white, blue. What will be the color of the 13th square in the fifth row?

Choose any strategy to solve.

8. Kareem is planting seedlings on a 30-ft square lot. He leaves 3 ft between rows of seedlings. He also leaves 3 ft between the outermost seedlings and the edge of the lot. How many seedlings will he plant?

9. Maral made cubes from blocks to display toys in her store.

   1st   2nd   3rd

   How many blocks will she use in the fifth display?

10. Akira made 9 baskets for a total of 25 points in a game. He made some 3-point goals and some 2-point goals, but no 1-point goals. How many of each type of goal did he make?

11. Maria bought 6 tulip bulbs for $5.00. At this rate, how much would 15 tulip bulbs cost?

CHAPTER 13   441

MIDCHAPTER CHECKUP

LANGUAGE & VOCABULARY

Explain the statement.
- **a.** Every number has an opposite.
- **b.** The opposite of 0 is 0.
- **c.** The sum of two opposites is 0.

QUICK QUIZ

Write the opposite. *(pages 432–433)*

1. 10 pounds gained
2. negative six
3. $^+24$

Order the integers from the least to the greatest. *(pages 434–435)*

4. 7, $^-18$, 10
5. 0, $^-5$, 1, $^-7$
6. $^-12$, $^-18$, $^-2$

Add. *(pages 438–439)*

7. $^-3 + {^-6}$
8. 14 + $^-9$
9. $^-11 + 11$

Solve. *(pages 440–441)*

10. Xanthe made a bead necklace with this repeating pattern.
    red-yellow-yellow-orange-brown-orange-yellow-yellow
    - **a.** What colors are the 3 beads after a red bead?
    - **b.** What colors are the 3 beads before a red bead?
    - **c.** If you start with a particular red bead, what color bead is 20 away from it in either direction?

Write the answer in your learning log.

1. Your friend thinks that ⁻6 is greater than ⁻3 because 6 is greater than 3. Explain why this thinking is wrong.

2. Describe the location of pairs of numbers such as 3 and ⁻3, 4 and ⁻4, 6 and ⁻6 on the number line.

DID YOU KNOW . . . ? The number of representatives each state sends to the House of Representatives in Washington is proportional to the state population. Find out the number of representatives and the population for your state. Estimate the ratio of representatives to people in your state.

Amazing Temperature Records

1. In 1916, the greatest one-day temperature change occured in Montana—a change of ⁻100°F from a high of 44°F. How cold did it get?

2. In 1943 in South Dakota, the temperature rose 49°F in 2 min. It started at ⁻4°F. How warm did it get?

3. In Siberia in the Soviet Union a temperature of ⁻94°F has been recorded. A temperature 192° warmer has also been recorded. What was the warmest Siberian temperature?

Exploring Subtracting Integers

You can use charged particles to subtract integers. Use a sheet of paper as a mat.

To find 8 − 3, place 8 ⊕ on your mat and then take away 3 ⊕.

1. What integer is left?

To find ⁻8 − ⁻3, place 8 ⊖ on your mat and then take away 3 ⊖.

2. What integer is left?

Now let's consider ⁻3 − ⁻8.

3. What do you place on your mat?

4. What do you need to take away?

5. You need to take away 8 ⊖, but you don't have that many charges. How can you get more ⊖ charges on the mat without changing the value of the integer?

6. How many zeros do you need to put on the mat before you can take out 8 ⊖?

7. Put in 5 zeros. Is the value of the integer on the mat still ⁻3? Explain.

8. Now take away ⁻8. What integer is left?

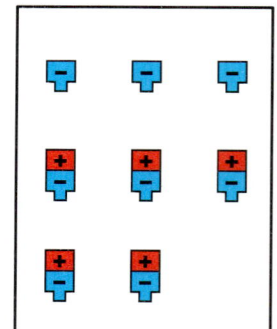

Now use charges to try to solve the problem:

$3 - 8 = $ ▨

9. How many ⊕ charges do you need to start with? How many ⊕ charges do you need to take away?

10. Do you need to put in any zeros? Explain.

11. Do the steps on your mat. What is your answer?

444   LESSON 13-6

Subtracting integers with different signs is the same process.

To find 3 − ⁻6, place 3 ➕ on your mat. You need to take away 6 ➖ charges.

12. How many zeros do you need to put in before you can take away 6 ➖?

Put in 6 zeros. Take out 6 ➖

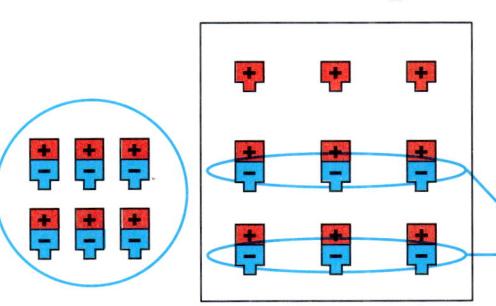

Taking away the negative charges from the zeros leaves more positive charges.

13. What charges are left? Name the integer.

14. CRITICAL THINKING What happens if you put in more than 6 zeros? less than 6 zeros?

Now use charges to solve the problem:
⁻6 − 2 = ▓

15. How many ➖ charges do you need to start with? How many ➕ charges do you need to take away?

16. Do you need to put in any zeros? Explain.

17. Do the steps on your mat. What is your answer?

Do the problems on your mat. Draw a picture of the integer that you start with and the integer that you end with.

18. ⁻10 − ⁻6   **19.** ⁻6 − ⁻8   **20.** ⁻4 − 2   **21.** 2 − ⁻4   **22.** 3 − 5

23. When you subtract a positive, you are left with fewer positive charges and/or more ___?___ charges.

24. When you subtract a negative, you are left with fewer negative charges and/or more ___?___ charges.

CHAPTER 13   445

Subtracting Integers

Ian has no money and owes his parents $8. We can say he has ⁻$8. His parents agree to cancel $5 of his debt if he cleans his little brother's room. How much money will Ian have then?

To find out, subtract ⁻8 − ⁻5.

You can use charged particles to help you subtract.

Start with 8 ⊖ and take away 5 ⊖. 3 ⊖ is left.

After cleaning the room, Ian will have ⁻$3, or a debt of $3.

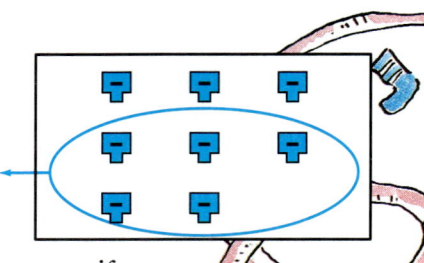

THINK ALOUD Would Ian have the same amount of money if his parents had paid him $5 instead of canceling $5 of his debt?

CRITICAL THINKING Study the pairs of equations. Instead of subtracting, what can you do to get the same answer?

⁻3 − 4 = ⁻7	4 − 9 = ⁻5	2 − ⁻5 = 7	⁻1 − ⁻6 = 5
⁻3 + ⁻4 = ⁻7	4 + ⁻9 = ⁻5	2 + 5 = 7	⁻1 + 6 = 5

To subtract an integer, add the opposite of that integer.

Other examples:

⁻5 − 6 → ⁻5 + ⁻6 = ⁻11   3 − ⁻6 → 3 + 6 = 9

THINK ALOUD How can you rewrite these examples as addition sentences? ⁻4 − ⁻8 and 7 − 11

GUIDED PRACTICE

Complete.

1. Subtracting ⁻6 is like adding __?__.
2. Subtracting 5 is like adding __?__.

Use charged particles to find the difference. Write an addition problem to find the difference. Do your answers agree?

3. 3 − ⁻1
4. ⁻3 − 1
5. 2 − ⁻4
6. 2 − 4

PRACTICE

7. 4 − ⁻1
8. 4 − 1
9. 5 − ⁻2
10. 5 − 2
11. ⁻5 − ⁻3
12. ⁻5 − 3
13. ⁻4 − ⁻2
14. ⁻4 − 2

15. $3 - {}^-2$ **16.** ${}^-6 - {}^-4$ **17.** ${}^-1 - 3$ **18.** $2 - {}^-5$

19. ${}^-1 - {}^-3$ **20.** $3 - 0$ **21.** $2 - 5$ **22.** ${}^-6 - 3$

MIXED REVIEW Find the difference.

23. $3{,}001 - 389$ **24.** $\frac{3}{8} - \frac{1}{5}$ **25.** $5.1 - 2.75$ **26.** $3\frac{3}{8} - 1\frac{5}{6}$

NUMBER SENSE Study the examples. When is the difference of two integers

27. 0?

28. positive?

29. negative?

$6 - 6 = 0$	$2 - {}^-4 = 6$	${}^-4 - 2 = {}^-6$
${}^-3 - {}^-3 = 0$	$7 - 4 = 3$	$4 - 7 = {}^-3$
${}^-5 - {}^-5 = 0$	${}^-1 - {}^-6 = 5$	${}^-6 - {}^-1 = {}^-5$

30. Write *positive, negative,* or *0.* Do not subtract.

  a. ${}^-2 - 5$ **b.** ${}^-4 - {}^-10$ **c.** ${}^-7 - {}^-7$ **d.** $6 - {}^-6$ **e.** $3 - 8$

MENTAL MATH Is the number sentence true if $n = {}^-4$?

31. $n - 4 = 0$ **32.** ${}^-5 < n + 1 < 1$ **33.** $3 - n < 3 + n$

PROBLEM SOLVING

34. Sue owes her sister Jean $5. She finds an IOU from Jean for $15. How much money or debt does Sue end up with?

35. Zack owes $23 to his father and $19 to his mother. If he pays his father $9 and his mother $4, to whom will he owe more money?

Critical Thinking

You can think about subtraction by picturing how far apart two numbers are on the number line.

3 is 8 units to the right of ${}^-5$, so $3 - {}^-5 = 8$.

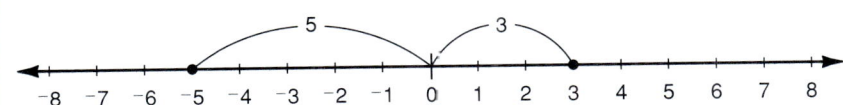

Use the number line to find the difference.

  1. $7 - {}^-2 = \blacksquare$ **2.** ${}^-1 - {}^-4 = \blacksquare$ **3.** $2 - {}^-8 = \blacksquare$

Coordinate Graphs and Ordered Pairs

Did you know that Reno, Nevada, is farther west than Los Angeles, California?

A **coordinate plane** can be used to show this. The horizontal line is the **x-axis** and the vertical line is the **y-axis**. The point where they meet is called the **origin** (0, 0).

Start at the origin and count 1 unit right and 11 units up. Where are you? The coordinates of Reno are the **ordered pair** (1, 11).

THINK ALOUD What ordered pair gives the coordinates of Los Angeles?

GUIDED PRACTICE

Name the ordered pair for the cities listed.

1. Santa Barbara (⁻1, ▓)
2. Bakersfield (▓, ⁻4)
3. San Francisco (▓, ▓)

Name the letter at these points.

4. (⁻6, 4)   5. (4, 4)   6. (⁻2, ⁻5)   7. (4, ⁻8)

8. **THINK ALOUD** Is (3, ⁻4) the same point as (⁻4, 3)? Explain.

PRACTICE

Write the letter of the point named by the ordered pair.

9. (⁻2, 1) 10. (⁻1, ⁻1) 11. (2, 0)
12. (0, ⁻3) 13. (4, ⁻3) 14. (⁻3, 4)
15. (1, 2) 16. (⁻3, ⁻3) 17. (⁻3, 0)

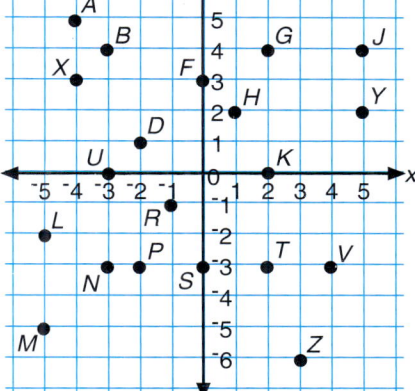

Write the coordinates of the point.

18. A 19. R 20. F 21. G
22. J 23. L 24. M 25. P
26. T 27. Z 28. V 29. X

PROBLEM SOLVING

Draw a coordinate plane on grid paper. Label each axis from ⁻10 through 10.

30. Find and label points A(⁻1, 5), B(8, 1), and C(2, 1). Connect A to B, B to C, and C to A. What type of triangle is made?

31. Find and label points D(3, ⁻3), E(10, ⁻7), and F(3, ⁻7). Connect D to E, E to F, and F to D. What type of triangle is made?

32. Find points M(⁻6, 1), N(⁻3, 1), P(⁻1, ⁻4), and Q(⁻8, ⁻4). Connect M to N, N to P, P to Q, and Q back to M. What type of quadrilateral is made?

33. **CREATE YOUR OWN** Draw a quadrilateral. Write the ordered pairs of the 4 vertexes. Have a friend locate and connect the points and name the type of quadrilateral.

Critical Thinking

The rectangle was slid 10 units to the right. Point A slid from (⁻6, 4) to (4, 4). What are the new coordinates for points B, C, and D?

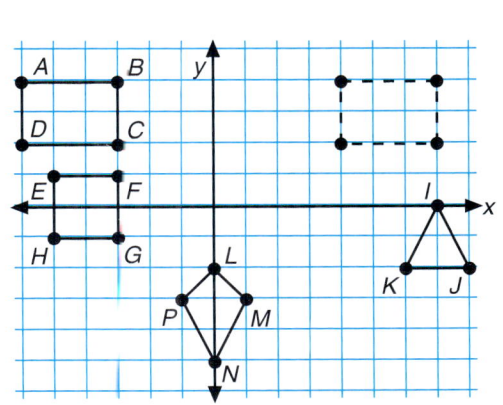

Write the new coordinates for each point of the figure.

1. Slide the square 3 units down.
2. Slide the diamond 5 units up.
3. Slide the triangle 6 units left.

Problem Solving Strategy: Working Backward

The stock market chart shows the gains (+) and losses (−) of three stocks for the week. The numbers show how much the stock changed by the end of each day.

If EEL stock ended the week at 7, what was its value on Tuesday night?

First find the value of the stock on Thursday night.

- How much did the stock change between Thursday night and Friday night?
- What was the value of the stock on Thursday night?
- Why did you add 1 to 7?

Find the value of the stock on Wednesday night and then on Tuesday night.

How can you check your answer?

- Which equation represents the problem? Why?
  - **a.** $n + 3 - 4 - 1 = 7$
  - **b.** $n - 3 + 4 + 1 = 7$

STOCK	M	T	W	Th	F
EEL	+2	+1	+3	-4	-1
NOD	-3	+1	-3	-2	+2
NAT	+1	-6	-4	-3	+5
LAM	-9	-1	-2	-9	-4
BEH	-1	-2	+1	+3	-1

GUIDED PRACTICE

THINK ALOUD Read the passage and use the chart above to answer the questions.

1. The value of NOD stock was 11 at the close of trading on Thursday.
   - **a.** How did the value of NOD stock change on Thursday?
   - **b.** How can you find out what the value of NOD stock was on Monday night?
   - **c.** What was the value of NOD stock on Monday night?

LESSON 13–9

PRACTICE

CHOOSE Choose mental math, pencil and paper, or calculator to solve.

2. Mr. Freedman is a stockbroker. He wants to be in his office by 9:00 A.M. His travel time is 35 min, and he needs 1 h 20 min to get ready at home. When does he need to get up?

3. Carmen bought some BEH stock Monday morning. The stock had these changes for the week: ⁻1, ⁻2, +1, +3, and ⁻1. The stock's value was 8 at the end of the week. At what price did she buy the stock?

4. Grandmother left her office and took the elevator up 7 floors. Then she took the elevator down 3 floors to the cafeteria. If the cafeteria is on the 6th floor, on what floor is Grandmother's office?

5. At the end of the day, Lee had $54.32. During the day he spent $15.56 on dry cleaning, bought 2 notepads at $2.59 each, and spent $6.79 on lunch. How much did he have at the beginning of the day?

6. A computer network charges $1.50 per month plus 21¢ for every minute on-line. Chris's bill was $25.65. How many minutes was she on-line last month?

7. Nita worked a total of 8 h. She began her day in a meeting. She worked $5\frac{3}{4}$ h on the computer and $1\frac{1}{2}$ h on a report. How long was the meeting?

CHOOSE Choose any strategy to solve. Fractions are used in writing stock values. A value of $7\frac{1}{2}$ means the stock sells for $7.50 per share.

8. Keiko and Ajani together own 1,200 shares of stock. Ajani owns twice as many shares as Keiko. How many shares does each of them own?

9. Of the stock that Sharon owns, 75% is NAT stock. If she owns 300 shares of NAT stock, what is the total amount of stock that she owns?

10. Ms. Martinez bought 200 shares of NAT stock. The following day NAT stock went down $1\frac{1}{2}$ and her 200 shares were worth $1,850. How much did Ms. Martinez spend for the stock?

11. Stock values and stock gains and losses are often reported in eighths.
    a. What is $\frac{1}{8}$ of $1?
    b. What is the cost of 100 shares of stock at $1\frac{1}{8}$? 200 shares? 300 shares? 400 shares?

CHAPTER 13   451

Investigating Opinion Polls

Some of the students at a school believed there should be more bicycle racks. They surveyed 30 students to get their opinions. You can present the results of the survey in different ways. Work with a partner to explore some of these ways.

Opinion	Number of Students
Strongly Agree	12
Agree	8
No Opinion	4
Disagree	3
Strongly Disagree	3

1. Give each response the following numerical values:

   strongly agree = 5 points
   agree = 4 points
   no opinion = 3 points
   disagree = 2 points
   strongly disagree = 1 point

   What is the average point score per student surveyed?

Opinion	Number of Students	Point Value	Total Points
Strongly Agree	12	5	60
Agree	8	4	
No Opinion	4	3	
Disagree	3	2	
Strongly Disagree	3	1	

2. Reverse the numerical values (*strongly agree* = 1 point, *agree* = 2 points, etc.) and find the average.

3. Now figure out the average based on these values:

   strongly agree = 2 points
   agree = 1 point
   no opinion = 0 points
   disagree = −1 point
   strongly disagree = −2 points

   What is the average?

LESSON 13–10

4. What percent of the students surveyed either agree or strongly agree?

5. Which of these ways to present the survey data would you use if you were in favor of having more bicycle racks?

6. Which would you use if you were against the idea?

7. Prepare a poll similar to the one on the previous page about a question facing your school or community.

8. Give the survey to at least 30 people. Organize the responses into a chart.

9. Present the results in at least four different ways. Which way is most helpful to each side of the question?

10. Compare these ways of reporting the surveys:
    - the scale from 1 to 5
    - the scale from -2 to $+2$
    - the percent that agrees or strongly agrees
    - showing the results in a chart

    Which way do you believe is the most accurate?

CHAPTER 13

CHAPTER CHECKUP

LANGUAGE & VOCABULARY

Explain why you can subtract an integer by adding its opposite.

TEST

CONCEPTS

Write the opposite. *(pages 432–433)*

1. 3 units to the left of 0
2. 25 ft below sea level
3. driving 10 mi east
4. losing $3.00
5. negative 2
6. ⁻31
7. 9
8. ⁻11

Write in order from least to greatest. *(pages 434–435)*

9. 3, 11, 8
10. ⁻3, 5, 1
11. ⁻7, 2, ⁻5, 5
12. ⁻12, ⁻9, ⁻10
13. 1, 0, ⁻4, ⁻5
14. ⁻5, ⁻11, ⁻6

SKILLS

Add or subtract. *(pages 438–439, 446–447)*

15. 20 + ⁻35
16. ⁻12 + 4
17. ⁻8 − ⁻12
18. 9 − 12
19. ⁻13 + ⁻7
20. ⁻10 − 17
21. 16 + 21
22. 12 − ⁻9
23. ⁻22 − 6

Write the coordinates of the point. *(pages 448–449)*

24. A
25. B
26. C
27. D
28. E
29. F

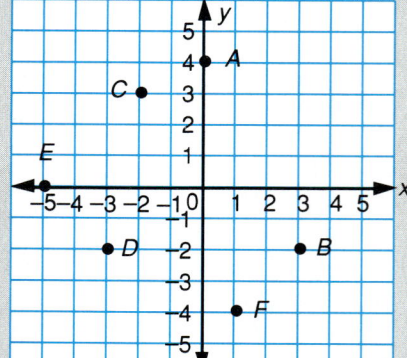

PROBLEM SOLVING

Work backward. *(pages 450–451)*

30. The chart shows the changes in Gwendolyn's bank account during 4 wk in January.

Week of	Change	Balance
Jan 7	+7	
Jan 14	−12	
Jan 21	+9	
Jan 28	−8	$29

  a. How much was in Gwendolyn's account at the end of the 4 wk?

  b. How much did she deposit during the 4 wk? How much did she withdraw?

  c. How much was in Gwendolyn's account at the beginning of the 4 wk?

Solve. *(pages 438–441)*

31. A hiker starting at sea level climbs a 20-m hill. She then descends 35 m into a valley. How many meters above or below sea level is she after climbing and descending?

32. Each rectangle fits inside the next one as shown. If the pattern continues, what would be the dimensions of the sixth rectangle?

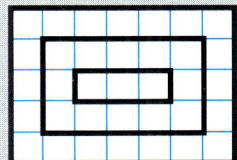

Write the answer in your learning log.

1. Your friend says that when you add integers you also use subtraction. Explain what your friend means.

2. Describe two integers that have a sum of zero.

EXTRA PRACTICE

Write the integer. Use the number line. *(pages 432–433)*

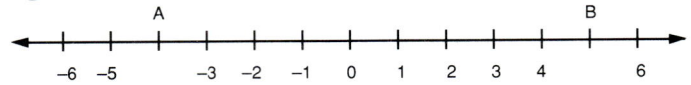

1. A
2. B
3. 3 units to the right of 0
4. 5 units to the left of 4
5. 2 units to the right of ⁻5

Write the opposite. *(pages 432–433)*

6. 75 ft north
7. 3°F below 0
8. 12 units to the left of 0
9. positive one
10. negative five
11. ⁻17
12. 10

Compare. Write >, <, or =. *(pages 434–435)*

13. 4 ▨ ⁻2
14. ⁻3 ▨ 1
15. ⁻7 ▨ ⁺7
16. ⁻10 ▨ ⁻2

Write in order from least to greatest. *(pages 434–435)*

17. 23, 7, 20
18. ⁻6, ⁻3, ⁻8
19. 4, ⁻8, ⁻1
20. 15, 27, ⁻35, ⁻20
21. 12, ⁻4, 0, ⁻8
22. 1, ⁻1, ⁻5, ⁻4

Add or subtract. *(pages 438–439, 446–447)*

23. 13 − 17
24. 6 + ⁻2
25. 21 + ⁻25
26. ⁻3 − 1
27. ⁻11 + ⁻16
28. ⁻9 − ⁻10
29. 0 − ⁻2
30. 20 − ⁻6
31. ⁻5 + 13
32. 9 + ⁻9
33. ⁻4 − 11
34. ⁻12 + 7

Write the letter of the point named by the ordered pair. *(pages 448–449)*

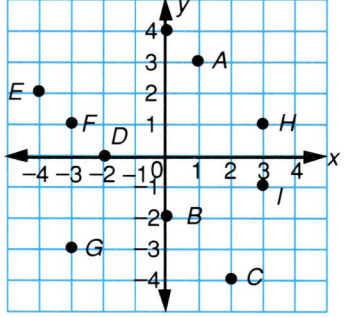

35. (3, ⁻1)
36. (⁻2, 0)
37. (⁻3, ⁻3)
38. (⁻3, 1)
39. (1, 3)
40. (2, ⁻4)

Solve. *(pages 440–441, 450–451)*

41. During a sale, the price of a coat was cut in half. It still didn't sell so another $20 was taken off the price. The coat finally sold for $65. What was its original price?

42. A company is putting prizes into some cereal boxes. There are prizes in the 1st, 7th, 13th, and 19th boxes. If the pattern continues, will there be a prize in the 40th box?

ENRICHMENT

A Game of Twenty

Play this game with a partner. The goal is to be the first to reach or exceed either $^+20$ or $^-20$. Use these patterns to make 2 number cubes.

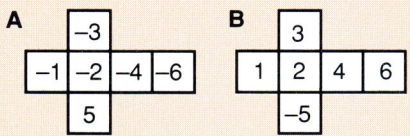

Each player rolls cube A. The player with the greater number goes first. The number rolled also becomes each player's first score.

Players take turns rolling the cube they choose. A table like the one below is used to keep a running total. In this game, player 1 tried for $^-20$ and player 2 tried for $^+20$.

Sample Game

Turn	Player 1		Player 2	
	Roll	Total	Roll	Total
1	−2		5		← goes first
2	−4	−6	1	6
3	−3	−9	3	9
4	−6	−15		

Patterns in Poetry

Some forms of poetry follow definite patterns. Haiku, for example, a Japanese form of poetry, has a pattern both in the number of lines and the number of syllables in each line.

Here is an example of Haiku:

    Silent sleek slider
    Soaring peacefully upward
    Floating on the clouds

Haiku always has three lines. The first line always has 5 syllables; the second, 7 syllables; and the third, 5 syllables.

Cinquain, a French form of poetry, and limericks, originally from Ireland also follow patterns. Find some examples of each type of poem. What patterns can you discover? Try writing a poem in one of these forms.

Coordinates on the Globe

DID YOU KNOW . . . ? The Aleutian Islands in Alaska extend so far west that they reach the eastern hemisphere.

Cape Wrangell, on Attu Island in Alaska, is the westernmost point in the United States. Its longitude is about $172\frac{1}{2}°$ East.

What are the latitude and longitude of a city or town near you?

CUMULATIVE REVIEW

Find the answer.

1. A pair of perpendicular lines form(s)
   a. 1 right angle
   b. 2 right angles
   c. 4 right angles
   d. none of these

2. The longest line segment in a circle is a(n)
   a. arc
   b. radius
   c. diameter
   d. none of these

Find the volume. Round to the nearest whole number.

3.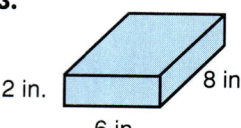
   a. 16 in.³
   b. 28 in.³
   c. 48 in³
   d. none of these

4.
   a. 150 ft³
   b. 2,826 ft³
   c. 471 ft³
   d. none of these

Find the greatest integer.

5. 2, ⁻9, ⁻17, ⁻3
   a. ⁻17
   b. ⁻9
   c. ⁻3
   d. none of these

6. 8, ⁻1, 6, ⁻12
   a. ⁻12
   b. 6
   c. 8
   d. none of these

7. 5, 1, 0, ⁻7
   a. 5
   b. 1
   c. ⁻7
   d. none of these

Find the sum or difference.

8. ⁻12 + 19
   a. ⁻7
   b. ⁻31
   c. 7
   d. none of these

9. 17 − ⁻8
   a. 25
   b. 9
   c. ⁻9
   d. none of these

10. ⁻6 − ⁻10
    a. ⁻16
    b. 4
    c. ⁻4
    d. none of these

Use the grid to name the coordinates of the point.

11. A
    a. (3, 3)
    b. (3, 2)
    c. (2, 3)
    d. none of these

12. B
    a. (⁻1, ⁻3)
    b. (1, ⁻3)
    c. (3, ⁻1)
    d. none of these

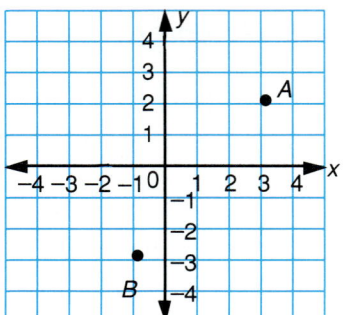

458   CUMULATIVE REVIEW

PROBLEM SOLVING REVIEW

Remember the strategies and types of problems you have had so far. Solve.

Problem Solving Check List
- Multistep problems
- Using estimation
- Using a pattern
- Choosing the operation
- Too little information
- Too much information

1. In one store, apples are sold in bags of 3 for $1.00. Oranges are sold in bags of 4 for $1.00.
   a. How many apples cost $1.00?
   b. In what amounts can you buy apples? oranges?
   c. What is the smallest number of each type of fruit that you can buy to get an equal number of apples and oranges? How much would you spend?

2. A small pizza is 8 in. in diameter and costs $5. A large pizza is 12 in. in diameter and costs twice as much. Is the large pizza at least twice as big as the small one?

3. A round-trip train ticket costs $35. The same trip costs $42 with one-way tickets. How much is saved by buying 10 round-trip tickets?

4. A chunk of Swiss cheese in the shape of a cylinder is 8 in. high with a radius of 10 in. What is the weight of the cheese?

5. A music store is offering a $1.50 discount on each CD sold. Jasmine is planning to buy 3 CDs, each of which usually costs $11.95. How much will she pay after the discount?

6. Chorus members stood on the stage in rows. There were 4 students in the front row, 6 in the second, 8 in the third and so on. If the pattern continues, how many students were in the seventh row?

7. For another concert, the pattern of chorus members was 6 in front, then 7, 8, and so on. If there were a total of 51 students, how many rows were there?

8. A spinner is divided as follows: $\frac{5}{8}$ red, $\frac{1}{8}$ blue, and $\frac{1}{4}$ green. In 1,000 spins, about how many of each color would you expect?

9. A stereo can be bought for a single payment of $229 or 6 payments of $41 each. How much is saved by paying all at once?

10. You started with $10. After buying lunch for $4.48, stamps for $2.90, and a card for $.85, do you have enough for a $2.00 bus fare?

TECHNOLOGY

PICTURE THIS

In the computer game "Sketch a picture," you will plot coordinates on a grid to draw a shape. This pencil and paper activity will sharpen your graphing skills.

Draw these axes on grid paper. Plot the coordinates below. When POINT appears before the coordinates, plot only the point. When DRAW appears before a group of coordinates, join those points in the order given. Compare your graph with those drawn by your classmates.

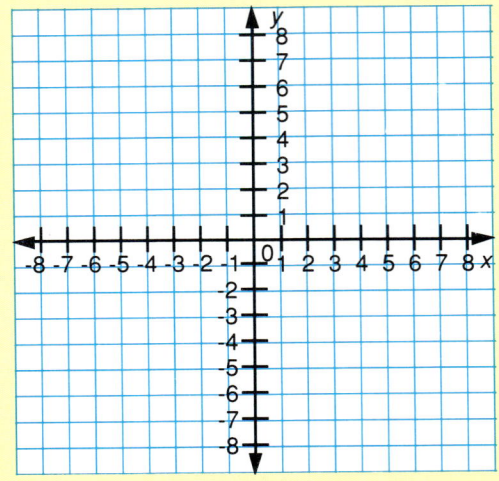

DRAW (5,5), (3,1), (2,4), (5,5)	POINT (⁻5, ⁻5)
DRAW (⁻3, ⁻1), (⁻2, ⁻4), (⁻5, ⁻5)	POINT (+5, ⁻5)

FOOTBALL FRENZY

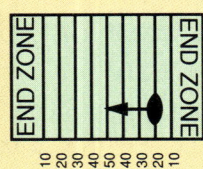

The Rangers have the football on the 20 yd line.

Down 1: gain of 2 yd

Down 2: 10 yd penalty against the Rangers

Down 3: 8 yd forward pass

Down 4: kick 38 yd; ball is caught by the Lions

The Lions run 8 yd and fumble. At what yard mark and at whose end of the field is the ball?

Alaska, with a population of 0.7 people per mi^2, is the least crowded state in the United States. How much greater is the population per square mile in your state than in Alaska?

MATH AMERICA

Pre-Algebra 14

DID YOU KNOW…?
It is estimated that at least 90 million people have thrown a Frisbee® at least once.

FRISBEE WORLD CHAMPIONSHIPS
(Ultimate Division)

Country	Teams (with an average 15 members)											
	Open Class				Women's Class				Junior Class			
	1983	1984	1986	1988	1983	1984	1986	1988	1983	1984	1986	1988
Australia												
Austria												
Belguim												
Canada												
Denmark												
England												
Finland												
France												
Holland												
Italy												
Japan												
Norway												
Poland												
Sweden												
Switzerland												
United States												
W. Germany												

About how many more women competed in 1988 than in 1983? (You may use a calculator if you wish.)

USING DATA
Collect
Organize
Describe
Predict

Explain how you found your answer.

Work with a partner. Take turns making up and answering 3 questions about the data in the table.

CHAPTER 14   461

Writing Expressions

While waiting to buy tickets, Vangy estimated that it takes 30 seconds for the Ferris wheel to make one revolution.

You can write an **expression** to show the number of seconds she waited in line.

- Let r be the number of revolutions she counted while waiting in line.
- Then $r \times 30$ is an expression for the number of seconds she waited.

THINK ALOUD How long would it take the Ferris Wheel to make 10 revolutions?

You can write expressions to represent other situations at the amusement park.

16 people in line increased by m people $\rightarrow$ 16 + m

x dollars for tickets shared among 4 people $\rightarrow \frac{x}{4}$

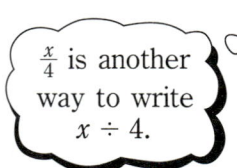

$\frac{x}{4}$ is another way to write $x \div 4$.

Other examples:

Word Phrase	Algebraic Expression
9 less than a number n	$n - 9$
the product of a number s and 8	$s \times 8$

CRITICAL THINKING Explain why you write *9 less than a number* as $n - 9$, not as $9 - n$.

GUIDED PRACTICE

Write as an expression.

1. the sum of a number c and 8
2. five times a number a
3. three less than a number b
4. 8 divided by a number d

Write as a word phrase.

5. $f + 6$
6. $3 \times m$
7. $r - 4$
8. $\frac{s}{5}$

462 LESSON 14–1

PRACTICE

Write as an expression. Let x be the unknown number.

9. four more than a number
10. a number multiplied by eight
11. a number decreased by three
12. twelve subtracted from a number
13. the product of nine and a number
14. ten divided by a number
15. a number increased by 100
16. a number added to itself

Write as a word phrase.

17. $x + 7$
18. 3×2
19. $8 - a$
20. $w \div 4$
21. $n - 5$
22. $11 \times q$
23. $3 \div a$
24. $5 \div p$
25. $\frac{m}{6}$
26. $d \times 15$
27. $\frac{12}{r}$
28. x^2

PROBLEM SOLVING

Work with a partner to write the expression.

29. Waiting to get on the roller coaster and then riding it took you 15 minutes. Let m represent the minutes you waited. Write an expression for the number of minutes you spent on the roller coaster.

30. Let s represent the number of seconds one bumper-car ride lasts. Write an expression for the number of seconds 8 bumper car rides would last.

31. Let h represent the total number of hours you spent in 5 trips to the carnival. Write an expression for the average number of hours you spent each trip.

Critical Thinking

Match each phrase with the expression that represents it.
(*Hint:* A decade is 10 years and a century is 100 years.)

a. $10 - n$
b. $\frac{n}{10}$
c. $100 - n$
d. $n + 10$
e. $10 \times n$
f. $100 \times n$

1. the number of decades in n years
2. the number of years in n decades.
3. Serena's age a decade from now if she is n years old now
4. Mr. Cecotti's age n years ago if he is a century old now.
5. the number of years left in the decade if n years have passed.
6. the number of years in n centuries.

CHAPTER 14

Order of Operations

Gudelia and Jim each used a different calculator to solve this problem:

$$17 + 38 \times 95 = \blacksquare$$

They both entered $17 + 38 \times 95$. Gudelia got 5,225 and Jim got 3,627. Why did they get different answers?

Mathematicians have agreed to follow a certain order when working with several operations. Some calculators do not follow this order.

The rules for the **order of operations** are:
- First do the operation(s) in parentheses.
- Next multiply and divide from left to right.
- Then add and subtract from left to right.

$$17 + 38 \times 95$$
$$17 + 3{,}610$$
$$3{,}627$$

No parentheses are in this problem.

THINK ALOUD Whose calculator followed the order of operations? What order did the other calculator use?

Other examples:

$(6.5 + 2) \times 4$
8.5×4
34.0

Do operations in parentheses. Multiply.

$8 - 8 \div 2 + 3 \times 4$
$8 - 4 + 12$
$4 + 12$
16

Multiply and divide from left to right. Subtract. Add.

THINK ALOUD Why did we subtract before adding in the second example?

CRITICAL THINKING Invent a way to help you remember the order of operations. Explain how it works.

GUIDED PRACTICE

1. Which expressions have the same value? Explain why.

   a. $(20 - 6) \times 2$ **b.** $20 - (6 \times 2)$ **c.** $20 - 6 \times 2$

464   LESSON 14-2

Use the rules for the order of operations to find the value.

2. $12 + 18.6 \div 6$
3. $(54 \div 9) - (2 + 3)$
4. $12 \div 2 \times (3 + 3)$

PRACTICE

Use the rules for the order of operations to find the value.

5. $7 \times (4 \div 2)$
6. $3 + 50.5 \div 5$
7. $2 \times (5 + 4)$
8. $25 - 4 \times 3$
9. $30 - 16 + {}^-10$
10. $6 \times 5 - 2 \times 3$
11. $7 \times 4 \div 2 - 1$
12. $4 + 5 \times (7 - 5)$
13. $27 \div 3 - 3 \times 2$
14. $1 + (3 \div 3) \times 8$
15. $66 \div (3 \times 2) - 3$
16. $16 + 12 - 8 \div 2$

The order of operations was not followed in the examples below. Find the correct answer and explain the mistake.

17. $12 + 18 \div 6 = 5$
18. $30 \div 3 \times 2 = 5$
19. $3 \times 11 - 4 \times 3 = 87$

NUMBER SENSE Use the symbols $+$, $-$, $\times$, and $\div$ to make the equation true.

20. $15 \;\blacksquare\; (5 \;\blacksquare\; 3) = 7$
21. $8 \;\blacksquare\; 5 \;\blacksquare\; 4 = 28$
22. $20 \;\blacksquare\; 8 \;\blacksquare\; 2 = 16$
23. $4 \;\blacksquare\; 8 \;\blacksquare\; 2 \;\blacksquare\; 9 = 14$

Use parentheses to make the equation true.

24. $19 - 8 + 5 = 6$
25. $8 + 3 \times 4 = 44$
26. $5 \times 16 \div 4 + 4 = 40$

ESTIMATE Use each of the digits **6**, **7**, **8**, and **9** once. Use the order of operations to write an expression with a value in the given range.

27. between 40 and 50
28. between 1 and 5
29. between 100 and 125

PROBLEM SOLVING

CHOOSE Choose mental math, pencil and paper, or calculator.

30. My numbers are 15, 13, 2, and 6. My value is 25. What is one way I can be written?

31. Use one of the properties you have learned to rewrite $83 \times (51 + 46)$ without changing its value.

32. **IN YOUR WORDS** I am written as $19 \times 21 \times 20 \div 10 = 798$. Can you add parentheses so that I have a different value? Explain.

33. My 4 numbers are all different. Each is a single-digit even number. My value is 36. What is one way I can be written?

Evaluating Expressions

The cheetah is the fastest land animal. It can run short distances as fast as 34.2 yd/s.

You can write an expression to find how far a cheetah can run in different amounts of time.

- Let s represent the number of seconds.
- Then $34.2 \times s$ represents the number of yards traveled.

- Substituting a value for the variable and finding the value of the expression is called **evaluating** the expression.

Distance (in yd) after 2 seconds:

$34.2 \times s$
34.2×2 ← Substitute 2 for s.
68.4 yd

THINK ALOUD Explain how you would find the distance a cheetah can run in 5 s.

A table can help you to evaluate an expression for several values of a variable.

n	$3 - n$
2	$3 - 2 = 1$
3	$3 - 3 = 0$
7	$3 - 7 = -4$
$^-1$	$3 - {^-1} = 4$

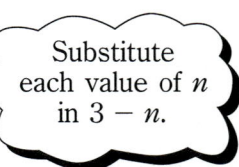

Substitute each value of n in $3 - n$.

To evaluate an expression that has two or more operations, you need to use the order of operations.

Evaluate $3 + 5 \times z$. Use $z = 4$.
$3 + 5 \times 4$ ← Substitute 4 for z.
$3 + 20$ ← Multiply before adding.
23

=== GUIDED PRACTICE ===

Evaluate $n + 5$ for the given value of n.

1. 6 **2.** 12 **3.** 5.5 **4.** $^-8$

Evaluate $6 + 6 \div z$ for the given value of z.

5. 2 **6.** 3 **7.** 6 **8.** 1.5

=== PRACTICE ===

Evaluate. Use $a = 2$, $b = 4$, $c = {^-6}$.

9. $b + 10$ **10.** $9 - a$ **11.** $\frac{a}{2}$ **12.** $5.3 \times b$

13. $12.8 + a$ **14.** $8 \div b$ **15.** $b - 1.2$ **16.** $a \times 13$

17. $b + b$ **18.** $b \times 1.5$ **19.** $\frac{a}{4}$ **20.** $c - 2$

21. $3 \times a - 2$ **22.** $3 \times (a - 2)$ **23.** $6 + b \div 2$ **24.** $\frac{a + 4}{3}$

25. $2 + 3 \times b$ **26.** $b - 0.5 + 2$ **27.** $20 \div a \times 5$ **28.** $10 \div (b + 1)$

29. $4 \times a - 2 \times 3$ **30.** $8 + b \times 4 \div 2$ **31.** $7 - a \div 8$ **32.** $6 - 2 \times 5 - c$

Complete the table.

33.

a	$a + 2$
1	$1 + 2 =$ ▨
2	▨
3.5	▨

34.

n	$3 \times n$
2	▨
4	▨
6	▨

35.

x	$x \div 2 - 1$
2	$2 \div 2 - 1 =$ ▨
6	▨
9	▨

36. CALCULATOR Find three values for m in $\frac{m}{8}$ so the value of the expression is between 1 and 3.

PROBLEM SOLVING

CHOOSE Choose calculator, mental math, or pencil and paper to solve.

Top Speed (in yd/s)	
Cheetah	34.2
Pronghorn antelope	29.3
Human	13.2

37. After running the first 350 yd, the cheetah's speed falls below that of the pronghorn antelope.
  a. Write an expression for the number of yards the pronghorn antelope can cover in s seconds.
  b. Evaluate it for $s = 2, 5,$ and 10.

38. Let y represent the speed in yards per second.
  a. We can represent the time it takes to run a distance of 50 yd as $\frac{50}{y}$. Explain why.
  b. Evaluate $\frac{50}{y}$ when y is the speed of a cheetah, a pronghorn antelope, a human.

CHAPTER 14   467

Exploring Equations: Addition and Subtraction

You already know how to solve some equations using mental math.

An equation is a mathematical sentence stating that two quantities are equal.

$6 + x = 11$

$6 + \square = 11$

Think: 6 plus what number equals 11?

$6 + 5 = 11$,

so $x = 5$ is the **solution** to the equation.

1. Now consider the equation $15 - y = 9$.
   a. What question do you ask yourself when you solve this equation mentally?
   b. What is the solution?

You can use a balance to show an equation.

This weight is unknown. We can call it x.

Each disk weighs 1 lb.

2. a. What expression represents the weight shown on the left side?
   b. What weight is on the right side?
   c. Write the equation shown.

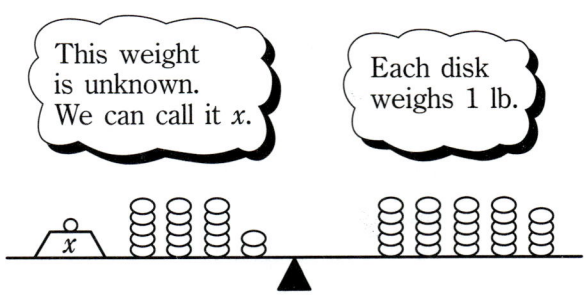

3. a. Is the scale still in balance if you add a 3-lb weight to both sides? Explain.
   b. Did you change the unknown weight x?
   c. Write the new equation.

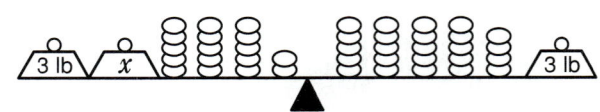

4. a. Is the scale still in balance if you subtract 5 lb from both sides? Explain.
   b. Did you change the unknown weight x?
   c. Write the new equation.

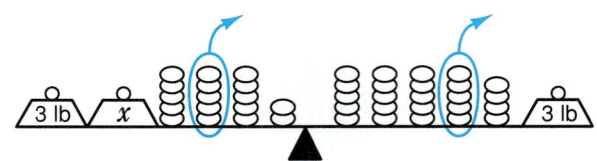

LESSON 14-4

> You can add or subtract the same number on both sides of an equation without changing the solution.

5. Now let's solve the equation you wrote in exercise 2c.

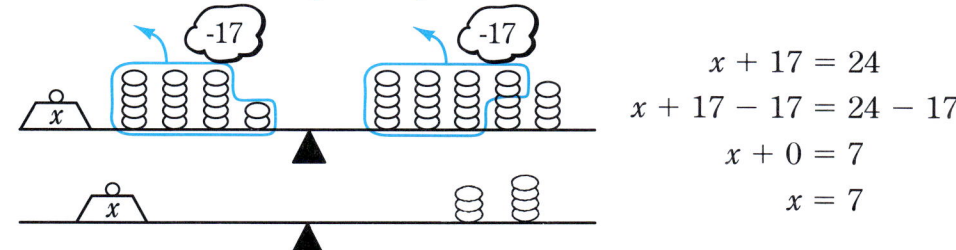

$$x + 17 = 24$$
$$x + 17 - 17 = 24 - 17$$
$$x + 0 = 7$$
$$x = 7$$

  a. What did you need to subtract from the left side to get x by itself?

  b. What did you need to subtract from the right side to keep the two sides equal? Why?

  c. What is the solution?

6. Since addition and subtraction are inverse operations, subtracting 17 undoes adding 17. How could you undo subtracting 12?

If you understand inverse operations, you can make a **MATH CONNECTION** to solve equations.

7. Solve the equation $x - 26 = 15$.

  a. What inverse operation do you need to use?

  b. What do you need to add to both sides to get the x by itself on the left?

  c. The equation becomes $x - 26 + 26 = 15 + 26$. So, $x = \underline{}$.

Use inverse operations to solve.

8. $b + 4 = 20$ **9.** $y - 13 = 7$ **10.** $z + 16 = 28$

11. $n - 17 = 39$ **12.** $c + 12 = 41$ **13.** $s - 9 = 54$

14. $r + 18 = 52$ **15.** $d - 25 = 76$ **16.** $g + 38 = 93$

17. $a - 83 = 151$ **18.** $w - 29 = 815$ **19.** $x + 159 = 204$

20. $m - 0.3 = 0.7$ **21.** $n + 1.5 = 4.2$ **22.** $h + \frac{3}{4} = 1\frac{1}{2}$

Solving Equations: Multiplication and Division

The highest waterfall in the world is Angel Falls in Venezuela. Its longest unbroken drop is about 810 m, which is 15 times as high as Niagara Falls in New York. How high is Niagara Falls?

> Niagara Falls has an average of 212,000 cubic feet of water flowing over it each second. Its water flow is the fourth greatest of any falls in the world.

Let n represent the height of Niagara Falls.

15 times what number is 810?
$15 \times n = 810$

THINK ALOUD What inverse operation would you use to solve for n?

$15 \times n = 810$

$\dfrac{15 \times n}{15} = \dfrac{810}{15}$ ← Divide both sides by 15.

$n = 54$ ← The solution is 54.

Niagara Falls is 54 meters high.

Check.

$15 \times n = 810$

$15 \times 54 = 810$ ✓

> **You can multiply or divide both sides of an equation by the same number without changing the solution. Remember, you cannot divide by zero.**

Another example:

The inverse of division is multiplication. Multiply both sides by 3.

$\dfrac{n}{3} = 12$

$\dfrac{n}{3} \times 3 = 12 \times 3$

$n = 36$

Check:

$\dfrac{n}{3} = 12$

$\dfrac{36}{3} = 12$ ✓

GUIDED PRACTICE

THINK ALOUD Explain how to solve and check the equation.

1. $\frac{x}{5} = 10$
2. $4 \times y = 20$
3. $\frac{z}{4} = 7$
4. $m + 3 = 11$

PRACTICE

Use inverse operations to solve. Check your answer.

5. $2 \times a = 6$
6. $\frac{p}{3} = 6$
7. $\frac{m}{5} = 25$
8. $4 \times n = 60$
9. $6 \times k = 42$
10. $7 \times d = 21$
11. $8 \times z = 108$
12. $\frac{s}{10} = 24$
13. $\frac{x}{16} = 4$
14. $\frac{n}{15} = 12$
15. $9 \times t = 117$
16. $9 \times f = 207$
17. $4 \times g = 10$
18. $2.5 \times r = 7.5$
19. $\frac{v}{4.5} = 2$
20. $a + a = 11.2$

MIXED REVIEW Solve for x. Check your answer.

21. $\frac{8}{12} = \frac{x}{9}$
22. $35\% = \frac{x}{20}$
23. 16% of $250 = x$.
24. $0.2 = \frac{x}{100}$

MENTAL MATH Use mental math to solve.

25. $y + 9 = 17$
26. $\frac{x}{11} = 8$
27. $r - 5 = 5$
28. $9 \times m = 72$

PROBLEM SOLVING

Let h represent the underlined fact. Select the equation that represents the problem and solve. Use the table below.

a. $h - 368 = 54$
b. $2 \times h = 979$
c. $h + 369 = 979$
d. $\frac{h}{2} = 54$

Total Height of Angel Falls: 979 m
Height of U.S. Niagara Falls: 54 m

29. The height of Niagara Falls is 368 m less than the <u>height of Gavarnie Falls,</u> found in France.

30. The total height of Angel Falls is 369 m more than the <u>height of Cuquenán Falls,</u> also in Venezuela.

31. Ribbon Falls in California is the highest waterfall in the U.S. The total height of Angel Falls is about twice the <u>height of Ribbon Falls.</u>

32. Victoria Falls, on the Zambia-Zimbabwe border, has the world's ninth greatest water flow. The <u>height of Victoria Falls</u> divided by 2 is the height of Niagara Falls.

Problem Solving Strategy: Writing and Solving Equations

The Ortiz family is planning a vacation. The cost to rent a cabin will be shared equally by 4 people. Each person will pay $21.25. What is the cost of the cabin?

You can write and solve an equation to answer the question.

- What are you asked to find? Let C be the variable to represent this.
- How many people will share the cost of the cabin?
- What expression represents one person's share?
- What amount is one person's share?
- How can you use the expression and the amount to write an equation?

cost of cabin divided by number of people is one person's share

$$C \div 4 = \$21.25$$

$$\frac{C}{4} = \$21.25$$

The equation balances. Both $\frac{C}{4}$ and $21.25 represent one person's cost.

- How can you use the equation to solve the problem?
- Solve the equation for C.
- Then look back and check your answer.

GUIDED PRACTICE

1. The distance from the Ortiz's home to the cabin is 270 miles. If their truck gets 15 miles per gallon, how many gallons of gas are needed to get to the cabin?
   a. What number tells you how many miles they need to travel to get to the cabin?
   b. Let g represent the number of gallons needed to get to the cabin. Write an expression that represents the number of miles they travel on g gallons of gas.
   c. Write an equation that balances. Solve it to find g.

PRACTICE

Decide on a variable to represent the unknown quantity. Then write an equation and solve.

2. The Ortiz family could have saved $12.50 per day by camping instead of staying in a cabin. Their total savings would have been $37.50. How many days was their trip?

3. While bird watching, the family saw a total of 71 birds. They saw 17 blue jays and 16 cardinals. The rest were robins. How many robins did they see?

4. Steve caught 8 fish. If the number of fish caught by Steve was 5 fewer than the number caught by Helene, how many fish did Helene catch?

5. While staying at the cabin, the Ortiz family spent $12.58 on gasoline. At $1.48/gal, how many gallons were used?

6. **CREATE YOUR OWN** Write a word problem that this equation describes. $\frac{x}{4} = 12$ Then solve the equation.

CHOOSE Choose any strategy to solve.

7. During the trip the 3 children hiked an average of 12 h. Each child hiked 4 h more than the next younger child. How many hours did each child hike?

8. Three people in the Jackson family will split the cost of a trip. Lodging cost $129, food cost $139, and gas cost $59. Will $100 from each be enough to cover the expenses?

9. One day the children spent $\frac{1}{8}$ of the day swimming, $5\frac{1}{3}$ h canoeing, 3 h 20 min horseback riding, 120 min eating, and the rest of the time sleeping. How many hours did they sleep?

CHAPTER 14

MIDCHAPTER CHECKUP

LANGUAGE & VOCABULARY

1. Explain why the expression $n - 8$ has a different meaning than the expression $8 - n$.

2. Explain why the value of the expression $4 + 5 \times 3$ is not equal to the value of the expression $(4 + 5) \times 3$.

QUICK QUIZ

Write as an expression. Let x be the unknown number. *(pages 462–463)*

1. six more than a number
2. a number divided by 10

Use the rules for the order of operations to find the value. *(pages 464–465)*

3. $30 - 6 \times 3$
4. $2 \times 8 + 12 \div 4$
5. $24 \div (6 + 2)$

Evaluate. Use $y = 4$. *(pages 466–467)*

6. $8 - y + 2$
7. $\dfrac{y + 5}{3}$

Use inverse operations to solve. Check your answer. *(pages 468–471)*

8. $14 + m = 32$
9. $\dfrac{x}{6} = 15$

Solve. *(pages 472–473)*

10. The Osbornes paid $8,100 for 12 months rent last year.

    a. What number tells how much rent money they paid for the year?

    b. Let m represent their monthly rent. Write an expression that represents the amount of rent they pay in 12 months.

    c. Write an equation that balances. Solve it to find m.

Write the answer in your learning log.

1. Explain how an equation is different from an expression.

2. Explain how a balance can help you understand the steps for solving an equation.

DID YOU KNOW . . . ? The largest island in the United States is Long Island, New York with an area of 1,396 mi². What is the area of the largest island near you?

Equations Puzzle
Solve for x, y, and z so that the equations balance both across and down.

$x =$ ▨ $y =$ ▨ $z =$ ▨

x	+	15	=	30
÷		÷		−
5	×	y	=	15
=		=		=
3	×	5	=	z

Exploring Functions

Tanya wanted to find out what flavors her friends liked. She had two conditions that her friends had to follow when choosing their favorite flavor.

Condition 1: You must choose a flavor.
Condition 2: You may not choose more than one flavor.

Her friends made these choices.

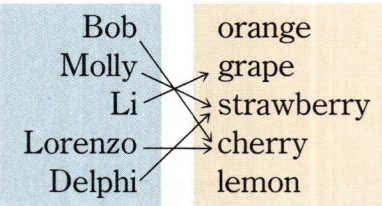

The pairing of students to favorite flavors is a **function** because one and only one flavor was chosen.

1. Notice that each friend can choose only one flavor, but two friends can choose the same flavor. Name two students who did.

2. This diagram does *not* show a function because not every person is assigned a flavor. How can you make it a function?

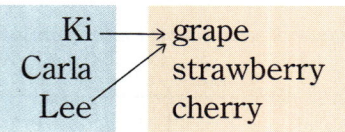

3. This diagram does *not* show a function because condition 2 does not hold. Which student was assigned 2 flavors?

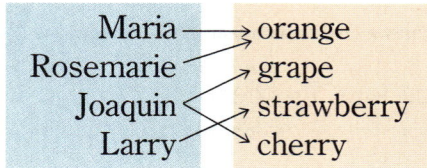

Which of these statements are true for a function?

4. Each student must choose one flavor.

5. Some students can choose 2 flavors.

6. No student can choose more than one flavor.

7. Two or more students can choose the same flavor.

The diagrams below show data about students.

8. One diagram does not show a function. Which one is it? Why?

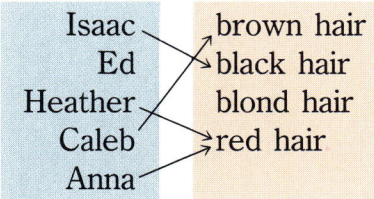
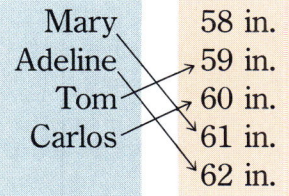

Does each diagram below show a function? If not, tell the condition that does not hold.

Condition 1: Each student has a hair color.

Condition 2: No student may have more than one hair color.

9. Pedro → brown hair
    Tara → black hair
    Hector → blond hair
    Rowana → red hair
    Jack

10. Jane → brown hair
    Carl → black hair
    Dahlia → blond hair
         → red hair

Many functions pair numbers with other numbers.

11. Sophie grows 2 inches per year from age 10 to age 15. Her height at age 10 was 56 in. Complete the diagram and the table to show the function for Sophie's age and height.

    10   56 in.
    11   58 in.
    12   ?
    13   ?
    14   ?
    15   ?

Sophie's age (a)	10	11	12	13	14	15
Sophie's height (h)	56	?	?	?	?	?

12. Make a table of the function for Lee's age and height. His height (h) increases $1\frac{1}{2}$ inches each year (y) from ages 11–16. Lee's height at age 11 is 60 in.

13. **CREATE YOUR OWN** Make up 2 functions of your own. Be sure to follow the two conditions for a function. Discuss them with a partner.

Exploring Function Rules

The sixth grade is going on a trip to Washington, D.C., by bus. Each bus holds 30 passengers.

The number of buses (*b*) and the number of passengers (*p*) that fit in each bus are shown below:

Buses	Passengers
1	30
2	60
3	90
4	120
5	150

Buses (*b*)	1	2	3	4	5
Passengers (*p*)	30	60	90	120	150

For each number of buses, there is one and only one number of passengers. This pairing of buses and passengers is a function.

1. How are the number of buses and passengers related?

2. You can find the number of passengers (*p*) by multiplying what number times the number of buses (*b*)?

You can write a **function rule** to show this relationship.

Number of passengers = 30 × number of buses
$p = 30 \times b$ ← function rule

You can use the function rule to find the number of people any number of buses will hold. How many passengers will 7 buses hold?

$p = 30 \times b$
$p = 30 \times 7$ ← Substitute 7 for *b*
$p = 210$

3. How many passengers will 10 buses hold?

4. How many buses will you need for 360 passengers?

The principal decides to put only 25 passengers on each bus.
The new function rule would be $p = 25 \times b$

5. Make a table to show the function for 1 to 8 buses.

6. What function rule would you write if 50 passengers fit in each bus?

Use the given function rule to complete each table.

7. Function rule: $l = 5 - t$

Hours traveled (t)	0	1	2	3	4	5
Hours left (l)						

8. Function rule: $p = \frac{s}{2}$

Suitcase (s)	4	8	12	16	20
Passengers (p)					

9. Function rule: $f = 3 \times y$

Number of yards (y)	2	4	6	8	9
Number of feet (f)					

10. Function rule: $h = \frac{m}{60}$

Hours (h)	1	2	3	4	5
Minutes (m)					

Write the rule for the function.

11.
Numbers of students (s)	2	4	6	8
Admission collected (a)	$30	$60	$90	$120

12.
Number of passengers (p)	1	2	3	4
Round trip fare (f)	$35	$70	$105	$140

13.
Pounds (p)	1	2	3	4	5
Ounces (o)	16	32	48	64	80

14.
Inches (i)	36	108	180	252	324
Yards (y)	1	3	5	7	9

15. Use the function in question 11 to find the admission collected from 14 students.

16. **CRITICAL THINKING** Is there a way of calculating someone's favorite color if you know his or her name? Does this function have a function rule?

CHAPTER 14   479

Problem Solving Strategy:
Guess and Check

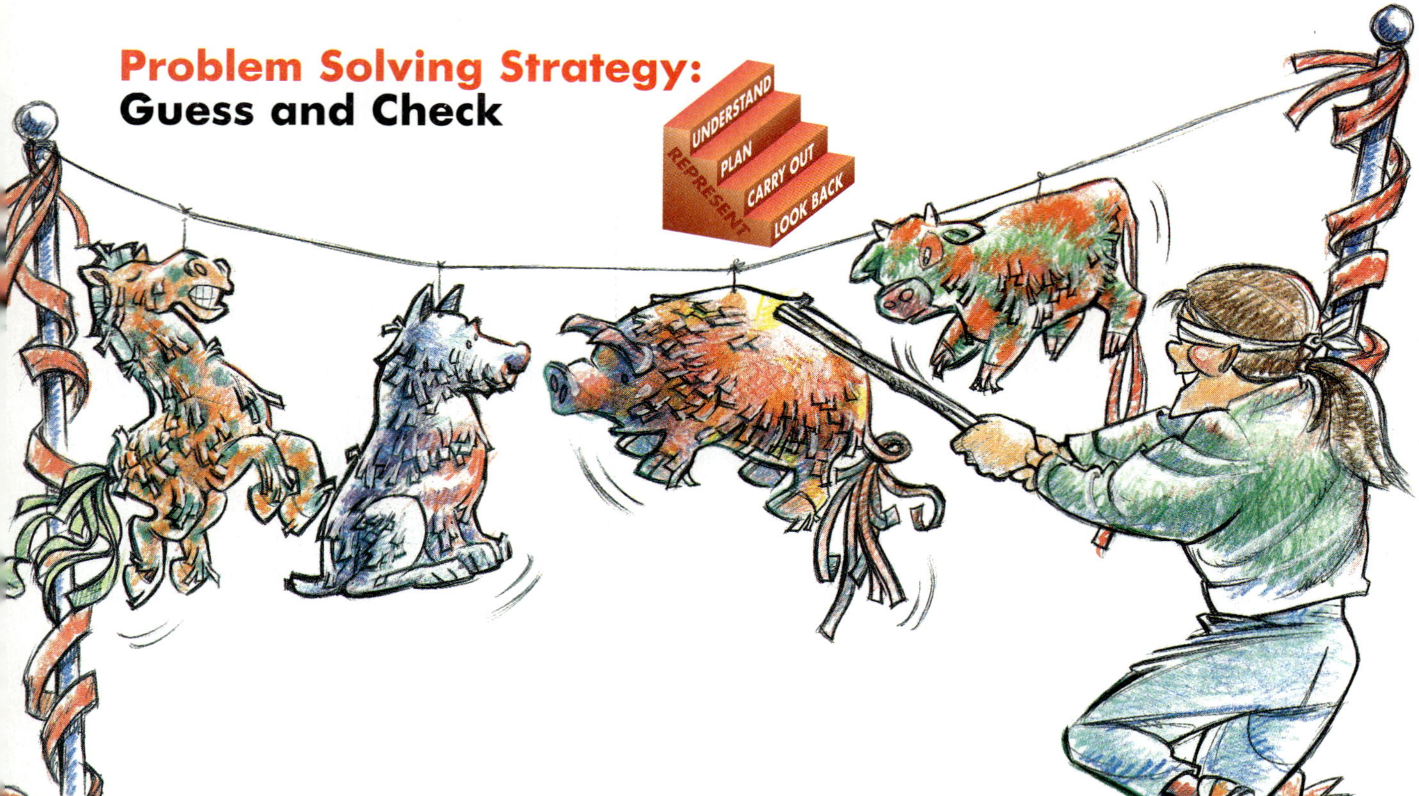

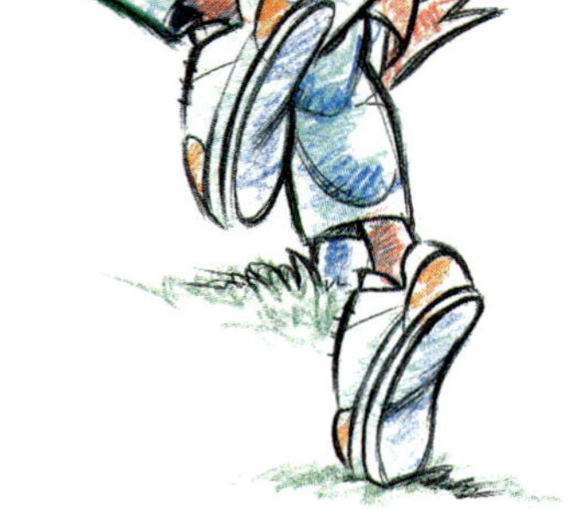

Luis made 4 piñatas for the fiesta. The piñatas will have a total of 52 toys. Each piñata has 2 more toys than the piñata to its left. How many toys are in each piñata?

Ask yourself questions to be sure you understand the problem.

- How many piñatas in all?
- How does the number in each piñata compare with the number in the piñata to its left?

Sometimes the best way to solve a problem is to guess and check. List your guesses in a table.

Try 14 toys in the horse

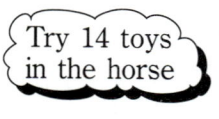

horse	dog	pig	cow	Total	Answer needed	Too high? Too low?
14	16	18	20	68	52	too high

Use your first guess to help you make a new one.

- How can you make your total lower?
- Keep using guess and check until you solve the problem.
- What would be your next guess?
- How can you check your answer?

GUIDED PRACTICE

Make a table and use guess and check to solve.

1. Suppose there are 110 toys in 5 piñatas. Each piñata contains 1 more toy than the piñata on its right.
   a. How many piñatas are there?
   b. Which piñata has the most toys?
   c. How many toys are there in piñata 1?

PRACTICE

Use the guess and check strategy to solve.

Refreshments	
Burritos	$1.25
Salsa	.75
Juice	.50

2. Elena bought some burritos, salsa, and juice. She spent $7.50. Name one group of items she could have bought.

3. Juan bought 8 items from the refreshment stand. He spent a total of $9.50. How many of each did he buy?

4. Alex bought 7 items, (burritos and juice only). If he bought 3 more burritos than juices, how much did he spend in all?

5. At the fireworks display, Suzi saw 13 rockets go up. Some rockets sent off 3 showers of light and others only 2 showers. If there were 34 showers in all, how many of each type of rocket were sent up?

PROBLEM SOLVING

 Choose any strategy to solve.

6. A souvenir cost $1 more than a trinket. Rosa bought 2 trinkets and 3 souvenirs. She spent a total of $15.50. What was the price of each?

7. The parade float is 6 ft by 10 ft. If crepe paper is put around the bottom, how many feet of crepe paper are needed?

8. A dance is done in this pattern of steps: 2 steps right, 2 steps left, 3 steps forward, 3 steps back. Where will the dancer be after 31 steps?

9. The fireworks begin at 8:00 p.m. Rosa wants to spend $\frac{3}{4}$ h sampling food, $\frac{3}{4}$ h playing games, and 25 min dancing before watching the fireworks. It takes her 15 min to get to the street where the festival is held. What time should she leave her house?

Mixed Review

Write in standard form.

1. thirty-one thousand, five
2. four thousandths
3. ten and thirteen hundredths
4. five million, six hundred

Round to the place of the underlined digit.

5. 7,8̲01
6. 8̲7,415
7. 9.01̲7
8. 0.07̲4
9. 4.1̲72
10. 14̲.89
11. $9\underline{\tfrac{5}{8}}$
12. $3\underline{\tfrac{5}{12}}$

Compare. Write >, <, or =.

13. 6.74 ▨ 6.8
14. 0.94 ▨ 0.904
15. 5.5 ▨ $5\tfrac{1}{2}$
16. 3.15 ▨ $3\tfrac{1}{10}$
17. $\tfrac{10}{3}$ ▨ $\tfrac{7}{2}$
18. $2\tfrac{2}{3}$ ▨ $2\tfrac{7}{12}$

Find the answer.

19. $125{,}074 + 89{,}654$
20. $19{,}123 - 8{,}974$
21. 324×28
22. $650 \div 27$
23. 1.4×0.04
24. $3.1 - 1.87$
25. $7.5 + 0.124$
26. $0.008 \div 0.2$
27. $6.89 + 0.987$
28. $8 - 2.78$
29. 3.42×1.05
30. $3.87 - 1.8$
31. $\tfrac{5}{12} + \tfrac{3}{8}$
32. $8\tfrac{1}{7} - 5\tfrac{6}{7}$
33. $2\tfrac{3}{4} \times 4$
34. $1\tfrac{3}{5} \div \tfrac{5}{8}$
35. $1\tfrac{2}{3} \times \tfrac{3}{5}$
36. $4\tfrac{1}{3} - 2\tfrac{7}{9}$
37. $9 \div 1\tfrac{1}{2}$
38. $2\tfrac{7}{9} + 3\tfrac{5}{6}$
39. $2\tfrac{1}{2} \times 1\tfrac{1}{4}$
40. 10% of 50
41. 15% of 40
42. 20% of 30

PROBLEM SOLVING

CHOOSE Choose estimation, mental math, calculator, or pencil and paper to solve.

43. Woo bought a bathing suit for $12.99, a snorkel for $19.95, and fins for $9.99. How much did he spend in all?

44. This year 120 children signed up for swimming lessons. There are 15 equal-sized classes. How many children are in each class?

LESSON 14–10

45. A swimming pool is 75 ft long by 48 ft wide. The water is $4\frac{1}{2}$ ft deep. How many cubic feet of water are in the pool?

46. Reiko left the house at 12:15 P.M. She swam for $1\frac{1}{2}$ h and returned home. The pool is 15 min from her house. When did she get home?

47. The number of cars in the beach parking lot on Monday through Friday was 1,089; 2,874; 1,879; 1,764; and 2,741. How many cars were in the lot on those days?

48. Luis is placing floats on the lane markers for a race. Each lane-marker rope is 50 ft long. If a float is placed every $2\frac{1}{2}$ ft, how many floats does he need for each rope?

49. Fred Baldasare completed the first underwater cross-Channel swim using scuba tanks in about 18 h. The distance covered was 42 mi. About how many miles did he swim per hour?

50. If Shani can swim $1\frac{1}{2}$ yd per second, how far can she swim in 1 min?

Solve. Use the chart.

51. How much faster is the sailfish than the wahoo?

52. How much greater or less is the speed of each fish than the 55 mi/h speed limit on a highway?

53. The sailfish's speed in miles per hour was calculated from how long it takes to swim 100 yd. The sailfish took 3 s to swim 100 yd. At this rate, how far can it swim in 1 s?

Speed of Fastest Fish

Fish	Speed in mi/h
sailfish	68.1
swordfish	57.6
wahoo	47.8

CHAPTER CHECKUP

LANGUAGE & VOCABULARY

Explain when you would solve an equation by
a. subtracting the same number from both sides.
b. multiplying both sides by the same number.

TEST

CONCEPTS

Write as an expression. Let x be the unknown number. *(pages 462–463)*

1. the product of 6 and a number
2. a number increased by 5.
3. 20 divided by a number
4. 8 less than a number

Write as a word phrase. *(pages 462–463)*

5. $12 - x$
6. $\frac{y}{2}$
7. $8 + z$
8. $15 \times m$

SKILLS

Use the rules for the order of operations to find the value. *(pages 464–465)*

9. $6 \times (8 + 3)$
10. $20 - 1 + (5 \times 2)$
11. $5 + 4 \times 3 + 2$

Evaluate. Use $m = 6$, $n = 8$. *(pages 466–467)*

12. $13 - m$
13. $2.5 \times n$
14. $4 + n \div 2$

Use inverse operations to solve. *(pages 468–471)*

15. $b - 19 = 52$
16. $r + 27 = 305$
17. $7 \times s = 35$
18. $\frac{x}{12} = 4$
19. $18 \times c = 90$
20. $p + 1.8 = 6.5$

Write the rule for the function. *(pages 478–479)*

21.
People (p)	1	2	3	4
Dollars (d)	8	16	24	32

22.
Pages read (r)	5	10	15	20
Pages left (l)	95	90	85	80

PROBLEM SOLVING

Solve using an equation. *(pages 472–473)*

23. Kurt bought some shirts for which he pays $23.50 each. The total cost was $141.
  a. What number tells you the amount he paid for all the shirts?
  b. Let s represent the number of shirts he bought. Write an expression that represents the amount of money he would spend on shirts.
  c. Write an equation that balances. Solve it to find s.

24. A fishing boat needs to catch 27 more fish to reach its limit of 305. How many fish have already been caught?

Solve using guess and check. *(pages 480–481)*

25. A farmer and two of his helpers are in a barn with some three-legged milking stools and some cows. There are 38 legs in the barn, including the people, the stools, and the cows. What is the largest number of cows that can be in the barn?

Write the answer in your learning log.
  1. Explain what it means when you solve an equation and get $x = 4$.

  2. Explain how you can tell whether a pairing of people to favorite songs is a function.

EXTRA PRACTICE

Write as an expression. Let x be the unknown number. *(pages 462–463)*

1. a number increased by three
2. a number added to nine
3. the product of two and a number
4. a number divided by eight
5. a number decreased by seven
6. ten less than a number

Write as a word phrase. *(pages 462–463)*

7. $p + 3$
8. $m \times 7$
9. $w - 1$
10. $r \div 2$
11. $2 \times b$
12. $6 + a$
13. $\frac{6}{x}$
14. $8 - e$

Use the rules for the order of operations to find the value. *(pages 464–465)*

15. $6 + 2 \times 5$
16. $21 - 6 \times 3$
17. $30 \div 6 + 5$
18. $2 \times (6 + 4)$
19. $8 - (3 \times 2) + 2$
20. $8 \div 2 \times 4 - 1$

Evaluate. Use $c = 8$, $d = 6$. *(pages 466–477)*

21. $c + 9$
22. $2 \times d$
23. $c + 8 - 1$
24. $8.3 - d$
25. $c \div 2 + 3$
26. $\frac{d}{3}$
27. $12 - c + 2$
28. $20 \div (10 - d)$

Use inverse operations to solve. Check your answer. *(pages 468–471)*

29. $m + 15 = 23$
30. $x - 115 = 97$
31. $y + 39 = 68$
32. $z - 2.5 = 11$
33. $\frac{n}{4} = 27$
34. $9 \times p = 72$
35. $7 \times r = 161$
36. $\frac{e}{10} = 12$

Solve. *(pages 472–473, 480–481)*

37. Ben made 3 trips last year. The total distance he traveled was 831 mi. If the first 2 trips were 268 mi and 359 mi, how far was the third trip?

38. The fencing that George needs comes in 5-ft and 9-ft sections. George bought 89 ft of fencing. How many sections of each length did he buy if he bought at least two sections of each size?

ENRICHMENT

FUNCTION MACHINES

A function machine operates on an input according to some rule. The machine gives the new value as an output.

You can have a 1- or 2-operation rule for a machine.

Play a game with a partner. Show at least 4 inputs and the corresponding outputs for a 1-operation machine. Challenge your partner to guess the rule. When you are successful with 1-operation machines, try some with 2 operations.

A GEOMETRY SCAVENGER HUNT

How many of these geometric figures can you find in your classroom?

- a triangle — either equilateral, isosceles, or scalene
- a quadrilateral that is not a rectangle
- any polygon with more than 4 sides
- an ellipse (oval)
- intersecting lines that are not perpendicular
- an obtuse angle
- an acute angle
- a prism that is not a rectangular prism
- any pyramid
- a space figure that is neither a prism nor a pyramid

CUMULATIVE REVIEW

Find the expression.

1. a number increased by three
   - a. $3 - n$
   - b. $n - 3$
   - c. $n + 3$
   - d. none of these

2. the product of ten and a number
   - a. $10 \times n$
   - b. $n + 10$
   - c. $n \div 10$
   - d. none of these

Find the word phrase.

3. $4 - m$
   - a. four less than a number
   - b. a number more than four
   - c. four decreased by a number
   - d. none of these

4. $\frac{m}{2}$
   - a. a number divided by two
   - b. two more than a number
   - c. two less than a number
   - d. none of these

Find the value. Use $x = 3$, $y = 4$.

7. $15 - x$
   - a. 18
   - b. 12
   - c. 5
   - d. none of these

8. $3 \times (x + 2)$
   - a. 15
   - b. 8
   - c. 11
   - d. none of these

9. $\frac{y + 2}{2}$
   - a. 2
   - b. 4
   - c. 6
   - d. none of these

Solve.

10. $c + 17 = 81$
    - a. 98
    - b. 74
    - c. 64
    - d. none of these

11. $d - 39 = 115$
    - a. 154
    - b. 86
    - c. 76
    - d. none of these

12. $\frac{e}{9} = 7$
    - a. 2
    - b. 16
    - c. 63
    - d. none of these

Find the function rule.

13.
a	1	2	3	4
b	2	4	6	8

   - a. $a = 2 \times b$
   - b. $b = 2 \times a$
   - c. $b = a + 2$
   - d. none of these

14.
m	1	2	3	4
n	4	4	6	7

   - a. $n = m + 3$
   - b. $m = n + 3$
   - c. $n + m = 3$
   - d. none of these

15.
x	$1.00	$2.00	$3.00
y	$1.50	$2.50	$3.50

   - a. $x - \$.50 = y$
   - b. $y + \$.50 = x$
   - c. $y = x + \$.50$
   - d. none of these

PROBLEM SOLVING REVIEW

Remember the strategies and types of problems you have had done so far. Solve.

Problem Solving Check List
- Using guess and check
- Working backward
- Multistep problems
- Choosing the operation
- Using proportion
- Drawing a diagram

1. The garden club had 34 members. Then 9 members left to join the sports club. This made the two clubs equal in size.
   a. How many members did the garden club originally have?
   b. Originally, which club had more members?
   c. How many members did the sports club originally have?

2. Kate is making a drink for a party. Her recipe uses 3 parts orange juice to 2 parts grapefruit juice. She has 9 pt of orange juice and 4 pt of grapefruit juice. If she uses all the orange juice, does she have enough grapefruit juice?

3. A dictionary weighs as much as 2 novels and 3 magazines. A novel weighs as much as 3 magazines. How many magazines weigh as much as a dictionary?

4. A group of friends ordered pizza slices. They ate half the slices, took a break to play a game, then ate 10 more slices. There were 2 slices left. How many slices did they order?

5. Michael wants to sleep $8\frac{1}{2}$ h tonight and wake up at 6:15 tomorrow morning. What is the latest he can go to sleep?

6. Andre has completed $\frac{1}{3}$ of a 150-mi bicycle race. If he averages 23 mi/h for the rest of the distance, can he complete the race in 4 more hours?

7. Sasha has test grades of 80, 82, and 76. What grade must she get on her next test in order to have an average of exactly 80 on all tests?

8. Doug is thinking of a number. If he multiplies it by 3, adds 3 to the result, and then divides by 3, he gets 8. What is his number?

9. A florist has fewer roses than tulips, more roses than irises, but no daffodils. She has fewer tulips than lilies. List the types of flowers the florist has from fewest to most.

10. It took Sam 15 min longer to do his math than his science homework. If the total time needed was 1 h 5 min, how long did he work on each assignment?

TECHNOLOGY

SLIDING HOME

In the computer game "Slide-Flip-Turn," you slide a figure on a coordinate graph. This activity will sharpen your graphing skills.

Copy the graph at the right on grid paper. Follow the directions to move the figure and record the coordinates on which the star on the figure rests. Make your moves from the point at which you stop. Record and compare your coordinates with your classmates.

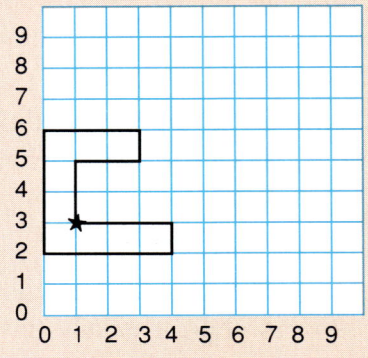

1. Up 2, Right 4, Stop.
2. Down 2, Left 3, Stop.
3. Up 3, Right 5, Stop.
4. Down 3, Left 6, Stop.

The first telephone book was published in New Haven, Connecticut in 1878. The book contained only 50 names. Use your calculator to estimate the number of names in the white pages of your local telephone book.

ON SALE

The Ski Shop is getting ready to put some items on sale. Use your calculator to determine the missing prices on the chart of sale items.

Item	Wholesale	Plus Profit	Less Discount	Sale Price
gloves	$36	$12	$4	
boots	$158	$25		$175
skis		$57	$25	$288
parka	$198		$15	$208

GAMES

This section contains some games you can play at school or at home with your friends or family. You will need only a few simple materials to play most of these games. This list tells you what these items are and how you can substitute if you don't have them.

Pencils, regular paper, graph paper
A calculator
Number cubes (1–6) or numbered slips of paper in a box
A spinner (0–6) or numbered slips in a box
Different color markers, buttons, or small circles cut
    from paper
Small index cards or pieces of construction paper

3 NUMBER CARD GAMES

The games on the next three pages are played with one set of number cards. There are ten groups of four cards in the set. Each group consists of a card with an answer to a multiplication or division problem with decimals and three cards showing problems with that answer.

To make this set of cards you will need 40 small index cards or slips of construction paper. Use a crayon or felt marker to write on each card. Be sure that the numbers do not show through the back of the cards.

Card 1	Card 2	Card 3	Card 4
7	14 ÷ 2	1.4 ÷ 0.2	0.14 ÷ 0.02
70	140 ÷ 2	14 ÷ 0.2	1.4 ÷ 0.02
15	10 × 1.5	100 × 0.15	1000 × 0.015
1.5	15 ÷ 10	150 ÷ 100	1500 ÷ 1000
20	60 ÷ 3	6 ÷ 0.3	0.6 ÷ 0.03
200	4 × 50	400 × 0.5	0.4 × 500
100	2300 ÷ 23	230 ÷ 2.3	23 ÷ 0.23
10	4 ÷ 0.4	0.4 ÷ 0.04	40 ÷ 4
2.1	6.3 ÷ 3	0.63 ÷ 0.3	63 ÷ 30
2.1	3 × 0.7	0.3 × 7	0.03 × 70

You can make different sets of cards to play these games. Make cards with basic facts or equivalent fractions or sides of polygons. See how many interesting ways you can find to play these games.

THE LAST CARD

Goal: To capture the last card
Number of Players: 2 or more
Skills: Chapters 3, 4
Materials: The set of 40 cards, a number cube

Rules
1. Choose 10 pairs of cards from the set. Any 2 cards with the same value are a pair. Mix up the cards, but keep them number side down.

2. Choose one card from these 20 cards. Put it number side up on the table. This is the Last Card.

3. Deal the cards to all players. One player may have an extra one.

4. All players sort their cards and place any pairs number side up on the table. Check that all the players have put down pairs.

5. Take turns rolling the number cube. The player with the least number goes first.

6. Take turns picking a card from the next player's hand. If the card you pick makes a pair with one of your cards, place the pair number side up on the table. If you can't make a pair, keep the card.

7. Play until all the pairs are made. The winner is the player who makes the last pair with the Last Card.

CATCH A PAIR

Goal: To make the most pairs
Number of Players: 2 or more
Skills: Chapters 3, 4
Materials: The set of 40 cards, a number cube (1–6)

Rules
1. Mix up all the cards, but keep them number side down. Give each player 6 cards. Put the rest number side down between the players.

2. Check your cards for pairs. Any 2 cards with the same value are a pair. Place any pairs number side up near you on the table. Check that all the players have put down pairs.

3. Take turns rolling the number cube. The player who rolls a number closest to 2 goes first.

4. On each turn ask the next player for a card with the same value as one in your hand.

   a. If the next player has a card with that value, she or he must give it to you. Put the pair number side up on the table and take another turn.

   b. If the next player doesn't have a card with that value, take the top card from the stack of cards. If that card has the value you asked for, put the pair number side up on the table and take another turn.

   c. If the card does not have the value you asked for, check to see whether it makes a pair with any card in your hand. If it does, put the pair number side up on the table. Then it is the next player's turn.

5. Play until all the pairs are made. The winner is the player with the most pairs.

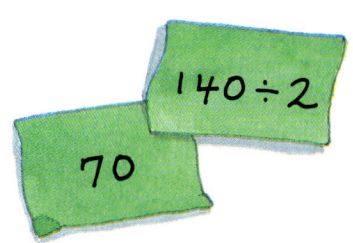

MAKE A MATCH

Goal: To make the most pairs
Number of Players: 2 or more
Skills: Chapters 3, 4
Materials: The set of 40 cards, a number cube

Rules
1. Choose 10 pairs of cards from the set. Any 2 cards with the same value are a pair. Mix up the cards, but keep them number side down.

2. Place the cards number side down in 4 rows with 5 cards.

3. Take turns rolling the number cube. The player with the greatest number goes first.

4. Take turns turning over 2 cards. If the cards you pick make a pair, take it, place it number side up near you on the table, and turn over 2 more cards. Continue if you turn over another pair. If the cards are not a pair, put them back number side down where they were.

5. Play until all the pairs are made. The winner is the player with the most pairs.

CONTACT 6

Goal: To get the most points for touching boxes
Number of Players: 2 (or 1)
Skills: Chapters 2, 3, 4
Materials: 24 different color markers for each player, 3 number cubes

Rules for Playing with a Partner

1. Take turns rolling 2 cubes to see who plays first. Multiply the 2 numbers. The player with the higher product plays first.

2. On each turn, roll 3 cubes. Make a number by adding, subtracting, multiplying, or dividing the numbers on the cubes. Put one of your markers on the box with your number, but don't cover another marker.

3. If the box you put the marker on touches other boxes with markers, you get a point for each box with a marker. The boxes may touch on a side or at a corner. After each turn, record the number of points you got on that turn.

4. Play until each player has had 8 turns.

5. The winner is the player who has more points. You may use a calculator to add your total points.

Rules for Playing Alone

1. Roll the 3 cubes. Make a number by adding, subtracting, multiplying, dividing or using as an exponent the numbers on the cubes. Put one of your markers on the box with your number, but don't cover another marker.

2. If the box you put the marker on touches other boxes with markers, you get a point for each box with a marker. The boxes may touch on a side or at a corner. After you put down a marker, record the number of points you got with that marker. Roll the cubes 8 times.

3. Find your total score. Use a calculator if you want. Play again. Try to get a higher total.

CONTACT 6

			0				

1	2	3	4	5	6	7	8
9	10	11	12	13	14	15	16
17	18	19	20	21	22	23	24
25	26	27	28	29	30	31	32
33	34	35	36	37	38	39	40
41	42	44	45	48	50	54	55
60	64	66	72	75	80	90	96
100	108	120	125	144	150	180	216

ACTION FRACTIONS (I AND II)

Goals: To obtain the most markers in Action Fractions I and the fewest markers in Action Fractions II.
Number of Players: 2 or 3
Skills: Chapters 6, 7
Materials: 90 markers, 2 number cubes (1–6)

Rules for Action Fractions I

1. Notice that the big circle is divided into sections containing small fraction circles. The fraction circles in each section add to 1. Cover each fraction circle with a marker.

2. Take turns rolling the 2 number cubes and making a fraction with the numbers you roll. You may make fractions greater or less than 1.

3. Pick up markers from fraction circles that make a sum equivalent to the fraction you made. If you can't make a sum equivalent to your fraction, your turn is over.

4. The person who has the most markers at the end of the game is the winner.

Rules for Action Fractions II

1. Remember that the small fraction circles in each section add to 1. Cover each fraction circle with a marker.

2. Take turns rolling the 2 number cubes and making a fraction with the numbers you roll. You may make fractions greater or less than 1.

3. Pick up markers from fraction circles that make a sum equivalent to the fraction you made. If you can't make a sum equivalent to your fraction, start your turn again.

4. The person who has the fewest markers at the end of the game is the winner.

ACTION FRACTIONS (I AND II)

FOUR IN A ROW

Goal: To make a line of four markers in a row, column, or diagonal
Number of Players: 2
Skills: Chapter 13
Materials: Graph paper, a spinner (0–6)

Rules

1. Make a grid that is six lines by six lines. Label the origin 0. See the sample game below.

2. Each player spins a number. The player with the greatest number goes first, and chooses whether to use X or O.

3. On each turn, the player spins 2 numbers. The first number tells how many lines to the right he or she moves from the origin. The second number tells how many lines up from the origin. The player plots the point with an X or an O. If a point is plotted incorrectly, it is changed to a mark for the other player.

4. The winner is the player who first plots 4 points consecutively, in a row, column, or diagonal.

Sample Game

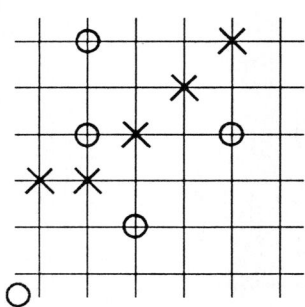

Player 1—X	Player 2—O
2,3	1,5
3,4	1,3
0,2	2,1
1,2	4,3
4,5 (winner)	

USING A CALCULATOR

Different Calculators Every calculator is different. Many of the keys that may appear on your calculator are pictured below, although the keys may be in different places. You should read the instructions for your calculator to learn how to use it.

The Display A calculator usually does not show commas. Usually 8 digits is the most the display will show. A display never shows a dollar sign. You must press the decimal point key to show a decimal number. You press the clear key to return the display to zero.

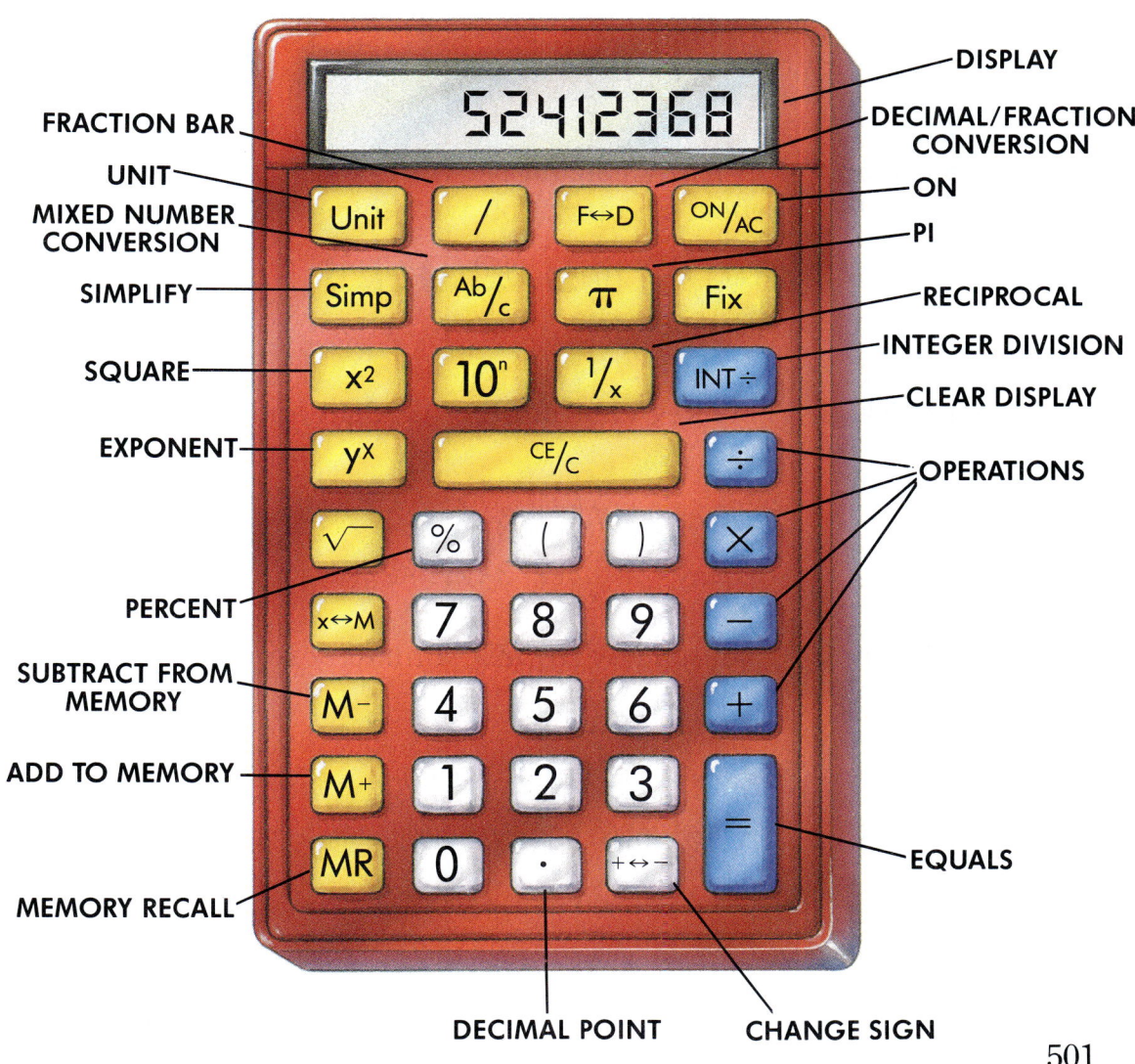

The Memory Feature   [M+] stores the displayed number in the memory or adds the displayed number to the memory.

[M–] subtracts the displayed number from the memory.

[MR] displays the number stored in the memory.

If you wish to clear the memory, press [MR] followed by [M–]. This subtracts the number stored in the memory. You can add a given number to several other numbers by using the memory feature. Add 36 to 9 and to 29.

Press   36 [M+] 9 [+] [MR] [=]

Display   36   M36   M9   M36   M45

Press   29 [+] [MR] [=]

Display   M29   M36   M65

So, $9 + 36 = 45$ and $29 + 36 = 65$.

The Constant Feature   On many calculators the [=] key can also be used to repeat an operation with the same number.

Press   10 [–] 2 [=] [=] [=] [=]

Display   10   2   8   6   4   2

Integer Division Key   This key can find the quotient and remainder of $17 \div 3$.

Press   17 [INT÷] 3 [=]

Display   17   3   Q5 R2

The Exponent Key   Use this key to raise a number to a power. Find 10^4.

Press   10 [y^x] 4 [=]

Display   10   4   10000

So, 10^4 is 10,000.

Operations Using Fractions and Mixed Numbers

The keys discussed here enable you to use your calculator to perform operations with fractions. If your calculator does not have them, change fractions to decimals before performing operations with them.

[Unit] enters the whole number part of a mixed number.

[/] fraction bar.

[Ab/c] converts a mixed number to a fraction.

[F↔D] converts a fraction to a decimal or vice versa.

[Simp] simplifies a fraction.

Add $1\frac{8}{10}$ and $3\frac{7}{10}$, and express the sum in simplest form.

Press 1 [Unit] 8 [/] 10 [+] 3 [Unit] 7 [/] 10   [=]

Display   1u 8/10   3u 7/10   4u 15/10

Press   [Ab/c] [Simp] [=]

Display   5u 5/10   5u 1/2

The result is $5\frac{1}{2}$.

The Percent Key

Many calculators have a percent key. You can use it to solve problems with percents. Find 50% of 80. On most calculators, enter 80 first.

Press   80 [×] 50 [%] [=]

Display   80   50 0.5 40

Other Keys

[π] lets you to calculate with π.

[1/x] gives you the reciprocal of a number.

[x^2] gives you the square of the number in the display.

[+↔−] can be used to enter a negative number or to change the sign of a number in the display.

GLOSSARY

A

actual probability (p. 381) The number of favorable outcomes divided by the total number of possible outcomes.
acute angle (p. 280) An angle that measures less than 90°.
acute triangle (p. 282) A triangle with all acute angles.
analogy (p. 73) A relationship between two pairs of words or numbers.
angle (p. 279) Two rays that have a common endpoint.
area (A) (p. 302) The number of square units that fit inside a figure.
Associative Property of Addition (p. 36) Changing the grouping of the addends does not change the sum.
Example: $(4 + 3) + 7 = 14$
$4 + (3 + 7) = 14$
so $(4 + 3) + 7 = 4 + (3 + 7)$
Associative Property of Multiplication (p. 68) Changing the grouping of the factors does not change the product.
Example: $(2 \times 5) \times 4 = 40$
$2 \times (5 \times 4) = 40$
so $(2 \times 5) \times 4 = 2 \times (5 \times 4)$
average (p. 124) The quotient found by dividing the sum of a group of numbers by the number of addends.

B

bar graph (p. 362) A graph that uses bars to compare data or show information.
base [of a space figure] (p. 408) The two congruent parallel faces of a prism or cylinder. The face of a cone or pyramid opposite the vertex.
base [of a power] (p. 180) One of equal factors in a product. The base in 10^3 is 10.

C

capacity (p. 144) The amount of fluid a container can hold.
Celsius (p. 434) The temperature scale with 0 degrees as the freezing point and 100 degrees as the boiling point.
center (pp. 300, 408) The point that is the same distance from all points on a circle or on a sphere.
centimeter (cm) (p. 152) A standard metric unit for measuring length. 100 cm = 1 m
circle (p. 300) A curved figure in a plane with all points an equal distance from a given point, called the center.
circle graph (p. 366) A circle divided into parts to show data.
circumference (C) (p. 300) The distance around a circle.
common denominator (p. 198) A common multiple of the denominators.
common factor (p. 192) A number that is a factor of two or more numbers. The common factors of 8 and 12 are 1, 2, and 4.
common multiple (p. 174) A number that is a multiple of two or more numbers. Some common multiples of 3 and 4 are 12 and 24.
Commutative Property of Addition (p. 36) Changing the order of the addends does not change the sum.
Example: $7 + 4 = 11 \quad 4 + 7 = 11$
so $7 + 4 = 4 + 7$
Commutative Property of Multiplication (p. 68) Changing the order of the factors does not change the product.
Example: $5 \times 8 = 40 \quad 8 \times 5 = 40$
so $5 \times 8 = 8 \times 5$
compatible numbers (p. 126) Numbers that divide easily that are used in estimating quotients.
composite numbers (p. 178) A number that has more than two factors.
cone (p. 408) A space figure with one circular face and one vertex.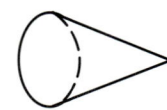

congruent figures (p. 296) Figures that have the same size and shape.

congruent segments (p. 278) Two segments that have the same length.
construction (p. 406) A figure made using only a compass and a straightedge.
coordinate plane (p. 448) A grid on a plane with two perpendicular number lines
corresponding parts (p. 296) The matching parts of congruent figures.
cube (p. 403) A special rectangular prism with all faces shaped like squares.

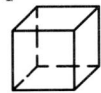

cup (c) (p. 144) A United States Customary unit of capacity. 2 c = 1 pt
cylinder (p. 408) A space figure with two parallel, circular, congruent bases.

D

data (p. 358) Numbers that give information.
decimal (p. 200) A number that shows tenths, hundredths, thousandths, and so on. 1.4 and 3.62 are decimals.
decimeter (dm) (p. 152) A metric unit of length. 1 dm = 10 cm
degree (p. 280) A unit for measuring angles.
degrees Celsius (°C) (p. 434) The metric unit for measuring temperature.
degrees Fahrenheit (°F) (p. 146) The U.S. Customary unit for measuring temperature.
denominator (p. 188) The bottom number of a fraction.
diagonal (p. 286) A segment that joins two vertexes of a polygon but is not a side.
diameter (d) (p. 300) The length of a segment through the center with endpoints on the circle.
discount (p. 490) A decrease in the price of an item.
distributive property (p. 77) The product of a factor and a sum is equal to the sum of the products.
Example: $5 \times (3 + 2) = (5 \times 3) + (5 \times 2)$
dividend (p. 102) The number being divided in division. In the equation $36 \div 4 = 9$, the dividend is 36.
divisible (p. 176) A number can be divided by another number without having a remainder.

Example: 72 is divisible by 3 because $72 \div 3 = 24$.
divisibility rules (p. 176) Ways to tell if one number is divisible by another without actually dividing.
divisor (p. 102) The number that divides the dividend in division. In the equation $30 \div 5 = 6$, the divisor is 5.
double bar graph (p. 364) A graph that uses bars to compare two sets of data simultaneously.

E

edge (p. 403) The line segment formed when two faces of a space figure meet.
elapsed time (p. 160) The amount of time from one point in time to another.
equation (pp. 46, 468) A statement using the equals sign to show that two quantities are equal
Example: $9 \times 6 = 54$
equilateral triangle (p. 282) A triangle with all sides congruent.
equivalent fractions (p. 190) Fractions that represent the same amount. $\frac{1}{3}$ and $\frac{2}{6}$ are equivalent fractions.
estimate (p. 12) An answer that is not exact.
evaluating (p. 466) Substituting a value for the variable to find a value of the expression.
expanded form (p. 4) The expanded form of 378 is $300 + 70 + 8$.
experimental probability (p. 380) The number of times the event occurs divided by the total number of trials in the experiment.
exponent (p. 180) Names that number of times a factor is used. The exponent in 10^3 is 3.
expression (p. 466) A way to write a relationship with numbers and symbols.

F

face (p. 402) A side of a space figure.

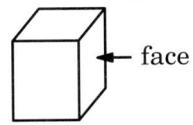

factor (p. 68) A number being multiplied to obtain a product.

factor tree (p. 178) A diagram showing the prime factorization of a product.
Fahrenheit (p. 146) The temperature scale with 32 degrees as the freezing point and 212 degrees as the boiling point.
flip (p. 295) A figure is reflected about a line.
foot (ft) (p. 142) A United States Customary unit for measuring length. 1 ft = 12 in.
formula (p. 418) A rule that is written using symbols. Example: $A = l \times w$
fraction (p. 188) A number that compares part of an object or a set with the whole.
front-end estimation (p. 40) Using the digits at the far left in numbers to guess a likely answer.
function (p. 476) A special relationship between the members of one group and another. For example, in the pairs (2, 5), (4, 7), and (5, 8) the first numbers are related to the second numbers by the rule "add 3."
function rule (p. 478) The rule that is used to make the pairs of a function.

G

gallon (gal) (p. 144) A United States Customary unit for measuring capacity. 1 gal = 4 qt
gram (g) (p. 158) A standard metric unit for measuring mass. 1,000 g = 1 kg
greatest common factor (GCF) (p. 192) The greatest number that is a factor of two or more given numbers. The GCF of 12 and 18 is 6.

H

hexagon (p. 286) A six-sided polygon.

I

inch (in.) (p. 144) A United States Customary unit for measuring length. 12 in. = 1 ft
independent events (p. 382) Events that have no effect on each other.
integers (p. 432) The positive numbers 1, 2, 3, . . ., the negative numbers −1, −2, −3, . . ., and zero.
intersecting lines (p. 278) Two or more lines that meet or cross at a common point.

inverse operations (p. 101) Two operations that undo each other, such as addition and subtraction, or multiplication and division.
isosceles triangle (p. 282) A triangle with two sides congruent.

K

kilogram (kg) (p. 158) A standard metric unit for measuring mass. 1 kg = 1,000 g
kiloliter (kL) (p. 156) A standard metric unit for measuring capacity. 1 kL = 1,000 L.
kilometer (km) (p. 152) A standard metric unit for measuring length. 1 km = 1,000 m

L

least common denominator (LCD) (p. 198) The least common multiple of the denominators of two or more given fractions. The LCD of $\frac{2}{3}$ and $\frac{1}{4}$ is 12.
least common multiple (LCM) (p. 174) The least multiple, excluding 0, of two or more numbers. The LCM of 6 and 10 is 30.
line (p. 278) A straight path of points that goes on forever in both directions.
line graph (pp. 186, 368) A graph with a line that shows changes over time.
line of symmetry (p. 298) A line through a figure so that if the figure were folded on the line, the two parts would match exactly.
liter (L) (p. 156) A standard metric unit for measuring capacity. 1 L = 1,000 mL
lowest terms (p. 194) When both terms of a fraction have no common factor greater than one. In lowest terms, $\frac{3}{6}$ is $\frac{1}{2}$.

M

mass (p. 158) The amount of matter in an object.
mean (p. 360) The quotient found by dividing the sum of a set of data by the number of items of data.
median (p. 360) The middle number in a set of data after the data is arranged in order from the least to the greatest.
meter (m) (p. 152) A standard metric unit for measuring length. 1 m = 100 cm

mile (mi) (p. 142) A U.S. Customary unit for measuring distance. 1 mi = 5,280 ft
milligram (mg) (p. 158) A metric unit for measuring mass. 1,000 mg = 1 g
milliliter (mL) (p. 156) A metric unit for measuring capacity. 1,000 mL = 1 L
millimeter (mm) (p. 152) A metric unit for measuring length. 10 mm = 1 cm
mixed number (p. 196) A whole number and a fraction. $2\frac{1}{2}$ is a mixed number.
mode (p. 358) The number that appears most often in a set of data.
multiple (p. 174) The product of a given number and any whole number. A multiple of 3 is 15.

N

negative number (p. 432) A number that is less than zero.
numerator (p. 188) The top number of a fraction.

O

obtuse angle (p. 280) An angle that measures more than 90° and less than 180°.
obtuse triangle (p. 282) A triangle with one obtuse angle.
octagon (p. 286) An eight-sided polygon.
opposites (p. 432) Two numbers that are the same distance from zero, but on opposite sides of zero.
ordered pair (p. 448) A pair of numbers in which the order shows the location of a point on a grid. (4, 3) is an ordered pair.
order of operations (p. 464) The rules that tell what operations to do before others if there are several operations to do in an expression.
origin (p. 448) The point (0, 0) where the axes of a coordinate plane meet.
ounce (oz) (p. 144) A United States Customary unit for measuring weight. 16 oz = 1 lb
outcome (p. 378) The result of a probability experiment.

P

palindrome (p. 63) A number that reads the same when written backward or forward.
parallel lines (p. 279) Lines that never intersect and are in the same plane.
parallelogram (p. 288) A quadrilateral with opposites sides parallel.
pentagon (p. 286) A five-sided polygon.
percent (p. 334) Hundredths written with a % sign. Percent means per hundred. Example: $0.33 = \frac{33}{100} = 33\%$
perimeter (P) (p. 154) The distance around a figure.
permutations (p. 386) The number of arrangements of objects when the order of the objects matters.
perpendicular (p. 280) Two lines that meet or cross to form right angles.
pint (pt) (p. 144) A United States Customary unit of capacity. 1 pt = 2 c
point (p. 278) An exact location.
polygon (p. 286) A simple closed figure formed by three or more segments.
positive number (p. 432) A number that is greater than zero.
pound (lb) (p. 144) A United States Customary unit of weight. 1 lb = 16 oz
power of 10 (p. 116) A number which can be written as a product of tens. For example, $10 \times 10 \times 10 \times 10 = 1,000$, so 1,000 is a power of 10.
prime factorization (p. 178) The product of prime numbers that names a given number, such as $2 \times 2 \times 3 \times 5$ to name 60.
prime number (p. 178) A number with exactly two factors, 1 and the number itself.
prism (p. 402) A space figure with five or more faces. Two of the faces, the bases, are parallel and congruent.
probability (p. 370) The number describing the chance that something will happen.
product (p. 68) The answer in multiplication. In the equation $8 \times 6 = 48$, the product is 48.

Property of One (p. 68) The product of one and any number is that number.
Example: 6 × 1 = 6
proportion (p. 326) Two equal ratios.
pyramid (p. 402) A space figure with four or more faces. The base can by any polygon. The other faces of the pyramid are triangles.

Q

quadrilateral (p. 288) A four-sided polygon.
quart (qt) (p. 144) A United States Customary unit of capacity. 1 qt = 2 pt
quotient (p. 102) The answer in division. In the equation 56 ÷ 7 = 8, the quotient is 8.

R

radius (r) (p. 300) The distance from the center of a circle to any point on the circle.

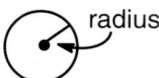

range (p. 358) The difference between the greatest and the least numbers of given data.
rate (p. 322) A special type of ratio that compares different kinds of quantities.
ratio (p. 320) The quotient of two numbers used to compare one number with the other.
Example: 2 bicycles to 3 cars = $\frac{2}{3}$.
ray (p. 278) A part of a line with one endpoint.

reciprocals (p. 255) Any two numbers whose product is 1; 2 × $\frac{1}{2}$ = 1, 2 and $\frac{1}{2}$ are reciprocals.
rectangle (p. 289) A special parallelogram with all angles right angles.
rectangular prism (p. 402) A space figure with six rectangular faces.
rectangular pyramid (p. 402) A space figure with four triangular faces and a rectangular base.
regular polygon (p. 286) A polygon in which all sides are congruent and all angles are congruent.

remainder (p. 102) In division, the dividend minus the product of the divisor and quotient.
rhombus (p. 289) A parallelogram with all sides congruent.
right angle (p. 280) An angle that measures 90°.
right triangle (p. 282) A triangle with a right angle.
round (p. 24) To replace a number by the nearest ten, hundred, thousand, and so on. 37 rounded to the nearest ten is 40.

S

sample (p. 374) A part of a large set of data that is used to make predictions.
sample space (p. 384) A list of all possible outcomes.
scale (p. 330) The ratio of the size in a drawing to the actual size.
scale drawing (p. 330) A drawing made using a scale.
scalene triangle (p. 282) A triangle with no sides congruent.
segment (p. 278) A part of a line with two endpoints.
similar figures (p. 296) Figures that have the same shape. They do not need to have the same size.
slide (p. 294) A figure is moved in one direction.
solution (p. 468) The answer to a problem or equation.
space figure (p. 402) A figure that has three dimensions, length, width, and height.
sphere (p. 408) A shape in space all points of which are the same distance from a point within, the center.

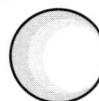

square (p. 289) A special rectangle with all sides congruent. It is also a rhombus with all angles right angles.

square root (p. 181) When you multiply a whole number by itself, the product is a

square number. The number you started with is called the square root.
standard form (p. 4) The usual short form of a number. The standard form of 5 hundreds, 7 tens, and 3 ones is 573.
straight angle (p. 280) An angle that measures 180°.
surface area (p. 404) The sum of the area of the faces of a space figure.
symmetrical figure (p. 298) A figure that can be folded so that both halves fit exactly on one another.
symmetry (p. 298) A figure has symmetry when it can be folded so that both parts match.

T

temperature (pp. 146, 434) Tells how hot or cold something is. Temperature is measured in degrees (°).
tessellation (p. 295) Uses figures that touch but do not overlap.
ton (t) (p. 144) A United States Customary unit of weight. 1 t = 2000 lb.
trapezoid (p. 289) A quadrilateral with exactly one pair of parallel sides.

tree diagram (p. 384) A picture showing outcomes of an activity.

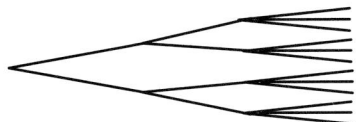

triangle (p. 286) A three-sided polygon.

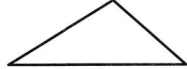

triangular prism (p. 402) A space figure with two triangular faces and three rectangular faces.
triangular pyramid (p. 402) A space figure with three triangular faces and a triangular base.
turn (p. 294) A figure is rotated about a point.

U

unit rate (p. 322) The ratio of a quantity to 1.

V

variable (p. 46) A letter that takes the place of a number.
Example: $4 + n = 7$ $n = 3$
Venn diagram (p. 284) A diagram to represent collections of people or objects and their relationships.
vertex (pp. 279, 403) The common endpoint of two rays of an angle or two segments of a polygon. The intersection of three or more edges of a space figure.
visualize (p. 410) To picture something in your head.
volume (V) (p. 414) The amount of space inside a space figure.

W

word form (p. 4) The word form for the number 4,832 is four thousand, eight hundred thirty-two.

X

x-axis (p. 448) The horizontal line in a coordinate plane.

Y

y-axis (p. 448) The vertical line in a coordinate plane.
yard (yd) (p. 144) A United States Customary unit for measuring length. 1 yd = 3 ft

Z

Zero Property of Addition (p. 36) The sum of any number and zero is that number.
Example: $4 + 0 = 4$
Zero Property of Multiplication (p. 68) The product of zero and any number is zero.
Example: $7 \times 0 = 0$

TABLE OF MEASURES

Time

60 seconds (s) = 1 minute (min)
60 minutes = 1 hour (h)
24 hours = 1 day (d)
7 days = 1 week

$\left.\begin{array}{r}\text{365 days}\\\text{52 weeks}\\\text{12 months}\end{array}\right\} = 1 \text{ year (y)}$

10 years = 1 decade
100 years = 1 century

Metric

LENGTH

10 millimeters (mm) = 1 centimeter (cm)
10 centimeters = 1 decimeter (dm)
$\left.\begin{array}{r}\text{10 decimeters}\\\text{100 centimeters}\end{array}\right\} = 1 \text{ meter (m)}$
10 meters = 1 dekameter (dam)
10 dekameters = 1 hectometer (hm)
$\left.\begin{array}{r}\text{10 hectometers}\\\text{1000 meters}\end{array}\right\} = 1 \text{ kilometer (km)}$

AREA

100 square millimeters = 1 square centimeter (mm^2) (cm^2)
10,000 square centimeters = 1 square meter (m^2)
10,000 square meters = 1 hectare (ha)

VOLUME

1000 cubic millimeters = 1 cubic centimeter (mm^3) (cm^3)
1,000,000 cubic centimeters = 1 cubic meter (m^3)

MASS

1000 milligrams (mg) = 1 gram (g)
1000 grams = 1 kilogram (kg)

CAPACITY

1000 milliliters (mL) = 1 liter (L)

United States Customary

LENGTH

12 inches (in.) = 1 foot (ft)
$\left.\begin{array}{r}\text{3 feet}\\\text{36 inches}\end{array}\right\} = 1 \text{ yard (yd)}$
$\left.\begin{array}{r}\text{5280 feet}\\\text{1760 yards}\end{array}\right\} = 1 \text{ mile (mi)}$

AREA

144 square inches (in^2) = 1 square foot (ft^2)
9 square feet = 1 square yard (yd^2)
4840 square yards = 1 acre (A)

VOLUME

1728 cubic inches = 1 cubic foot (ft^3)
27 cubic feet = 1 cubic yd (yd^3)

WEIGHT

16 ounces (oz) = 1 pound (lb)
2000 pounds = 1 ton (t)

CAPACITY

8 fluid ounces (fl oz) = 1 cup (c)
2 cups = 1 pint (pt)
2 pints = 1 quart (qt)
4 quarts = 1 gallon (gal)

INDEX

A
Addition
  addend, 36–37, 60
  calculator activities, 38–39, 49, 52–53, 66, 138
  column, 49, 52
  of decimals, 52–53, 87, 93
  estimating with, 38–39, 40, 41, 57, 60, 232–233
  of fractions, 216–217, 218–219, 228–229, 230–231, 232–233
  integers, 436–437, 438–439
  mental math and, 39, 46–47
  mixed numbers, 218–219, 232–233, 241
  with money, 40, 41, 47, 52, 53, 56–57, 77, 89, 110–111, 237
  properties, 36–37, 48
  units of measure, 142–143
  of units of time, 481
  of whole numbers,
    four through six digits, 38–39, 43, 66
    three addends, 38–39
Angle
  acute, 280–281
  congruent, 278–279, 280–281
  measuring, 280–281, 283
  obtuse, 280–281
  right, 280–281
Applications, *see* Problem solving
Arc of a circle, 406–407, 424
Area
  counting square units, 302–303
  of a circle, 306–307
  estimating with, 303, 306
  of a rectangle, 304–305
  of a square, 304–305
  of a triangle, 302–303, 304–305
Arrangements, 386–387
Associative property
  of addition, 36–37, 48
  of multiplication, 68–69, 83
Average, 124–125, 135, 273, 360–361, 398, 452–453

B
Bar graph, 11, 35, 91, 99, 139, 147, 215, 335, 362–363, 364–365, 370–371, 373, 392–394, 431
Base five, 33
Base ten, 33
Better buy, 323
Billions, 14

C
Calculator activities
  addition, 49, 66, 138
  area, 305
  average, 125, 452–453
  change-sign key, 439
  checking work, 75, 98, 138
  decimals, 34, 55, 66
  division, 125, 138
  fractions, 214, 246
  geometry, 301, 305, 318, 430
  integers, 439
  measurement, 155, 318, 430
  multiplication, 75, 79, 98, 116, 138
  percent, 339, 343, 356
  statistics, 398
  subtraction, 55, 66
  volume, 430
  when to use, 50–51, 155, 165, 201, 202–203, 207, 265, 301, 311, 337, 348–349, 390–391, 422–423, 451, 465, 467, 482–483
Capacity
  metric units of, 156–157
  U.S. customary units of, 144–145
Celsius, 435
Centimeter, 152–153
Circle
  arc, 406–407
  area, 306–307
  center of, 300–301, 406
  circumference, 300–301, 355
  diameter, 300–301
  drawing a, 406–407
  estimated area of, 306
  radius, 300–301

Circle graph, 173, 199, 230, 355, 366–367
Circumference, 300–301, 355
Combinations, 386–387, 395
Common denominator, 198–199, 228–229, 230–231, 232–233, 243
Common factors, 192–193, 194–195
Common multiples, 174–175
Communication
  creating problems, 51, 73, 81, 97, 137, 147, 149, 161, 165, 175, 213, 217, 221, 235, 243, 245, 253, 363, 365, 369, 377, 385, 387, 449, 461, 473, 477
  speaking, 3, 6, 8, 10, 13, 20, 28, 38, 40, 44, 51, 52, 56, 74, 78, 81, 102, 106, 108, 121, 122, 129, 142, 144, 149, 154, 179, 182, 186–187, 190, 192, 194, 195, 196, 198, 200, 216, 220, 230, 250, 255, 264, 284, 290, 296, 298, 304, 320, 322, 334, 336, 344, 361, 362, 364, 366, 368, 370–371, 374, 376, 382, 383, 384, 386, 401, 432, 434, 438, 446, 448, 450, 462, 464, 466, 470, 471
  writing, 3, 9, 13, 19, 23, 29, 31, 49, 51, 61, 75, 81, 83, 93, 113, 129, 133, 141, 147, 149, 151, 153, 167, 185, 187, 189, 191, 193, 195, 197, 209, 217, 227, 233, 241, 254, 259, 260, 267, 271, 288, 289, 293, 313, 325, 329, 333, 337, 344, 351, 365, 367, 369, 371, 373, 381, 393, 401, 404–405, 413, 425, 435, 437, 443, 455, 465, 475, 485
Commutative property
  of addition, 36–37, 48
  of multiplication, 68–69
Comparing
  decimals, 26–27, 34
  estimating with, 40–41, 53
  fractions, 198–199, 222–223
  integers, 434–435
  mental math with, 87

511

using the number line, 8,
222–223
whole numbers, 6–7, 8, 14–15
Compass, 406–407
Compatible numbers, 126–127,
263, 343
Composite numbers, 178–179
Computer activities, 34, 66, 98,
138, 172, 214, 246, 276, 318,
356, 398, 430, 460, 490
Cone, 408–409, 420–421
Congruence
of angles, 280–281
of line segments, 278–279
of triangles, 302–303
Congruent figures, 278–279,
280–281, 296–297, 302–303,
403
Connections
to other disciplines, *see*
Mathematics across the
curriculum
mathematical, 6, 24, 72, 76,
119, 152, 200, 202, 228, 232,
234, 252, 261, 338, 469
real-life, *see* Problem solving
applications
Constructions, 406–407
Consumer, 7, 11, 15, 33, 40–41,
52, 53, 69, 71, 77, 79, 110–111,
123, 130–131, 256–257, 323
Cooperative learning, 13, 17, 23,
27, 33, 34, 51, 55, 63, 73, 81,
87, 95, 121, 129, 130–131, 135,
138, 149, 155, 162–163, 169,
179, 211, 215, 222–223,
238–239, 243, 265, 267, 273,
278–279, 288–289, 299, 307,
308–309, 315, 324–325, 344,
353, 358, 371, 387, 388–389,
395, 398, 407, 420–421, 427,
431, 452–453, 457, 461, 463
Coordinate grid, 448–449, 456,
460, 490
Creative problem solving,
130–131, 162–163, 238–239,
308–309, 388–389, 420–421,
452–453
Critical thinking, *see* Logical
thinking
Cross multiplication, 326–327,
328–329
Cube, 403
Cubic numbers, 180–181
Cubic units, 414–415, 416–417
Cumulative review, *see* Review

Cup, 144–145
Customary units, *see* U.S. cus-
tomary units
Cylinder, 408–409, 416–417

D

Data collection, *see* Statistics

Decimals
addition of, 52–53, 87, 93
calculator activities, 34, 55,
66, 98
comparing, 26–27, 34
dividing, 114–115, 117,
118–119, 120–121, 122–123
estimating products, 86, 98
estimating quotients,
126–127
estimating sums and differences, 52–53, 54–55
expanded form, 22–23
and fractions, 200–201,
202–203, 338–339
hundredths, 20–21, 24–25,
84–85
mental math with, 87,
116–117, 121
and metric measurements,
152–153
and money, 52–53, 54, 88–89,
122–123
multiplying, 84–85, 86–87,
88–89, 92–93, 116
ordering, 26–27
and percents, 336–337,
338–339
place value, 20, 31
problem solving with, *see*
Problem solving applica-
tions
rounding, 24–25, 122–123
subtraction of, 54–55, 66
tenths, 20–21, 24–25,
122–123
thousandths, 22–23, 84–85,
122–123
Decimeter, 152–153
Degree, 280
Denominator, 188–189
Diagonal of polygon, 286–287
Diameter, 300–301
Discount, 490
Distributive property, 77, 253
Divisibility rules, 176–177
Division
calculator activities, 106–107,

108–109, 114–115, 124–125,
138
checking, 102, 108, 138
compatible numbers, 126–127
with decimals, 99, 114–115,
117, 118–119, 120–121,
122–123, 133
equations, 47, 260–261,
470–471
estimating with, 107, 115,
126–127, 138
with fractions, 202–203,
260–261, 262–263, 264–265,
266–267
mental math with, 105, 107,
109, 117, 121
with mixed numbers,
266–267
of money, 110–111, 122–123
powers of ten, 116–117, 133
rules of divisibility, 176–177
with whole numbers,
102–103, 104–105, 106–107,
108–109, 114–115, 117, 118
zeros in the quotient,
108–109, 114–115, 120–121
Double bar graphs, 364–365

E

Elapsed time, 59, 160–161, 311
Enrichment, 33, 63, 95, 135, 169,
211, 243, 273, 315, 353, 395,
427, 457, 487
Equations
addition, subtraction, 45–47,
97, 269, 468–469, 472–473
calculator activities, 55
decimals, 469, 471
fractions, 268–269
multiplication, division,
46–47, 261, 265, 268–269,
470–471, 472–473, 475
mental math with, 46–47, 471
problem solving with, *see*
Problem solving strategies
using inverse operations to
solve, 468–469, 470–471
writing, 80–81, 472–473
Equivalent fractions, 190–191,
198
Equivalent ratios, 320–321,
324–325, 326–327, 328–329,
330–331, 353
Estimating
area, 302–303
decimals, 86–87, 98

512

differences, 44–45, 54, 60,
    232–233
fractions, 232–233, 339
measures, 55, 143, 153, 156,
    159, 223, 302–303
money, 40–41, 53
percent, 339, 343, 356
problem solving and, 13, 25,
    37, 41, 55, 57, 91, 125, 143,
    157, 165, 256–257, 348–349,
    367, 390–391, 415, 422–423,
    473, 482–483
products, 74–75, 76–77, 78,
    98, 138
quotients, 107, 115, 126–127,
    138
statistics, 368–369
strategies
    clustering, 257
    comparisons, 231
    compatible numbers,
        126–127, 263, 343
    front-end, 40–41, 49
    proportional thinking,
        324–325, 327
    range, 75, 127, 233
    reading graphs, 368–369
    reasonable answers,
        38–39, 44–45, 232–233
    rounding, 38–39, 44–45,
        49, 54, 74–75, 232–233
    sampling and estimating,
        374–375
    using measurement, 143,
        153, 156, 159, 223,
        302–303
sums, 38–39, 40–41, 52–53, 60
volume, 415
when to use, 12–13, 57, 59,
    125, 165, 201, 207, 301,
    310–311, 348–349, 390–391,
    422–423, 482–483
Expanded form
    decimals, 22–23
    whole numbers, 4–5
Exploring, 84–85, 100–101,
    116–117, 118–119, 140–141,
    176–177, 202–203, 222–223,
    260–261, 278–279, 288–289,
    294–295, 302–303, 324–325,
    340–341, 358–359, 380–381,
    386–387, 402–403, 404–405,
    406–407, 436–437, 444–445,
    468–469, 476–477, 478–479
Exponents, 180–181

Expressions, 462–463, 466–467,
    472, 475
Extra practice, 32, 62, 94, 134,
    168, 210, 242, 272, 314, 352,
    394, 426, 456, 486

F
Factor
    common, 192–193, 194–195
    greatest common factor
        (GCF), 192–193, 194–195
    prime, 178–179
    tree, 178–179
Fahrenheit, 146–147
Foot, 142–143
Formulas
    area
        of a circle, 306–307
        of a rectangle, 304–305
        of a square, 304–305
        of triangles, 302–303,
            304–305
    circumference, 300–301
    perimeter, 154–155, 418–419
    problem solving with, *see*
        Problem solving applications
    volume
        of a cylinder, 416–417
        of a rectangular prism,
            414–415
Fractions
    addition of
        different denominators,
            228–229, 230–231,
            232–233
        same denominators,
            216–217, 218–219,
            230–231, 232–233
    comparing, 198–199, 222–223
    and decimals, 200–201,
        202–203, 338–339
    dividing, 202–203, 260–261,
        262–263, 264–265, 266–267
    equivalent, 190–191, 198
    estimation with, 199,
        218–219, 231, 232–233, 253,
        263, 339
    lowest terms, 194–195,
        320–321, 338–339
    meaning of, 188–189,
        196–197, 334–335
    and measurement, 222–223
    mental math with, 217, 219,
        221, 229, 235, 251, 265
    mixed numbers, 196–197,
        198–199, 200–201, 216,
        218–219, 222–223, 232–233,
        234–235, 252–253, 254–255,
        266–267, 417
    multiplying, 248–249,
        250–251, 252–253, 254–255
    ordering, 198–199
    and percents, 334–335,
        338–339
    problem solving, *see* Problem
        solving applications
    rounding, 218, 232–233, 253
    subtraction of
        different denominators,
            228–229, 230–231,
            232–233, 234–235
        same denominators,
            216–217, 218–219,
            230–231, 232–233,
            234–235
Front-end estimation, 40–41, 49
Functions, 79, 105, 229, 265, 439,
    467, 476–477, 478–479, 487

G
Gallon, 144–145
Geometry
    angle
        acute, 280–281
        obtuse, 280–281
        right, 280–281
    area
        circle, 306–307
        rectangle, 302–303,
            304–305
        triangle, 302–303, 304–305
    calculator activities, 301, 305,
        318, 419, 430
    circumference, 300–301
    classifying triangles, 282–283
    classifying quadrilaterals,
        288–289, 290–291
    congruent figures, 278–279,
        280–281, 296–297, 302–303,
        403
    constructions, 406–407
    cylinder, 408–409, 416–417
    diagonal, 286–287
    diameter, 300–301, 306–307
    formulas, *see* Formulas
    line, 278–279
    line segment, 278–279
    parallel lines, 278–279
    perimeter, 154–155, 418–419
    perpendicular lines, 280–281,
        407

513

plane, 278–279, 448–449
point, 278–279
polygon, 286–287, 406–407
problem solving with, see Problem solving applications
radius, 300–301, 306–307
ray, 278–279, 280–281
similar figures, 296–297, 315
slide, flip, turn, 294–295, 449
surface area, 404–405
symmetry, 298–299
three-dimensional figures, see Space figures
two-dimensional figures, see Plane figures
vertex, 278–279, 280–281, 403
volume, 414–415, 416–417
    cylinder, 416–417
    rectangular prism, 414–415
Glossary, 504–509
Gram, 158–159
Graph
    bar, 11, 35, 91, 99, 139, 147, 215, 335, 362–363, 364–365, 370–371, 392–394, 431
    circle, 173, 199, 230, 355, 366–367
    coordinate, 448–449, 457, 460, 490
    line, 186–187, 368–369, 372
Greatest common factor (GCF), 192–193, 194–195
Grid, 448–449, 457

H
Hexagon, 286–287
Hundreds, 4–5
Hundred thousands, 4–5, 10–11
Hundredths, 20–21, 24–25, 84–85

I
Inch, 142–143, 223
Integers
    adding, 436–437, 438–439
    calculator activities, 439
    comparing, 434–435
    graphing ordered pairs, 448–449, 456, 460
    on the number line, 432–433
    opposites, 432–433
    ordering, 434–435
    problem solving with, see Problem solving applications
    subtracting, 444–445, 446–447

K
Kilogram, 158–159
Kilometer, 152–153

L
Language review, see Review
Least common denominator (LCD), 198–199, 228–229, 230–231, 232–233, 243
Least common multiple (LCD), 174–175, 198–199
Length
    metric units of, 152–153
    U.S. customary units of, 142–143
Let's talk math, 2–3, 12–13, 28–29, 50–51, 80–81, 128–129, 148–149, 186–187, 344–345, 362–363, 370–371, 400–401
Line
    identifying, 278–279, 280–281
    intersecting, 278–279
    parallel, 278–279
    perpendicular, 280–281
    segment, 278–279
    of symmetry, 298–299
Line graphs, 186–187, 368–369, 372
Liquid measure, see Capacity
Liter, 156–157
Logical thinking
    analogies, 73
    analysis, synthesis, questioning, 27, 95, 105, 118–119, 265
    classifying, 23, 287, 288–289, 290–291, 408
    cues and clues, 55, 63, 214
    deductive and inductive reasoning, 63, 75, 95, 98, 100–101, 105, 118–119, 154, 174, 188, 228, 237, 267, 283, 290–291, 305, 313, 360, 377, 384, 437, 441, 445, 462, 464, 479
    error analysis, 54, 76, 102
    flow charts, 100–101
    generalizations, 154, 174, 188, 228, 267, 283, 290–291, 360, 377, 384, 437, 445, 462, 464, 478–479
    inferences, 176–177, 374–375, 380–381
    non-routine problems, 16–17, 182–183, 204–205, 224–225, 237, 256–257, 269, 290–291, 331, 347, 386–387, 440–441, 450–451, 473, 480–481
    non-unique solutions, 130–131, 162–163, 238–239, 308–309, 388–389, 420–421, 452–453
    observing, 37, 72, 293, 295, 299, 403, 407, 410–411
    patterns, 16–17, 80–81, 97, 118, 243, 440–441, 457
    predictions and outcomes, 27, 75, 87, 95, 98
    problem solving with, see Problem solving applications
    process of elimination, 224–225, 480–481
    reasonable answers, 38–39, 44–45, 58, 129, 232–233
    spatial visualization, 299, 403, 407, 410–411, 425, 427, 449
    statements, 2–3, 7, 77, 145, 176–177, 201, 231, 249, 291, 316, 327, 358, 364–365, 392, 435
    tree diagram, 384–385
    Venn diagrams, 284–285, 287, 290–291
    verifying, 71, 75, 128–129, 147, 155, 181, 191, 193, 194, 203, 204–205, 249, 253, 261, 307, 327, 364, 446
Longitude, latitude, 457
Lowest terms, 194–195, 320–321, 338–339

M
Mass
    estimating with, 55, 159
    metric units of, 158–159
Math America, 19, 49, 83, 113, 151, 185, 227, 259, 293, 315, 333, 373, 413, 443, 457, 475
Mathematics across the curriculum
    art, 328–329, 414–415
    ecology, 70–71
    geography, 38–39, 44–45, 198–199, 232–233, 432–433, 448–449
    history, 90–91, 126–127, 194–195, 228–229, 298–299

literature, 218–219
music, 148–149, 250–251
science, 14–15, 16–17, 22–23,
  50–51, 58–59, 72–73,
  108–109, 114–115, 142–143,
  158–159, 200–201, 254–255,
  330–331, 336–337, 348–349,
  388–389, 408–409
social studies, 4–5, 36–37,
  104–105, 124–125, 324–327,
  362–363, 390–391, 422–423,
  470–471
Mean, 124–125, 360–361
Measurement
  addition and subtraction,
    142–143
  of angles, 280–281, 283
  area, 302–303, 304–305,
    306–307, 308–309, 404–405,
    418–419
  calculator activities, 155, 301,
    305, 318, 419, 430
  decimals and, 86–87, 152–159
  estimation with, 143, 153,
    156, 159, 223, 302–303
  fractions and, 222–223
  mental math with, 145, 159
  metric system of
    capacity, 156–157
    length, 152–153
    mass, 158–159
    temperature, 434–435
  perimeter, 154–155, 418–419
  problem solving with, see
    Problem solving applications
  with a protractor, 280–281
  ratios in, 326–327, 328–329,
    330–331, 353
  with a ruler, 222–223
  surface area, 404–405
  Table of measures, 142, 144,
    152, 158, 510
  of time, 34, 160–161
  U.S. customary system of
    capacity, 144–145
    length, 142–143
    temperature, 146–147
    weight, 144–145
Median, 360–361
Mental math
  addition, 39, 46–47
  decimals, 87, 116–117, 121
  division, 105, 107, 109, 117,
    121

fractions, 217, 219, 221, 229,
  235, 251, 265
measurement, 145, 159
multiplication, 69, 72–73, 77,
  107, 116, 251, 253
percent, 337, 339
ratio and proportion, 327
statistics, 125
strategies
  changing numbers and operations, 43, 46–47, 49,
    69, 72–73, 77, 87, 105,
    107, 109, 116, 121, 125,
    145, 159, 175, 179, 193,
    217, 219, 221, 229, 235,
    251, 253, 265, 327, 339,
    439, 447, 471
  comparing and ordering,
    87
  counting on, 56–57, 219
  using properties, 69, 77,
    253
  using visual images, 251
  subtraction, 43, 49, 56–57
  when to use, 13, 50–51, 57,
    59, 155, 201, 207, 229, 265,
    337, 348–349, 390–391,
    422–423, 451, 465, 467,
    482–483
Meter, 152–153, 416–417
Metric system, see Measurement
Midchapter checkups, see
  Review
Mile, 106, 120–121, 142–143
Milliliter, 156–157
Millimeter, 152–153
Millions, 14
Mixed numbers
  addition of, 218–219, 232–233
  comparing, 198–199, 222–223
  and decimals, 200–201
  dividing with, 266–267
  estimating with, 218,
    232–233, 253
  and fractions, 196–197,
    198–199
  meaning of, 222–223
  mental math with, 219, 221,
    235
  multiplication of, 252–253,
    254–255
  subtraction of, 218–219,
    220–221, 232–233, 234–235
Mixed review, see Review
Mode, 358–359

Modeling
  hands-on activities, 20,
    84–85, 140–141, 202–203,
    222–223, 260–261, 278–279,
    288–289, 294–295, 302–303,
    308–309, 340–341, 381,
    388–389, 402–403, 404–405,
    406–407, 436–437, 444–445
  making representations, 73,
    80–81, 111, 139, 182–183,
    204–205, 239, 249, 268–269,
    284–285, 290–291, 440–441,
    453, 472–473, 480
Money
  addition, 40, 41, 47, 52, 53,
    56–57, 77, 89, 110–111, 237
  and decimals, 52, 122–123
  division, 110–111, 122–123
  estimating, 40, 41
  interest, 53, 344–345
  making change, 56–57
  multiplying with, 47, 57, 69,
    77, 79, 89, 237
  subtraction, 43, 44–45, 54,
    56–57
Multiple
  common, 174–175
  least common, 174–175,
    198–199
Multiplication
  calculator activities, 75, 79,
    116–117, 130–131, 138
  cross, 326–327, 328–329
  of decimals, 84–85, 86–87,
    88–89, 92–93, 98, 116
  of decimals and whole numbers, 88–89
  equations, 46–47, 470–471
  estimating with, 74–75,
    76–77, 79, 98, 138
  factor, 74–75, 93
  of fractions, 248–249,
    250–251, 252–253, 254–255
  of greater numbers, 50–51,
    79, 87, 89
  mental math with, 69, 72–73,
    77, 107, 116, 251, 253
  of mixed numbers, 252–253,
    254
  by powers of ten, 116–117
  properties of, 68–69, 77, 83,
    253
  of whole numbers, 70–71,
    72–73, 76–77, 78–79, 88–89,
    107, 116

515

by multiples of 10, 100, 1,000, 72–73
by one-digit number, 70–71
by two-digit number, 76–77, 107
with money, 76–77, 78–79, 88–89

N

Non-numerical graphs, 186–187
Number line
    decimals, 20
    fractions, 199
    integers, 432–433, 434, 447
    whole numbers, 8
Number sense
    magnitude of numbers, 2–3, 7, 11, 14–15, 21, 69, 71
    and measurement, 2–3, 147, 157, 307, 323
    number meaning, 2–3, 5, 21
    number relationships, 25, 37, 41, 45, 75, 115, 125, 181, 191, 199, 335, 337, 343, 377, 433, 447, 465
    operations sense, 84–85, 87, 89, 121, 249, 251, 253, 263, 265, 267, 447
Numbers
    comparing, 6–7, 8, 14–15, 26–27, 198–199, 434–435
    divisibility, 176–177
    integers, 432–433, 434–435
    opposites, 432–433, 439
    ordering, 6–7, 14, 26–27, 34, 198–199, 434–435
    prime, composite, 178–179
    prime factorization, 178–179
    rounding, 8–9, 10–11, 24–25
    square, 180–181, 440–441
    triangular, 440–441
Numerator, 188–189

O

Obtuse angles, 280–281
Octagon, 286, 312, 314
Opposite(s) property
    of addition and subtraction, 101, 468–469
    of multiplication and division, 101, 470–471
Order of operations, 464–465
Ordered pairs, 448–449
Ordering
    decimals, 26–27
    fractions, 198–199
    integers, 434–435
    whole numbers, 6–7, 8, 14–15, 34
Ounce, 144–145
Outcomes, 378–379, 381, 384–385

P

Palindromes, 63
Parallel lines, 278–279
Parallelograms, 288–289, 290, 303
Patterns, 16–17, 72–73, 95, 98, 116–117, 118–119, 318, 440–441, 457
Pentagon, 286–287, 292, 314
Pentominoes, 427
Percent
    calculator activities, 356
    comparing, 335, 337, 340–341, 343
    and circle graphs, 367
    and decimals, 336–337
    discounts, 347
    and fractions, 334–335, 338–339
    estimating with, 339, 343, 347, 356
    gain or loss, 344–345, 347
    meaning of, 334–335
    number sense and, 335, 337, 343
    of a number, 342–343
    problem solving, *see* Problem solving applications
    sales tax and, 346–347
Perimeter, 154–155, 418–419
Perpendicular lines, 280–281
Pi (π), 300–301, 306
Pint, 144–145
Place value
    addition and, 38–39
    bases other than ten, 33
    billions, 14
    decimals, 20–21, 22–23, 31
    expanded form, 4, 18, 22–23
    exponents, 180–181
    hundredths, 20–21, 24–25, 84–85
    hundred thousands, 4, 10, 14
    millions, 14–15
    multiplication and, 70–71, 76–77
    standard form, 4–5, 14, 18, 22–23
    subtraction and, 42–43
    tenths, 20–21, 24–25, 122–123
    thousands, 4–5, 6–7, 8–9, 45
    word form, 4, 18, 22–23, 30
    whole numbers, 4–5, 14–15
Plane, 448–449
Plane figures
    circle, 300–301
    hexagon, 286–287
    octagon, 286, 312
    parallelogram, 288–289, 290, 303
    pentagon, 286–287, 292, 314
    polygon, 286–287, 406–407
    quadrilateral, 288–289, 449
    rectangle, 288–289, 290–291
    rhombus, 288–289, 290–291
    square, 288–289, 290–291
    trapezoid, 288–289, 290, 303
    triangle, 282–283
Point, 278–279
Polygon, 286–287, 406–407
Pound, 144–145
Powers of ten, 116–117, 133
Pre-algebra, *see* Equations, Exponents, Expressions, Formulas, Integers
Prime factorization, 178–179, 180–181
Prime numbers, 178–179, 211
Prism
    rectangular, 402–403, 414–415
    surface area of, 404–405
    volume of, 414–415
    triangular, 402–403
Probability
    arrangements, 386–387
    chance, 376–377
    combinations, 386–387, 395
    equally likely, 378–379, 380
    experimental and actual, 380–381, 398
    independent events, 382–383
    of one, 376–377
    outcomes, 378–379, 384–385
    predictions, 376–377, 378–379, 380, 395, 398
    problem solving with, *see* Problem solving applications
    sampling, 380–381
    tree diagram, 384–385
    of zero, 376–377
Problem solving applications
    consumer, 7, 11, 15, 47, 52,

53, 57, 69, 71, 77, 79, 111, 123, 130–131, 237, 256–257, 441
decimals, 21, 25, 26–27, 41, 52, 55, 61, 71, 87, 89, 103, 107, 115, 121, 123, 133, 237, 482–483
equations and inequalities, 47, 80–81, 97, 245, 268–269, 418–419, 472–473
fractions, 191, 195, 197, 199, 217, 219, 229, 237, 241, 249, 311, 339, 348–349, 417, 422–423, 481
geometry, 155, 283, 297, 299, 305, 307, 308–309, 415, 417, 418–419, 420–421, 425, 483
integers, 433, 435, 439, 447, 455
measurement, 55, 75, 87, 89, 105, 109, 115, 125, 127, 143, 147, 153, 155, 157, 159, 197, 221, 236–237, 269, 283, 305, 307, 308–309, 415, 417, 418–419, 425, 481, 483
percent, 335, 337, 339, 343, 346–347, 367, 452–453
probability, 377, 379, 383, 385, 386–387
ratio and proportion, 197, 321, 323, 327, 329, 330–331, 441, 451
scale drawing, 330–331
statistics, 11, 90–91, 111, 123, 147, 182–183, 199, 205, 207, 231, 233, 238–239, 321, 323, 335, 337, 348–349, 361, 362–363, 365, 367, 369, 370–371, 375, 393, 452–453, 477
visual perception, 407, 410–411, 420–421
whole numbers, 7, 9, 25, 38, 41, 43, 45, 50–51, 61, 75, 77, 79, 87, 89, 103, 107, 143, 433, 483
Problem solving skills
four part process, 5, 7, 9, 11, 13, 21, 23, 26–27, 28–29, 37, 38, 41, 43, 45, 47, 52, 53, 55, 57, 58–59, 61, 65, 69, 71, 80–81, 87, 90–91, 93, 97, 103, 105, 109, 110–111, 115, 123, 125, 127, 128–129, 130–131, 133, 137, 143, 145, 147, 148–149, 153, 155, 157, 159, 160–161, 162–163, 165, 167, 175, 179, 181, 189, 191, 193, 195, 197, 199, 201, 207, 209, 217, 219, 221, 229, 231, 233, 235, 238–239, 241, 245, 249, 251, 253, 256–257, 263, 265, 267, 269, 281, 283, 284–285, 287, 291, 297, 299, 301, 305, 308–309, 311, 313, 321, 323, 327, 329, 330–331, 335, 337, 339, 343, 347, 348–349, 351, 361, 362–363, 365, 367, 369, 370–371, 375, 377, 379, 383, 385, 388–389, 390–391, 393, 409, 410–411, 415, 417, 418–419, 420–421, 422–423, 425, 433, 435, 441, 447, 449, 451, 452–453, 455, 463, 465, 467, 471, 473, 477, 481, 482–483
looking back, 58, 128–129
reading for understanding, 28–29
representing a problem, 80–81
writing problems, 17, 81, 101, 137, 147, 148–149, 161, 165, 217, 221, 235, 253, 363, 365, 387, 449, 473
Problem solving strategies
choosing the operation, 5, 29, 73, 77, 79, 87, 89, 90–91, 97, 105, 111, 121, 125, 133, 161, 167, 205, 245, 257
choosing a strategy, 28–29, 50–51, 59, 73, 77, 79, 89, 90–91, 109, 111, 125, 161, 171, 182–183, 205, 213, 224–225, 237, 245, 257, 269, 291, 317, 331, 347, 355, 397, 419, 429, 441, 451, 459, 473, 481
drawing a picture, diagram, graph, 80–81, 90–91, 99, 137, 139, 161, 204–205, 209, 213, 269, 291, 308–309, 441, 453
estimating, 13, 25, 37, 57, 97, 125, 137, 143, 153, 157, 159, 165, 256–257, 347, 348–349, 367, 390–391, 415, 422–423, 473, 482–483
logical thinking, 205, 237, 285
non-routine, 16–17, 23, 25, 39, 53, 73, 80–81, 89, 91, 111, 155, 161, 175, 182–183, 197, 204–205, 224–225, 229, 237, 256–257, 269, 290–291, 331, 347, 384–385, 386–387, 419, 440–441, 450–451, 473, 480–481
open-ended, 130–131, 162–163, 238–239, 308–309, 388–389, 420–421, 452–453
reasonable answers, 128–129, 379
simplifying the problem, 236–237, 241, 317, 473
too much or little information, 28–29, 90–91, 93, 111, 137, 245, 459
trial and error, 224–225, 291, 451, 473, 480–481
two or multi-step problems, 97, 110–111, 137, 237
using data from a graph, list, chart, or table, 11, 26–27, 41, 43, 55, 80–81, 90–91, 103, 111, 123, 147, 165, 182–183, 189, 199, 205, 207, 225, 231, 233, 238–239, 251, 263, 267, 291, 311, 321, 323, 335, 337, 348–349, 362–363, 365, 367, 369, 370–371, 393, 422–423, 441, 450–451, 452–453, 455, 467, 471, 477, 480–481
using equations, 47, 80–81, 97, 245, 268–269, 418–419, 472–473, 481
using generalizations, 290–291, 305, 313
using patterns, 16–17, 80–81, 97, 137, 281, 440–441, 451, 455, 481
using proportions, 327, 329, 330–331
working backwards, 161, 257, 311, 450–451, 455, 481
Product, 68–69, 74–75, 98
Properties
associative, 36–37, 48, 68–69, 82, 83
commutative, 36–37, 48, 68–69, 82
distributive, 77, 253
of one, 68–69
opposites, 101, 468–469, 470–471
of zero, 36–37, 48, 68–69, 78–79, 82, 88–89

517

Proportion, *see* Equivalent ratios
Proportional thinking, 315, 324–325, 327, 330–331
Protractor, 280–281
Pyramid, 402–403

Q

Quadrilaterals, 288–289, 449
Quart, 144–145
Quotient, 102–103

R

Radius, 300–301
Range, 75, 127, 233, 358–359
Rates, 322–323
Ratios
   equivalent, 320–321, 324–325, 326–327, 328–329, 330–331, 353
   mental math with, 327
   problem solving, *see* Problem solving applications
   unit rates and, 322–323
Rays, 278–279, 280–281
Reasonable answers, 38–39, 44–45, 58, 128–129, 232–233
Reciprocal, 255, 261, 262–263, 266
Rectangle, 288–289, 290–291
Rectangular prism, 402–403, 414–415
Remainder, 102, 122–123
Review
   cumulative review, 64, 96, 136, 170, 212, 244, 274, 316, 354, 396, 428, 458, 488
   extra practice, 32, 62, 94, 134, 168, 210, 242, 272, 314, 352, 394, 426, 456, 486
   language and vocabulary, 18, 30, 48, 60, 82, 92, 112, 132, 150, 166, 184, 208, 226, 240, 258, 270, 292, 312, 332, 350, 372, 392, 412, 424, 442, 454, 474, 484
   midchapter checkup, 18–19, 48–49, 82–83, 112–113, 150–151, 184–185, 226–227, 258–259, 292–293, 332–333, 372–373, 412–413, 442–443, 474–475
   mixed review, 71, 75, 77, 79, 87, 89, 103, 107, 115, 121, 145, 157, 164–165, 206–207, 217, 219, 221, 229, 231, 233, 235, 251, 253, 263, 265, 267, 310–311, 348–349, 390–391, 422–423, 471, 482–483
   problem solving review, 65, 97, 137, 171, 213, 245, 275, 317, 355, 397, 429, 459, 489
Rhombus, 288–289, 290–291
Right angle, 280–281
Right triangle, 282–283
Roman numerals, 33
Rounding
   decimals, 24–25, 122–123
   mixed numbers, 218–219, 232–233, 253
   money, 24–25, 122–123
   whole numbers, 8–9, 10–11, 39

S

Sales tax, 346–347
Sampling, 374–375
Scale drawing, 315, 330–331, 353
Segments, 278–279
Sides, 282–283, 286–287
Similarity, 296–297, 315
Slide, flip, turn, 294–295, 449
Solution, 468–469
Space figures
   cone, 408–409, 420–421
   cube, 403, 414–415
   cylinder, 408–409, 416–417
   rectangular prism, 402–403, 414–415
   rectangular pyramid, 402–403
   sphere, 408–409
   triangular prism, 402–403
   triangular pyramid, 402–403
Sphere, 408–409
Square, 288–289, 290–291, 304–305
Square number, 180–181, 440
Square root, 181
Square unit, 303–307
Standard form
   decimals, 22–23
   whole numbers, 4–5, 14
Statistics
   average, 124–125, 135, 139, 273, 360–361, 398, 452–453
   bar graph, 11, 35, 91, 99, 139, 147, 215, 335, 362–363, 364–365, 370–371, 392–393, 431
   calculator activities, 398
   circle graph, 173, 199, 230, 355, 366–367
   data
      collecting, 73, 95, 130–131, 135, 169, 173, 215, 238–239, 247, 273, 319, 357, 359, 369, 388–389, 395, 399, 427, 431, 452–453
      describing, 67, 99, 173, 215, 277, 357, 388–389, 431, 461
      organizing, 1, 35, 73, 80–81, 99, 130–131, 139, 182–183, 238–239, 277, 357, 358–359, 366–367, 371, 395, 431, 452–453, 477, 479, 480–481
      predicting, 27, 247, 319, 369, 374–375, 395, 427
      using in problem solving, 11, 23, 25, 26–27, 37, 41, 43, 53, 55, 80–81, 90–91, 103, 123, 145, 147, 157, 165, 175, 181, 189, 193, 199, 207, 233, 251, 253, 256–257, 263, 267, 268–269, 290–291, 311, 313, 321, 323, 329, 335, 337, 347, 348–349, 361, 365, 367, 369, 393, 422–423, 450–451, 452–453, 455, 467, 471
   estimating with, 91, 368–369, 423
   line graph, 186–187, 368–369, 372
   mean, 124–125, 360–361, 452–453
   median, 360–361
   mental math with, 125
   mode, 358–359, 360–361
   pictograph, 362–363
   problem solving with, *see* Problem solving applications
   range, 358–359
   sampling, 374–375
   surveys, 135, 365, 379
   time lines, 1, 277
   tree diagram, 384–385
Subtraction
   calculator activities, 42–43, 44–45, 54–55, 66
   checking by adding, 42, 44
   with decimals, 54–55

equations, 47, 60, 468–469
estimating with, 44–45, 54, 60, 232–233
of fractions, 216–217, 218–219, 220–221, 228–229, 230–231, 232–233, 234–235
of integers, 444–445, 446–447
of measures, 142–143
mental math with, 43, 49, 56–57
of mixed numbers, 218–219, 220–221, 232–233, 234–235
of money, 53, 56–57, 110–111
of whole numbers
    of four through six digits, 42–43, 44–45, 66
    of two- and three-digit numbers, 42–43
    zeros in, 44–45, 54, 66
Sum, 36, 60
Surface area, 404–405
Symbols to show relations, 6–7, 14–15, 26–27, 198–199, 434–435
Symmetry, 298–299

T

Table of measures, 510
Table of numbers, *xvi*
Temperature
    Celsius, 434–435
    Fahrenheit, 125, 146–147
Tenths, 20–21, 24–25, 122–123
Ten-thousands, 4–5, 10–11
Tests, Chapter, 30–31, 60–61, 92–93, 132–133, 166–167, 208–209, 240–241, 270–271, 312–313, 350–351, 392–393, 424–425, 454–455, 484–485
Thousands, 4–5, 6–7, 10–11, 14
Thousandths, 22–23, 84–85, 122
Three-dimensional figures, *see* Space figures
Time
    addition and subtraction, 473, 481
    comparing, 34, 463
    elapsed time, 59, 160–161, 311
    twenty-four hour clock, 169
Ton, metric, 158–159
Trapezoid, 288–289, 290, 303
Tree diagram, 384–385
Triangles, *see also* Area
Triangular numbers, 440

Triangular prism, 402–403
Triangular pyramid, 402–403
Two-dimensional figures, *see* Plane figures

U

Unit rate, 322–323
U.S. customary units, *see* Measurement

V

Variables, 37, 46–47, 53, 60, 462–463
Venn diagrams, 284–285, 287, 290–291
Vertex
    of an angle, 278–279, 280–281
    of a polygon, 286
    of space figures, 403
Vocabulary review, *see* Review
Volume
    of a cylinder, 416–417
    of a rectangular prism, 414–415

W

Weight, 140–141, 144–145, 158–159
Whole numbers
    adding, 38–39, 43, 66
    comparing, 6–7, 8, 14–15
    dividing, 102–103, 104–105, 106–107, 108–109, 117, 118
    expanded form, 4, 18
    multiplying, 70–71, 72–73, 76–77, 78–79, 107, 116
    on a number line, 8
    ordering, 6–7, 8, 14–15, 34
    place value, 4–5, 6, 14
    problem solving, *see* Problem solving applications
    rounding, 8–9, 10–11, 39
    standard form, 4–5, 14
    subtracting, 42–43, 44–45, 66

Y

Yard, 120–121, 142–143

Z

Zero
    in decimal products, 85, 86–87
    in the multiplier, 78–79
    in the quotient, 108–109, 114–115, 120–121
    in subtraction, 44–45, 54

properties of, 36–37, 48, 68–69, 82
Zero property
    of addition, 36–37, 48, 82
    of multiplication, 68–69, 82

Acknowledgments

HOMESICK MY OWN STORY by Jean Fritz. Cover-illustration copyright © 1982 by Margot Tomes. Reprinted by permission of Dell a division of Bantam, Doubleday, Dell Publishing Group, Inc.

Credits

Series Design Pronk & Associates

Cover and Title page design Sheaff Design, Inc.

Cover and title page photography Schlowsky Photography

Art Production of openers for Chapters 2, 3, 5–14 Kirchoff/Wohlberg

Technical Art Morgan Slade & Associates

ILLUSTRATION
Meg Kelleher Aubrey 65, 135
Rudi Backart iv, 22–23, 154
Karen Bell 35
Kristine Bergenheim 148–149
Eliot Bergman 357
Nancy E. Bernard 7 (lettering), 24–25 (lettering), 28, 51, 68 (border), 97, 122, 156–157, 181, 196 (left), 222, 251 (music staff), 252, 253, 266–268, 346–347, 383
Annie Bissett 21, 67 (middle), 215
Rob Brooks 174–175, 218, 298, 299
Brian Callanan 277 (spots)
Wendy Caporale 139
Patrick O. Chapin 24–25, 157 (inset)
Chris Demarest 63, 214, 475
Julie Durrell 95, 245
Ruth Flanigan 171 (bottom), 497, 499
Gary Fujiwara 434, 435
Cameron Gerlach 10
David Graves 27, 160–161, 186, 187, 235, 281, 382, 470–471
Annie Gusman 58, 59, 368 (top left), 369 (top right), 441
Dave Joly ix, 131, 358, 359, 378 (right), 380, 381 (top and bottom), 436–437
Roger Jones 99
Piotr Kaczmarek 200, 201
Beth Krommes 173
Joseph LeMonnier 117, 188 (map), 199, 232, 296, 331, 337, 401, 432–433
Karin Lidbeck 70, 216–217, 452–453
Valerie McKeown 3, 19
Scott MacNEILL 38, 39, 137 (bottom), 315, 373, 430, 487
Valerie Marsella 74, 142–143, 164, 342, 349, 376 (top), 377, 384–385, 480, 481
Paul McCusker xii–xv
Margery Mintz 100, 119, 124 (chart), 248
Debby Morse 501
Sal Murdocca 137 (top), 171 (top), 332
Andy Myer 472–473, 478–479
Cheryl Kirk Noll p. 495
Thomas R. Krepcio/Olive Jar Studios viii, 294, 295
Jennie Oppenheimer vi, 44, 152, 153
Lori Osiecki 15 (bottom right)
Diane Palmisciano 5, 102–103, 140, 145, 182, 305, 321, 366, 367, 376 (bottom), 378 (left), 379 (top), 381 (middle), 408 (bottom), 420
Cyndy Patrick 33, 172, 211 (bottom)
Deborah Perugi 8, 9, 11, 26, 27 (chart), 43, 46 (top), 52, 53, 55, 91, 108–109, 123, 127 (map), 129, 146, 147, 229, 238, 284, 335, 336, 362–365, 368–371 (charts), 379 (bottom), 406 (bottom), 407 (middle), 416 (top), 444–445
Linda Phinney 1, 16, 76, 80, 81, 84, 85, 88–89, 176, 254, 255, 450–451
Andrew Shiff 178–179, 192, 193, 207, 290, 291 (bottom), 345, 374, 375, 419
Stuart Siegal 40, 41, 158, 263, 322–323, 330, 404–405, 438–439, 483
Valerie Spain 6, 7, 12, 13, 15 (bottom left), 110, 111, 165, 196 (right), 197, 204, 256, 360, 391, 446–448
Brad Teare 114, 162, 163, 230–231, 302, 309, 402
Gary Torrisi 36 (map)
Sally Vitsky v, 56, 57, 189, 289, 300, 324, 325

PHOTOGRAPHY
Sam Gray 182, 183, 334.
Richard Haynes/RM International   141, 311, 491.
Ken Karp   118, 260, 464, 476, 477.
Steve Nelson/Fay Foto   225.
David Shopper   14–15, 130, 252, 253, 261, 264, 308–309, 387, 411

vii, Traditional Origami models folded by Gay Merrill Gross. Photo by Ellen Silverman;   **4** Mud Island (top);   **4** Allen Mims (bottom);   **8** Sand Sculptures International;   **9** Chad Slattery;   **14** NASA;   **15** Steve Rosenthal;   **16** Vu Cabisco/Visuals Unlimited (bottom);   **16** Jeff Rotman (top);   **29** Joe Carini/Pacific Stock;   **36** Breck P. Kent (left);   **36** Stan Osolinski/The Stock Market (right);   **42** Mark M. Walker/Nawrocki Stock Photo (middle left);   **42** The Image Works (middle right);   **42** Ken Regan/Camera 5 (left);   **42** Ken Regan/Camera 5 (right);   **45** Susan I. Cunningham.;   **46** Lawrence Migdale;   **49** Michele Burgess/The Stock Market;   **50** Bruce Roberts/Photo Researchers, Inc.;

50 Breck P. Kent (inset); **52** Chris Jones/The Stock Market; **53** Andrew Sacks/Tony Stone Worldwide; **54** Timothy Ross/Picture Group; **67** Comstock; **68** Jim Arndt/Visual Images West; **72–73** The Stock Market; **76** Sam C. Pierson/Photo Researchers, Inc. (top); **76** Larry Lee/H. Armstrong Roberts (middle); **76** Seth H. Goltzer/The Stock Market (bottom); **78** Show America, Inc.; **79** David Parker/Photo Researchers, Inc.; **83** Luis Villota/The Stock Market; **86** Library of Congress (top); **86** Rick Browne/Stock Boston (bottom); **90** William Williams Collection. United States History, Local History & Geneology Division. New York Public Library, Aster, Lenox and Tilden Foundations (top); **90** Library of Congress (bottom); **102–103** Ironworld, USA; **104** The Bettmann Archive; **104–105** Bernard Hoffman, LIFE Magazine, © Time Warner Inc.; **106** Daniel Nouvel/Gamma Liaison (top); **106** Jy Ruszhiewski/Sportschrome (bottom); **113** The Bettmann Archive; **120** Stickland/Allsport; **124** Alice K. Taylor/Berg and Associates (top); **124** Steve Solum/Third Coast Stock Source (bottom); **126** The Granger Collection; **126–127** The Thomas Gilcrease Institute of American History and Art, Tulsa, Oklahoma.; **128** Houston Chronicle; **145** Philip Little/APA Photo Agency; **151** The Granger Collection; **172** Superstock; **185** Bob Woodward/The Stock Market; **188** David Weintraub/Photo Researchers, Inc.; **190–191** Traditional origami models folded by Gay Merrill Gross, Photo by Ellen Silverman; **194** Robert Caputo/Stock Boston; **198** NASA; **206** Harvey Lloyd/The Stock Market; **214** Mark E. Gibson; **220** Philippa Scott/Photo Researchers, Inc.; **220–221** Randa Bishop/Dupont Corporation; **224** National Baseball Library, Cooperstown, NY; **227** National Baseball Library, Cooperstown, NY; **228** American Numismatic Society (top); **228** Eric Fordham (bottom) **234** Steve Cram/Focus on Sports; **236** Alan Smith/Tony Stone Worldwide; **239** Jon Riley/Tony Stone Worldwide; **239** Stacy Pick/Stock Boston (inset); **247** Animals Animals/©Margot Conte (top); **247** Tanaka Associates/FPG International (bottom); **250** John P. Endress/The Stock Market; **254** Animals Animals/©E.R. Degginger (top); **254** M. Abbey/Visuals Unlimited (bottom); **255** Stanley Flegler/Visuals Unlimited (top); **255** Animals Animals/©E.R. Degginger (bottom); **259** Pete Saloutos/The Stock Market; **262** Robert Frerck/Odyssey; **268** Bob Daemmrich/Stock Boston; **277** Superstock (left); **277** Mary Evans Picture Library (right); **279** Bill Reaves/Viesti Associates, Inc.; **283** Ted Strechinsky/Photo 20–20; **284–285** Superstock; **286** Tony Stone Worldwide; **288** Museum Ludwig, Rheinisches Bildarchiv Koln; **293** Pete Saloutos/The Stock Market; **294** Jim Berman/Olive Jar Studios, Inc.; **298** Dallas and John Heaton/Stock Boston (top); **298** Superstock (bottom); **300** Private Collection; **301** SPNEA; **304** Judith Larzelere; **306** Kindra Clineff; **315** Superstock; **319** Index Stock International (top); **319** Ken Karp/Omni Photo Communications (6); **320** Culver Pictures; **322** Kindra Clineff; **326** Henry Meyer/Berg and Associates; **328–329** Paul T. Goodnight;

333 North Wind Picture Archives; **338** The Granger Collection (top left); **338** North Wind Picture Archives (top right); **357** Robert Frerck/Tony Stone Worldwide; **370–371** Steve Elmore; **388** The Watch and Clock Museum; **389** Terry Wild (bottom); **389** Stu Rosner/Stock Boston (top); **399** Fred McConnaughey/Photo Researchers, Inc. (top left); **399** Allen Power/Photo Researchers, Inc. (top right); **399** Animals Animals/©Z. Leszcynski (middle left); **399** Animals Animals/©W. Gregory Brown (bottom left); **399** Animals Animals/©Z. Lesczynski (bottom right); **400** Ben Simmons/The Stock Market (top left); **400** ©RAGA/The Stock Market (top right); **400** Joe Sohn, Chromosohn/Stock Boston (bottom left); **400** Claudia Parks/The Stock Market (bottom right); **401** Cameramann International, LTD (left); **401** Loren McIntyre/Woodfin Camp & Associates, Inc. (right); **408** NASA (top right); **408** Museum of Science (left); **409** NASA; **414** Private Collection, photo by Geoffrey Clements, courtesy Paula Cooper Gallery, NYC.; **415** Kasimir Meduniezky, "Construction No. 557" Yale University Art Gallery, Gift of Collection Societe Anonyme; **416** David Ryan/Photo 20/20; **421** Kindra Clineff (top); **421** R. Ian Lloyd/The Stock Market (bottom left); **421** Kindra Clineff (bottom right); **423** National Archives Pacific Sierra Region; **427** Kindra Clineff; **431** M. Yazaki/FPG International; **431** Chris Hackett/The Image Bank (inset); **433** Jeff Gnass/The Stock Market; **443** Paul Conklin/Monkmeyer Press; **460** John W. Warden/Superstock; **461** Craig Aurness/West Light; **462** Kindra Clineff; **467** Animals Animals/©Terry Murphy

Table of Numbers

	O	P	Q	R	S	T	U	V	W
A	50	32	73	15	61	92	28	86	47
B	568	240	809	485	714	123	926	679	391
C	82	325	146	65	508	237	78	444	99
D	6,385	3,672	5,127	8,996	2,001	9,131	1,518	4,280	7,324
E	2,430	767	6,420	3,518	942	5,291	895	4,713	1,727
F	0.7	0.3	0.9	0.1	0.5	0.8	0.2	0.6	0.4
G	0.48	0.99	0.10	0.66	0.83	0.26	0.50	0.72	0.36
H	5.4	3.1	1.6	4.3	3.9	3.2	2.8	2.5	7.7
I	3	2.18	3.5	2.19	7.42	1.38	4.2	4.02	5.08
J	$\frac{4}{5}$	$\frac{1}{5}$	$\frac{2}{2}$	$\frac{1}{10}$	$\frac{7}{10}$	$\frac{2}{5}$	$\frac{3}{10}$	$\frac{1}{2}$	$\frac{3}{5}$
K	$1\frac{1}{2}$	$3\frac{1}{4}$	$4\frac{1}{3}$	$2\frac{1}{4}$	$5\frac{5}{6}$	$1\frac{1}{3}$	$3\frac{1}{2}$	$2\frac{2}{3}$	$1\frac{3}{4}$
L	$3\frac{1}{8}$	$6\frac{1}{2}$	$4\frac{3}{4}$	$5\frac{7}{8}$	$7\frac{2}{4}$	$1\frac{1}{4}$	$7\frac{7}{16}$	$2\frac{2}{4}$	$8\frac{4}{8}$
M	$\frac{1}{4}$	3.5	$3\frac{5}{10}$	2.8	5.2	$8\frac{3}{4}$	0.75	0.15	4.25
N	5%	25%	100%	30%	10%	75%	1%	50%	80%

Ideas for using this table for estimation, mental math, and for calculator activities are found periodically in the Two Minute Math sections of the Teacher's Edition.